Edited by
Fabrizio Cavani, Stefania Albonetti,
Francesco Basile, and Alessandro Gandini

**Chemicals and Fuels from Bio-Based
Building Blocks**

Volume 2

Edited by Fabrizio Cavani, Stefania Albonetti,
Francesco Basile, and Alessandro Gandini

Chemicals and Fuels from Bio-Based Building Blocks

Volume 2

WILEY-VCH

Verlag GmbH & Co. KGaA

Editors

Prof. Fabrizio Cavani
Dipartimento di Chimica Industriale
Viale Risorgimento 4
40136 Bologna
Italy

Prof. Stefania Albonetti
Dipartimento di Chimica Industriale
Viale Risorgimento 4
40136 Bologna
Italy

Prof. Francesco Basile
Dipartimento di Chimica Industriale
Viale Risorgimento 4
40136 Bologna
Italy

Prof. Alessandro Gandini
Universidade de Sao Paulo
PB 780
13560-970 Sao Carlos, SP
Brazil

■ All books published by **Wiley-VCH** are carefully produced. Nevertheless, authors, editors, and publisher do not warrant the information contained in these books, including this book, to be free of errors. Readers are advised to keep in mind that statements, data, illustrations, procedural details or other items may inadvertently be inaccurate.

Library of Congress Card No.: applied for

British Library Cataloguing-in-Publication Data
A catalogue record for this book is available from the British Library.

Bibliographic information published by the Deutsche Nationalbibliothek
The Deutsche Nationalbibliothek lists this publication in the Deutsche Nationalbibliografie; detailed bibliographic data are available on the Internet at <http://dnb.d-nb.de>.

© 2016 Wiley-VCH Verlag GmbH & Co. KGaA, Boschstr. 12, 69469 Weinheim, Germany

Print ISBN: 978-3-527-33897-9
ePDF ISBN: 978-3-527-69819-6
ePub ISBN: 978-3-527-69822-6
Mobi ISBN: 978-3-527-69821-9
oBook ISBN: 978-3-527-69820-2

Cover Design Formgeber, Mannheim, Germany
Typesetting SPi Global, Chennai, India
Printing and Binding Markono Print Media Pte Ltd, Singapore

Printed on acid-free paper

Contents

List of Contributors

Stefania Albonetti
Alma Mater
Studiorum – University of
Bologna
Dipartimento di Chimica
Industriale "Toso Montanari"
Viale Risorgimento 4
40136 Bologna
Italy

Aristos Aristidou
Cargill Biotechnology R&D
2500 Shadywood Road
Excelsior, MN 55331
USA

Udo Armbruster
Leibniz-Institut für Katalys e.V.
Albert-Einstein-Street 29a
18059 Rostock
Germany

Mehrdad Arshadi
Swedish University of
Agricultural Sciences
Department of Forest
Biomaterials and Technology
Umea
Sweden

Francesco Basile
Alma Mater
Studiorum – University of
Bologna
Dipartimento di Chimica
Industriale "Toso Montanari"
Viale Risorgimento 4
40136 Bologna
Italy

Anna Katharina Beine
RWTH Aachen University
Institut für Technische und
Makromolekulare Chemie
Lehrstuhl für Heterogene
Katalyse und Technische Chemie
Worringerweg 2
52074 Aachen
Germany

Giuseppe Bellussi
ENI S.p.A
Downstream R&D Development
Operations and Technology
Via Maritano 26
20097 S. Donato Milanese
Italy

Daniele Bianchi
Eni S.p.A.
Renewable Energy and
Environmental R&D
Center – Istituto eni Donegani
Via G. Fauser 4
28100 Novara
Italy

Eric Black
Cargill Corn Milling North
America
15407 McGinty Road West
Wayzata, MN 55391
USA

Thomas Bonnotte
Univ. Lille
CNRS, Centrale Lille, ENSCL
Univ. Artois
UMR 8181 – UCCS – Unité de
Catalyse et Chimie du Solide
F-59000 Lille
France

Thomas R. Boussie
Rennovia Inc.
3040 Oakmead Village Drive
Santa Clara California 95051
USA

Massimo Bregola
Cargill Starches & Sweeteners
Europe
Divisione Amidi
Via Cerestar 1
Castelmassa
RO 45035
Italy

Pieter C.A. Bruijnincx
Utrecht University
Faculty of Science
Debye Institute for
Nanomaterials Science
Inorganic Chemistry and
Catalysis
Universiteitsweg 99
3584 CG Utrecht
The Netherlands

Tuong V. Bui
University of Oklahoma
Chemical, Biological, and
Materials Engineering
100 East Boyd Street
Norman, OK 73019
USA

Vincenzo Calemma
ENI S.p.A
Downstream R&D Development
Operations and Technology
Via Maritano 26
20097 S. Donato Milanese
Italy

Federico Capuano
Eni S.p.A.
Refining and Marketing and
Chemicals
Via Laurentina 449
00142 Roma
Italy

Alfred Carlson
BioAmber, Inc.
3850 Annapolis Lane North
Plymouth, MN 55447
USA

Roberto Werneck do Carmo
BRASKEMS.A.
Chemical Processes from
Renewable Raw Materials
Renewable Technologies
Rua Lemos Monteiro 120
05501-050 São Paulo, SP
Brazil

Fabrizio Cavani
Alma Mater
Studiorum – University of
Bologna
Dipartimento di Chimica
Industriale "Toso Montanari"
Viale Risorgimento 4
40136 Bologna
Italy

Annamaria Celli
University of Bologna
Department of Civil, Chemical,
Environmental and Materials
Engineering
Via Terracini 28
40131 Bologna
Italy

Gabriele Centi
University of Messina
ERIC aisbl and CASPE/INSTM
Department DIECII
Section Industrial Chemistry
Viale F. Stagno D'Alcontras 31
98166 Messina
Italy

Sanjay Charati
Solvay R&I
Centre de Lyon
Saint Fons 69190
France

Alessandro Chieregato
Alma Mater
Studiorum – Università di
Bologna
Dipartimento di Chimica
Industriale "Toso Montanari"
Viale Risorgimento 4
40136 Bologna
Italy

James H. Clark
University of York
Green Chemistry Centre of
Excellence
York
YO10 5DD
UK

Corine Cochennec
Solvay R&I
Centre de Lyon
Saint Fons 69190
France

Bill Coggio
BioAmber, Inc.
3850 Annapolis Lane North
Plymouth, MN 55447
USA

Martino Colonna
University of Bologna
Department of Civil, Chemical,
Environmental and Materials
Engineering
Via Terracini 28
40131 Bologna
Italy

Steven Crossley
University of Oklahoma
Chemical, Biological, and
Materials Engineering
100 East Boyd Street
Norman, OK 73019
USA

Manilal Dahanayake
Solvay R&I
Centre de Lyon
Saint Fons 69190
France

Paulo Luiz de Andrade Coutinho
BRASKEM S.A.
Knowledge Management
Intellectual Property and
Renewables
Corporative Innovation
Rua Lemos Monteiro 120
05501-050 São Paulo, SP
Brazil

Jean-Claude de Troostembergh
Cargill Biotechnology R&D
Havenstraat 84
Vilvoorde 1800
Belgium

Gary M. Diamond
Rennovia Inc.
3040 Oakmead Village Drive
Santa Clara California 95051
USA

Eric Dias
Rennovia Inc.
3040 Oakmead Village Drive
Santa Clara California 95051
USA

Jean-Luc Dubois
ARKEMA France
420 Rue d'Estienne d'Orves
92705 Colombes
France

Franck Dumeignil
Univ. Lille
CNRS, Centrale Lille
ENSCL, Univ. Artois
UMR 8181 – UCCS – Unité de
Catalyse et Chimie du Solide
F-59000 Lille
France

and

Maison des Universités
Institut Universitaire de France
IUF
103 Bd St-Michel
75005 Paris
France

Alan Barbagelata El-Assad
BRASKEM S.A.
Innovation in Renewable
Technologies
Corporative Innovation
Rua Lemos Monteiro 120
05501-050 São Paulo, SP
Brazil

Mihaela Florea
University of Bucharest
Department of Organic
Chemistry
Biochemistry and Catalysis
4-12 Regina Elisabeta Boulevard
030016 Bucharest
Romania

Alessandro Gandini
University of São Paulo
São Carlos Institute of Chemistry
Avenida Trabalhador
São-carlense 400
CEP 13466-590
São Carlos, SP
Brazil

Nicholas Gathergood
Tallinn University of Technology
Department of Chemistry
Tallinn
Estonia

Patrick Gilbeau
Solvay R&I
Centre de Lyon
Saint Fons 69190
France

Claudio Gioia
University of Bologna
Department of Civil, Chemical,
Environmental and Materials
Engineering
Via Terracini 28
40131 Bologna
Italy

Gianni Girotti
Versalis S.p.A.
Green Chemistry R&D Centre
Via G. Fauser 4
28100 Novara
Italy

Peter J.C. Hausoul
RWTH Aachen University
Institut für Technische und
Makromolekulare Chemie
Lehrstuhl für Heterogene
Katalyse und Technische Chemie
Worringerweg 2
52074 Aachen
Germany

Cheng-Jyun Huang
Industrial Technology Research
Institute
Material and Chemical Research
Laboratories
321 Kuang Fu Road
Hsinchu 30011
Taiwan

Ying-Ting Huang
Industrial Technology Research
Institute
Material and Chemical Research
Laboratories
321 Kuang Fu Road
Hsinchu 30011
Taiwan

Thuan Minh Huynh
Leibniz-Institut für Katalys e.V.
Albert-Einstein-Street 29a
18059 Rostock
Germany

and

4 Nguyen Thong Street
District 3
Ho Chi Minh City
Vietnam

Alberto Iaconi
SPIGA BD S.r.l.
Via Pontevecchio 55
16042 Carasco, GE
Italy

Selma Barbosa Jaconis
BRASKEM S.A.
Knowledge Management and
Technology Intelligence
Corporative Innovation
Rua Lemos Monteiro 120
05501-050 São Paulo, SP
Brazil

Guang-Way Bill Jang
Industrial Technology Research
Institute
Material and Chemical Research
Laboratories
321 Kuang Fu Road
Hsinchu 30011
Taiwan

Ruben Jolie
Cargill Biotechnology R&D
Havenstraat 84
Vilvoorde 1800
Belgium

Benjamin Katryniok
Univ. Lille
CNRS
Centrale Lille
ENSCL, Univ. Artois
UMR 8181 –UCCS –Unité de
Catalyse et Chimie du Solide
F-59000 Lille
France

and

Ecole Centrale de Lille
ECLille 59655
Villeneuve d'Ascq
France

Apostolis Koutinas
Agricultural University of Athens
Department of Food Science and
Human Nutrition
Athens
Greece

Marie-Pierre Labeau
Solvay R&I
Centre de Lyon
Saint Fons 69190
France

Talita M. Lacerda
University of São Paulo
São Carlos Institute of Chemistry
Avenida Trabalhador
São-carlense 400
CEP 13466-590
São Carlos, SP
Brazil

Paola Lanzafame
University of Messina
ERIC aisbl and CASPE/INSTM
Department DIECII
Section Industrial Chemistry
Viale F. Stagno D'Alcontras 31
98166 Messina
Italy

Philippe Lapersonne
Solvay R&I
Centre de Lyon
Saint Fons 69190
France

Susanna Larocca
SO.G.I.S. S.p.A.
Via Giuseppina 132
26048 Sospiro, CR
Italy

Kit Lau
BioAmber, Inc.
3850 Annapolis Lane North
Plymouth, MN 55447
USA

Chia-Ling Li
Industrial Technology Research
Institute
Material and Chemical Research
Laboratories
321 Kuang Fu Road
Hsinchu 30011
Taiwan

Alice Lolli
Alma Mater
Studiorum – Università di
Bologna
Dipartimento di Chimica
Industriale "Toso Montanari"
Viale Risorgimento 4
40136 Bologna
Italy

Erica Lombardi
Alma Mater
Studiorum – Università di
Bologna
Dipartimento di Chimica
Industriale "Toso Montanari"
Viale Risorgimento 4
40136 Bologna
Italy

Mateus SchreinerGarcez Lopes
BRASKEM S.A.
Innovation in Renewable
Technologies
Corporative Innovation
Rua Lemos Monteiro 120
05501-050 São Paulo, SP
Brazil

Rafael Luque
University of Cordoba
Department of Organic
Chemistry
E14014 Cordoba
Spain

Rodolfo Mafessanti
Alma Mater
Studiorum – Università di
Bologna
Dipartimento di Chimica
Industriale "Toso Montanari"
Viale Risorgimento 4
40136 Bologna
Italy

Philippe Marion
Solvay R&I
Centre de Lyon
Saint Fons 69190
France

Andreas Martin
Leibniz-Institut für Katalys e.V.
Albert-Einstein-Street 29a
18059 Rostock
Germany

Sergio Martins
Solvay R&I
Centre de Lyon
Saint Fons 69190
France

Christopher Mercogliano
BioAmber, Inc.
3850 Annapolis Lane North
Plymouth, MN 55447
USA

Jim Millis
BioAmber, Inc.
3850 Annapolis Lane North
Plymouth, MN 55447
USA

François Monnet
Solvay R&I
Centre de Lyon
Saint Fons 69190
France

Piergiuseppe Morone
University of Rome
Department of Law and
Economics
Unitelma-Sapienza
Rome
Italy

Vince Murphy
Rennovia Inc.
3040 Oakmead Village Drive
Santa Clara California 95051
USA

Ronaldo Nascimento
Solvay R&I
Centre de Lyon
Saint Fons 69190
France

Juliana Velasquez Ochoa
Alma Mater
Studiorum–Università di
Bologna
Dipartimento di Chimica
Industriale "Toso Montanari"
Viale Risorgimento 4
40136 Bologna
Italy

Regina Palkovits
RWTH Aachen University
Institut für Technische und
Makromolekulare Chemie
Lehrstuhl für Heterogene
Katalyse und Technische Chemie
Worringerweg 2
52074 Aachen
Germany

Vasile I. Parvulescu
University of Bucharest
Department of Organic
Chemistry
Biochemistry and Catalysis
4-12 Regina Elisabeta Boulevard
030016 Bucharest
Romania

Sébastien Paul
Univ. Lille
CNRS
Centrale Lille
ENSCL, Univ. Artois
UMR 8181 –UCCS –Unité de
Catalyse et Chimie du Solide
F-59000 Lille
France

and

Ecole Centrale de Lille
ECLille
59655 Villeneuve d'Ascq
France

Siglinda Perathoner
University of Messina
ERIC aisbl and CASPE/INSTM
Department DIECII
Section Industrial Chemistry
Viale F. Stagno D'Alcontras 31
98166 Messina
Italy

Carlo Perego
Eni S.p.A.
Renewable Energy and
Environmental R&D
Center – Istituto eni Donegani
Via G. Fauser 4
28100 Novara
Italy

Paolo Pollesel
ENI S.p.A
Downstream R&D Development
Operations and Technology
Via Maritano 26
20097 S. Donato Milanese
Italy

Katie Privett
University of York
Green Chemistry Centre of
Excellence
York
YO10 5DD
UK

Daniel E. Resasco
University of Oklahoma
Chemical, Biological, and
Materials Engineering
100 East Boyd Street
Norman, OK 73019
USA

Marco Ricci
Versalis S.p.A.
Green Chemistry R&D Centre
Via G. Fauser 4
28100 Novara
Italy

Giacomo Rispoli
Eni S.p.A.
Refining and Marketing and
Chemicals
Via Laurentina 449
00142 Roma
Italy

Matthias N. Schneider
Baerlocher GmbH
Freisinger Straße 1
85716 Unterschleissheim
Germany

Franco Speroni
Solvay R&I
Centre de Lyon
Saint Fons 69190
France

Everton Simões Van-Dal
BRASKEM S.A.
Innovation in Renewable
Technologies
Corporative Innovation
Rua Lemos Monteiro 120
05501-050 São Paulo, SP
Brazil

Micaela Vannini
University of Bologna
Department of Civil, Chemical,
Environmental and Materials
Engineering
Via Terracini 28
40131 Bologna
Italy

Bert M. Weckhuysen
Utrecht University
Faculty of Science
Debye Institute for
Nanomaterials Science
Inorganic Chemistry and
Catalysis
Universiteitsweg 99
3584 CG Utrecht
The Netherlands

Sophie C.C. Wiedemann
Utrecht University
Faculty of Science
Debye Institute for
Nanomaterials Science
Inorganic Chemistry and
Catalysis
Universiteitsweg 99
3584 CG Utrecht
The Netherlands

and

Croda Nederland
B.V. Buurtje 1
2800 BE Gouda
The Netherlands

Jinn-Jong Wong
Industrial Technology Research
Institute
Material and Chemical Research
Laboratories
321 Kuang Fu Road
Hsinchu 30011
Taiwan

Yu Zhang
Alma Mater
Studiorum – Università di
Bologna
Dipartimento di Chimica
Industriale "Toso Montanari"
Viale Risorgimento 4
40136 Bologna
Italy

Preface

Today the biorefinery concept is being applied to integrate the production of chemicals, fuels, and materials from renewable resources and wastes. But the concept is not new, since the industry of oils and fats, among others, has already for some time been transforming by-products or co-products received from other industries into chemicals for several diverse sectors while combining this production with that of fuels, such as biogas, obtained from organic residues.

However, the sector is rapidly evolving, and new concepts and ideas are setting the scene. It is impressive to see how the scientific and technical advancements in this field have been, and still are, changing the scenario at an unprecedented pace.

This book was conceived with the ambitious aim of preparing something different from the information already available. In other words, for the authors it was not only a matter of updating the panorama with the latest developments in technologies and transformation processes but also of offering readers the possibility to view the "world of renewables" from a more rational perspective. Instead of contributions focusing on how a certain biomass or bio-based building block can be transformed into specific products, we decided to offer an overview from the standpoints of *reaction* and *products*. This led to the organization of the book into different sections in which the classes of reactions (oxidation, hydrodeoxygenation, C–C bond formation) are examined or the different types of monomers (succinic acid, adipic acid, furandicarboxylic acid, glycols, and acrylic acid), polymers, and drop-in chemicals (olefins, aromatics, and syngas, produced from renewables) are discussed. A closing section of the book contains several contributions from chemical industries operating in the field of biorefinery development. Throughout the book it is possible to find points of intersection between the chapters on products and those on reactions that are finding a place in biorefinery models. This approach gives a three-dimensional perspective on the production of bio-based building blocks and, looking at the future, facilitates the development of new processes and placement of new products in an increasingly integrated context.

This book is therefore organized into three main sections. The first section (15 chapters) is devoted to a discussion of the main products attainable from renewable raw materials. The different types of products have, in turn, been further separated into three sections: (i) chemicals and fuels from bio-based building blocks (B⁴), (ii) B⁴ monomers, and (iii) polymers from B⁴. To provide a reliable description of the state of research and development and of industrial implementations of the production of different molecules, we asked several experts from universities and industry to present the most recent results obtained in their respective field, together with their ideas on the topic. Given the structure of the book, the chapters on polymers concentrate exclusively on the use of bio-based building blocks as potential monomers and on the most recent studies dealing with their polymerizations and copolymerizations, as well as on the properties of the ensuing materials.

The second part of the book focuses on the class of reactions needed to obtain chemicals and fuels from bio-based building blocks. Hydrogenation, deoxygenation, C–C bond formation, and oxidation are key reactions in the upgrading of these compounds, and Chapters 14–17 provide the reader with a basic understanding, offering an overview of the possibilities offered by these tools using different raw materials. Several examples of homogeneous and heterogeneous catalysis are discussed, with an emphasis on the industrial aspects, providing a comprehensive picture and addressing the main issues associated with biomass transformations.

The book concludes with several contributions from scientists working in the industry. This part is introduced by Centi's chapter on "A vision for future biorefineries." In recent years, numerous bio-based companies, seeing new technologies as ways to gain market positions through innovation, have been created all over the world. However, they have often appeared to be driven primarily by the need to raise funds, rather than by a true desire for innovation and a clear analysis of the market perspectives. For example, the US government's push to develop hydrocarbons from biomass led to the creation of over 100 venture-capital companies, most of which have now closed or markedly changed their strategies. Hence there is the trend to reconsider the biorefinery model and analyze the new directions and scenarios to evaluate possibilities and anticipate needs for research. Companies themselves are involved in this process, and the last part of the book presents the different models of biorefineries they have developed in recent years.

SOGIS SpA and Baerlocher GmbH (Chapter 19) analyze the possibility of fitting biorefinery concepts into the well-established oleochemistry value chains. Indeed, oleochemistry has always used renewable raw materials (plant oils) or side streams from food production (animal fats) and may therefore be considered a model for green chemistry. Specifically, the emerging opportunities of food supply chain waste form the topic covered by R. Luque and coworker (Chapter 26). Researchers around the world are working to find innovative solutions to the food waste problem through valorization, and interdisciplinary collaborative approaches are crucial for creating a viable set of solutions in chemical, material, and fuel production.

Large chemical companies, such as Arkema (Chapter 20) and Eni (Chapter 25), highlight the necessity to facilitate the transition to bio-based raw materials, creating existing fossil-based molecules from renewables (benefiting from the existing market and regulations) and using existing infrastructure to lower capital costs. In 2014, renewable products generated around €700 million of the turnover of Arkema; today this company has several plant-based factories operating worldwide, demonstrating the economic feasibility of bio-based production when the products have a technical advantage over fossil-based materials. Similar considerations have also been reported by Solvay (Chapter 24), describing several case studies that provide a good picture of the issue of introducing new sustainable chemical processes into the various sectors of chemical production. Moreover, they have shown the wide range of benefits that the use of renewable raw materials can bring to the chemical industry, taking into account the real sustainability of the processes applied.

The keys to the current and future success of a bioeconomy are analyzed by Cargill in Chapter 21, describing the concept of colocation as a model for the production of bio-based chemicals from starch, while Braskem reports on the economic viability of sugarcane biorefineries in Brazil (Chapter 22), highlighting the importance of integration in value chain models as a key element for the future of chemical production from renewables. Moreover, they underscore the importance of industrial biotechnology as the basis of these industries in the future.

Versalis, the major Italian chemical company belonging to the Eni group (Chapter 23), discusses the company's strategy in the field, describing several projects mainly related to the synthetic elastomer business. Particular attention is focused on the production of natural rubber from guayule plants and the development of a process for the production of butadiene from bio-based raw materials.

This book is aimed at R&D engineers and chemists in chemical and related industries, chemical engineering and chemistry students, and chemical, refining, pharmaceutical, and biotechnological industry managers. Moreover, it can be useful for decision-making managers in funding institutions, providing them with an overview of new trends in the bio-based economy.

In closing, we would like to express our sincere thanks to all the authors who contributed to the compilation of this book, without whom this project would never have seen the light. We are also grateful to all the people at Wiley for their patience and professionalism.

1 December 2015
Bologna

Fabrizio Cavani, Stefania Albonetti,
Francesco Basile, and Alessandro Gandini

Part IV
Reactions Applied to Biomass Valorization

Chemicals and Fuels from Bio-Based Building Blocks, First Edition.
Edited by Fabrizio Cavani, Stefania Albonetti, Francesco Basile, and Alessandro Gandini.
© 2016 Wiley-VCH Verlag GmbH & Co. KGaA. Published 2016 by Wiley-VCH Verlag GmbH & Co. KGaA.

14
Beyond H$_2$: Exploiting H-Transfer Reaction as a Tool for the Catalytic Reduction of Biomass

Alice Lolli, Yu Zhang, Francesco Basile, Fabrizio Cavani, and Stefania Albonetti

14.1
Introduction

The last century has witnessed a dramatic growth of the energy and chemical industry, fostered by a steep rise in population. In the coming decades, further growth and changes are expected, brought about by the sustainability imperative, progress in science, and a gradual switch from fossil-based raw materials to renewable ones [1, 2]. In contrast to oil-based resources – which are characterized by high H/C ratio, chemical stability, and low/medium molecular weight – renewable feedstocks contain highly functionalized carbohydrates with abundant oxygen [3]. As a consequence, the change to a biomass-based economy will face several challenges, as the chemistry involved in the conversion of new platform molecules may significantly differ from the processes based on petroleum feedstock.

Since biomass-derived molecules are generally highly oxidized compounds, their essential transformation to alternative fuels and chemicals requires hydrogen. For reduction processes, catalytic hydrogenation is still the main technique in use today [4]. Nevertheless, the reduction of carbonyl groups using alcohols as hydrogen sources, that is, the Meerwein–Ponndorf–Verley (MPV) reaction, offers an alternative approach to working under H$_2$ pressure using supported precious metal catalysts [5]. This reaction is a well-known process which involves hydrogen transfer from alcohol to carbonyl carbon [6]. The reaction is chemoselective. Indeed, under the conditions of the MPV reaction, unsaturated aldehydes and ketones are not reduced to saturated alcohols [7].

The MPV reaction is usually carried out through homogeneous catalysis by using Lewis acids such as Al, B, or Zr alkoxides. However, with these materials, the reaction requires an alkoxide excess of at least 100–200% and the subsequent need to neutralize the residual alkoxide in the medium by means of a strong acid [8].

Conversely, the use of heterogeneous catalysts has been extensively explored. Solid catalysts can be generally classified into two types. In one case, metal cations,

Chemicals and Fuels from Bio-Based Building Blocks, First Edition.
Edited by Fabrizio Cavani, Stefania Albonetti, Francesco Basile, and Alessandro Gandini.
© 2016 Wiley-VCH Verlag GmbH & Co. KGaA. Published 2016 by Wiley-VCH Verlag GmbH & Co. KGaA.

such as Al^{3+}, Zr^{4+}, and Sn^{4+}, are either incorporated as Lewis acidic sites in a zeolite framework [9–11] or supported on oxides [12]. In this case, the reaction mechanism is similar to that observed with classical alkoxide catalysts [13]. In the second case, some solid bases, such as MgO or amphoteric oxides, serve as catalysts. The involvement of both acidic and basic sites is often suggested with these materials.

In this chapter, we summarize the potential of homogeneous and heterogeneous catalysis in the H-transfer reaction while paying specific attention to the hydrogenation of bio-based molecules.

14.2
MPV Reaction Using Homogeneous Catalysts

The MPV reaction is a catalytic carbonyl reduction in the presence of an alcohol as a hydrogen donor. Many elements belonging to the second transition series in the periodic table – such as salts and complexes of Pd, Pt, Ru, Ir, Rh, Fe, Ni, and Co – have been used as catalysts for hydrogen transfer from alcohols. The general principle is to transfer a hydride and a proton from the hydrogen donor AH_2, typically an alcohol, to an acceptor B, such as a ketone or an aldehyde, which is then reduced (Scheme 14.1).

$$AH_2 + B \rightleftarrows A + BH_2$$

Scheme 14.1 Main concept for the hydrogen transfer process (AH_2 hydrogen donor, B hydrogen acceptor).

In the 1920s, Meerwein, Ponndorf, and Verley independently found that carbonyl compounds were reduced in the presence of 2-propanol [9]. Later, Oppenauer found that secondary alcohols were oxidized in the presence of acetone [10] (Scheme 14.2).

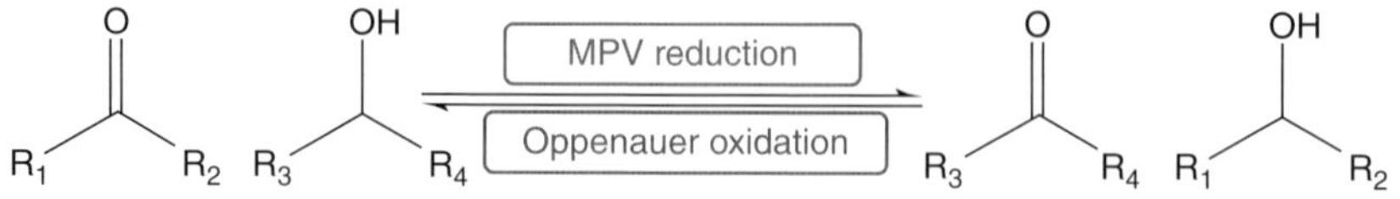

Scheme 14.2 Meerwein–Ponndorf–Verley reduction and Oppenauer oxidation.

The efficiency of the reduction is determined, in part, by thermodynamic considerations. Indeed, secondary alcohols are stronger reductants than primary ones, and aldehydes are more easily reduced than ketones. As a consequence, *iso*propanol is the most commonly used H-donor. The reaction is an equilibrium and is usually driven to the right by using a large excess of reductant; this is the reason why, in liquid phase processes, alcohols are often used as both H-donor and solvent.

In the original hydrogen transfer reaction, $Al(Oi\text{-}Pr)_3$ was used in a stoichiometric amount to mediate the reaction. Later it was found that transition metals could catalyze hydrogen transfer. Since then, transfer hydrogenation has been used successfully for both imines and ketones [11].

Transfer hydrogenation has the advantages of operational simplicity and environmental friendliness: no hydrogen pressure is used and no special equipment is required. In addition, no hazardous waste is produced, as in the case of the stoichiometric reduction by borane reagents. When 2-propanol is used as the hydrogen donor, the only side product formed is acetone, which is easily removed by distillation during workup. 2-Propanol and formic acid (FA), which are both often used as hydrogen donors, are nontoxic and also inexpensive reactants. One drawback of the use of this reaction is that each step in the cycle is reversible and the selectivity is driven by the thermodynamic properties of products and intermediates. This disadvantage can be overcome by the use of a formate, since the reaction is irreversible in this case, with the consequent loss of CO_2.

Transfer hydrogenation has been studied in detail [12]: two main reaction mechanisms have been suggested depending on the type of metal used, namely, the direct hydrogen transfer and the hydridic route. In general, direct hydrogen transfer is suggested for main group elements, whereas the hydridic route is considered to be the major pathway for transition metals:

1) *Direct hydrogen transfer*
   In direct hydrogen transfer, hydrogen is transferred directly from the donor to the acceptor, without any involvement of metal hydrides. The mechanism is thought to proceed through a cyclic six-membered transition state (Scheme 14.3) [14]. This mechanism was originally proposed for the MPV reduction, where $Al(Oi\text{-}Pr)_3$ was used as the catalyst.
2) *Hydridic route*
   The hydridic route proceeds stepwise through a metal hydride formation. Typically, the metal catalyst removes a hydrogen from the donor such as 2-propanol, through β-hydride elimination (Scheme 14.4, step 2) [12]. Hydrogen is then transferred from the metal to the acceptor, for example, a ketone. Transition metal catalysts usually operate via this kind of mechanism.

One of the main advantages in the use of the MPV reaction for the reduction of carbonyl compounds is its high selectivity for the C=O double bond. As a matter of fact, α,β-unsaturated carbonyl compounds can be selectively reduced leaving C=C untouched.

In addition to aluminum complexes, the zirconium complex bis(cyclopentadienyl) zirconium dihydride has also been reported to be active in the MPV reaction for several carbonylic compounds such as benzaldehyde, acetone, and benzophenone. Different linear (e.g., butanol, octanol, dodecanol, allyl alcohol) and cyclic

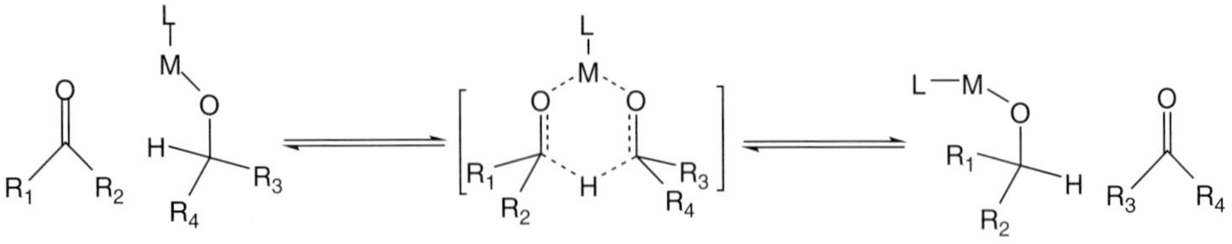

Scheme 14.3 Direct hydrogen transfer processes via six-membered ring intermediate.

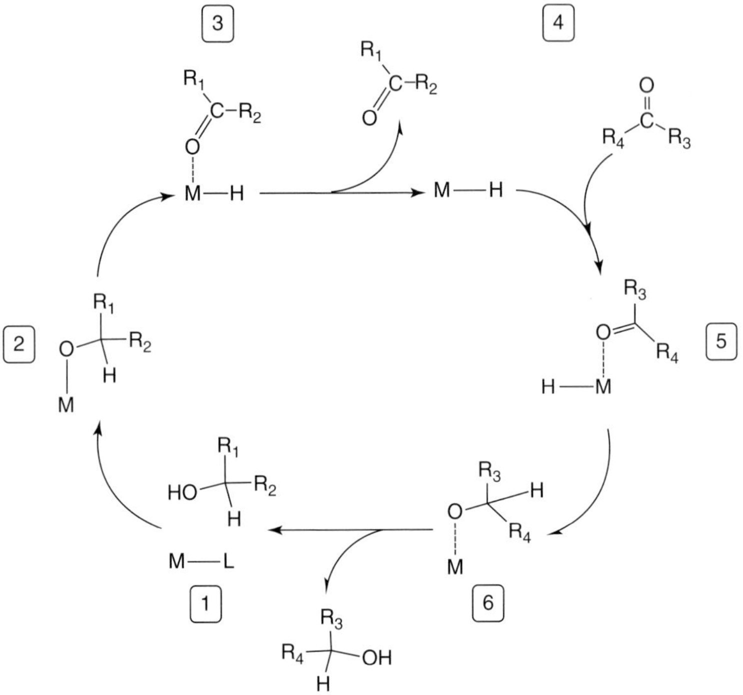

Scheme 14.4 Hydrogen transfer via metal hydride pathway.

(cyclohexanol, benzyl alcohol) alcohols have been used as hydrogen donors and were all active in the hydrogen transfer reaction [15].

Shvo's catalyst has also been demonstrated to be a powerful catalyst for transfer hydrogenation [16]. 5-Hydroxymethylfurfural (HMF) reduction to 2,5-bishydroxymethylfuran (BHMF) has occurred under mild conditions (10 bar H$_2$, 90 °C) and with complete selectivity toward the formation of the desired product. The catalyst has been successfully recycled, after removing the product from the reaction mixture by precipitation and filtration.

In addition to these homogeneous catalysts, some other alternative materials have also been reported, based on either nonprecious metals, such as Fe [17–20], or alkali metal ions, such as Li alkoxides [21, 22] or KOH [23].

14.3
MPV Reaction Using Heterogeneous Catalysts

The need for a large amount of reagents and the presence of some undesired side reactions are the main drawbacks of MPV reduction using homogeneous catalysts. To overcome these problems, the reaction has been investigated on different heterogeneous catalysts, including metal oxides, hydrous zirconia, hydrotalcites, zeolites, and even metal alkoxides immobilized on mesoporous materials. As a matter of fact, H-transfer reduction by means of hydrogen donors on heterogeneous catalysts has many interesting advantages, compared to the classical reduction procedure, which requires molecular hydrogen. Using the H-transfer reaction, no gas tank is necessary, thus avoiding any possible hazards connected to the presence of a gas with a low molecular weight and high diffusibility in a chemical plant. Therefore, over the past two decades, an increasing number of reports on heterogeneous catalysts for the MPV reaction have been published [24].

Most of the solid catalysts reported in literature are based on alkali and alkaline earth oxides, which are sometimes characterized by the insertion of metal atoms, bringing Lewis acidic properties. The solid catalysts studied for this kind of reaction may be divided into three main groups: Lewis base, Lewis acid, and metal-supported catalysts. Table 14.1 lists the most frequently used catalysts.

The H-transfer method can improve product selectivity, depending on the hydrogen donor chosen. Each alcohol can affect the reaction differently because of the different ways of absorption onto the catalyst surface; this is mainly due to the fact that the MPV reaction on the solid catalyst surface seems to proceed via a hydrogen transfer mechanism, from the alcohol to the carbonyl compound adsorbed in the vicinity of alcohol [82].

Metal-doped zeolites are included in Lewis acid-type catalysts together with Zr-containing systems. In particular, β zeolites have been found to be highly stereoselective catalysts in MPV reduction, especially when some guest element such as Ti, Sn, or Zr was introduced into the structure. For example, zeolites

Table 14.1 List of the most important H-transfer catalysts, with relevant references.

	Type of catalyst	References
Lewis acid type	Zr-based catalysts	[25–30]
	Metal-doped zeolites (with Ti, Sn, Zr)	[31–34]
	Alumina	[35]
	Metal-doped mesoporous silica	[36–38]
Lewis base type	Mg-based mixed oxide	[39–63]
	CuO	[64, 65]
	MnO_x	[66]
Metal-supported catalysts	Pt, Au, Pd, Ir	[67–72]
	Co, Cu, Ni	[73–77]
	Ru, RuO_x	[78–81]

such as H-β [31–33] and alumina [35] were used in the H-transfer reaction using isopropanol, ethanol, or cyclopentanol as the donor. Corma and coworkers [34] developed a totally water-resistant Lewis acid catalytic system based on a Sn-β zeolite. Compared to Ti-β and Al-β zeolites, Sn-β exhibited an improved stereoselectivity to the thermodynamically less favorable *cis*-alcohol isomer, when alkyl-cyclohexanones were used as substrates. It should be stressed that, in addition to Lewis acidic strength, the surface hydrophobicity/hydrophilicity is an important characteristic; Sn-β, being hydrophobic, can be a good catalyst even in the presence of water.

Furthermore, high-surface-area mesoporous materials should be included: Ti or Sn incorporated into mesoporous silica [36–38] was used as catalysts with isopropanol or 2-butanol as the reducing agents.

As far as the Zr-based systems are concerned, the most significant results were obtained with ZrO$_2$ and hydrous zirconia, either doped or as is [25, 26], anchored/grafted Zr over supports [27, 28] and Zr-β [29, 30].

In the field of solid base materials, magnesium oxide has been widely used in H-transfer reactions. More specifically, some examples include MgO, either doped or as is, and Mg/M mixed oxides, which have been used for the reduction of substrates such as citral [39, 40], cyclohexanone [41–43], acetophenone [44, 45], hexenone [46–48], acetone [49], benzaldehyde [50, 51], crotonaldehyde [52], and in general, for various aliphatic aldehydes and ketones [53–59] or aralkylketones [60, 61]. In most of the cases, isopropanol was used as the H-transfer reactant, with a few exceptions in which ethanol [49, 51] and other C$_4$ alcohols [50] were taken into consideration. The majority of these processes were carried out in the liquid phase. Some gas phase reactions over MgO were reported by Di Cosimo and coworkers [62]. The author concluded that the unsaturated ketone conversion pathways toward unsaturated alcohols and other compounds also depended on the ketone structure. Unsaturated alcohols were formed on MgO as a primary product from both 2-cyclohexenone and mesityl oxide. However, saturated alcohol may be produced by a consecutive reaction of unsaturated alcohols in the presence of 2-cyclohexenone but not with mesityl oxide. The reduction of the C=C bond was negligible regardless of the reactant structure, whereas competing reactions such as the C=C bond shift were more likely to contribute during the reduction of the acyclic reactant. Another work concerning MgO in a gas phase was carried out by Gliński [63]. Jyothi *et al.* [56] investigated the catalytic transfer reduction of alkyl, alkyl aryl, cyclic, and unsaturated ketones over calcined Mg–Al hydrotalcites in a gas phase. Some catalysts based on alkali oxides were also reported to be efficient catalysts for processing the catalytic reduction of aldehydes; in particular, Chuah and coworkers [83] investigated benzaldehyde reduction over potassium phosphate with 2-propanol as a hydride source. Transition metal oxides such as CuO- [64, 65] and MnO$_2$-based catalysts [66] with cyclohexanol, 1,4-butanediol, or ethanol as the H-donor were investigated; these are examples of soft basic sites which are able to catalyze the MPV reaction.

Among the heterogeneous catalytic systems, supported noble metal catalysts such as Pt [67, 68], Au [69, 70], Pd [71], and Ir [72] acted as effective and reusable heterogeneous catalysts for the same reaction. Au supported on different supports (TiO_2, Fe_2O_3, carbon) and Pd/C were used as catalysts for the H-transfer hydrogenation of acetophenone and many other carbonyl compounds, using 2-propanol as the H-donor. Au catalytic performances were much higher than the Pd ones, in the presence of a base [69]. $Pt-TiO_2$ was tested in acetophenone reduction in the presence of KOH, achieving a high product yield in a short reaction time [67]. The cheap noble metal Ru was chosen as the active phase; in order to study the role of the support in this reaction when using Ru-based catalysts, the same active phase $RuOH_x$ was supported on different materials. Mizuno and coworkers [78] reported a strategy for designing efficient heterogeneous catalysts for various functional group transformations, mainly based on the creation of metal hydroxide species on the appropriate support. Metal hydroxide species possess both Lewis acidic and Brønsted basic sites, while various functional group transformations are likely to be promoted through the concerted activation of these groups. After this case, Mizuno and coworkers [79] continued to develop the $Ru(OH)_x$ catalytic system and applied it in allylic alcohol reduction, using Al_2O_3 as the support. Various allylic alcohols were tested, and all the results showed higher chemoselectivity using 2-propanol as the H-donor. The mechanism studied indicated that the initial ruthenium monohydride species is formed from $Ru(OH)_x$ and that this hydride species further reacts with unsaturated compounds. In addition to conventional metal oxide supports, some novel carbon supports were also explored in the hydrogen transfer reaction. Kim and coworkers [80] used nanostructured RuO_2 on Multiwall Carbon Nanotubes (MWCNTs) to carry out the hydrogen transfer hydrogenation of carbonyl compounds. The main feature of the system with Ru supported on carbon is a tolerance of a wide range of functional groups. As a result, various carbonyl-containing compounds were tested and a high chemoselective carbonyl reduction was observed. Chuah and coworkers [81] reported on an AlO(OH)-entrapped ruthenium catalytic system for the transfer hydrogenation of aldehydes and ketones. One of the advantages of the use of AlO(OH) as a support is the precise control of the Ru nanoparticle size; indeed, particles were narrowly distributed with mean diameters of $1.5-1.8$ nm. In this example, due to the utilization of potassium formate as the H-donor, the hydrogen transfer was obtained in mild conditions. Formate seemed to be a good hydrogen donor; however, the reaction is believed to occur via *in situ* generation of H_2 by the decomposition of FA or butyl formate, rather than by transfer hydrogenation. Therefore, further study for an in-depth understanding of this mechanism must still be carried out. As cost-competitive alternatives, nonnoble metal heterogeneous catalysts such as transition metal catalysts Co [73], Cu [74, 75], and Ni [76, 77] were also indicated.

14.3.1
Mechanism and Path on Heterogeneous Catalysts

A reaction mechanism similar to that of the classical alkoxide catalysts has been suggested for heterogeneous processes [84]. In the reported hydrogen transfer path, both the alcohol (donor) and the carbonylic compound (acceptor) coordinate to the same metal center. Then a six-membered ring intermediate is formed. Lastly, the hydrogen transfer takes place, forming new products (Scheme 14.5).

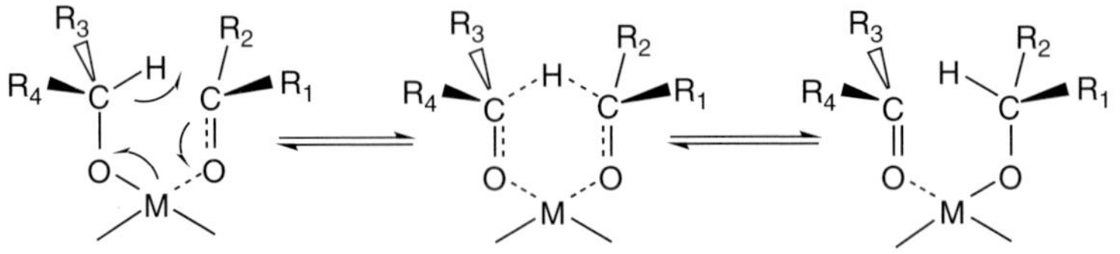

Scheme 14.5 H-transfer mechanism over metal oxide.

This mechanism has been accepted for most of the heterogeneous catalysts, especially for those which are characterized by Lewis acidic sites such as Al^{3+}, Zr^{4+}, and Sn^{4+}. Amphoteric oxides may have different reaction mechanisms, often involving both acidic and basic sites. In the second case, the mechanism (Scheme 14.6) includes the heterolytic dissociation of the alcohol followed by the transfer of alkoxide α-hydrogen to the absorbed carbonylic compound. In this mechanism, the rate-limiting step is the hydrogen abstraction from the α-carbon of the alcohol.

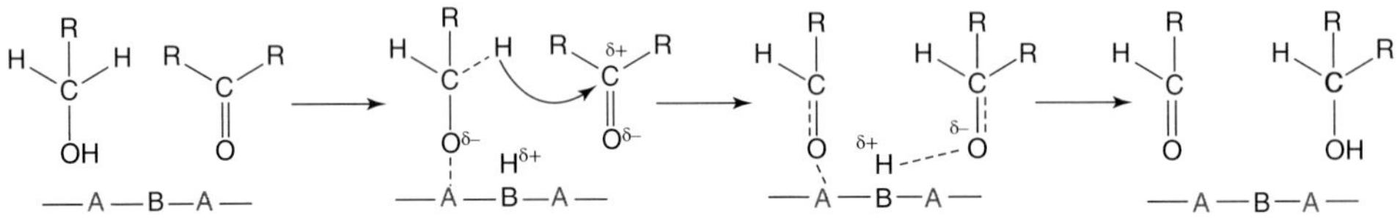

Scheme 14.6 Hydrogen transfer mechanism between an alcohol and a ketone for catalysts carrying both acidic (**A**) and basic (**B**) sites.

Recently, the computational density functional theory (DFT) method was applied to investigate the mechanism of H-transfer. In particular, a complete computational study of the mechanism of the MPV reduction was reported by Corma and coworkers [85]. Theoretical calculations were carried out on the way the substrate (cyclohexanone and 2-butanol) can absorb on Sn-β and Zr-β, and it has been pointed out that the C=O of the substrate can only be activated if it is coordinated to the Lewis acid center. Taking this into account, the reaction mechanism is considered to be the same as that of the classical MPV reaction, where $Al(Oi\text{-}Pr)_3$ is used as the catalyst. Therefore, the reaction takes place through 2-butanol deprotonation, the formation of a six-membered cyclic intermediate, and the H-transfer from the catalyst to the cyclohexanone. The role of the Sn−OH group in the MPV reaction is that of making possible the deprotonation of the alcohol, by forming an alcoholate intermediate bonded both to the tin center and to a water molecule. Zr-β was demonstrated to follow the same reaction mechanism as Sn-β, being a good catalyst for the MPV reaction. However, different active phases may have totally different catalytic paths;

therefore, with other metals, the typically six-membered ring intermediate was not formed during this reaction.

14.4
H-Transfer Reaction on Molecules Derived from Biomass

Current interest is mainly focused on the use of biomass as an alternative source for the production of fuels and chemicals. In order to upgrade biomass feedstock, it is necessary to set up an oxygen removal step, due to the high oxygen content which is typical for these compounds. In this section, we systematically summarize the various catalytic systems and the latest research progress for the selective hydrogenation of biomass-derived platform molecules by using H-transfer reaction with different hydrogen donors. A list of the most important heterogeneous catalysts used for H-transfer processes of biomass derivatives appears in Table 14.2. This table is not exhaustive but provides the references of most of the available literature.

14.4.1
Levulinic Acid Hydrogenation

The acid hydrolysis of lignocellulosic materials leads to levulinic acid (LA) formation. The hydrogenation of LA then yields γ-valerolactone (GVL), which is a sustainable fluid for energy- and carbon-based chemicals. Figure 14.1 shows the main products obtained from GVL, with their relevant application in fuels, fuel additives, and chemical production.

Many processes have been reported for the hydrogenation of LA to GVL using gaseous H_2 and metal-based catalysts (Ru, Pd, Pt), in either homogeneous or heterogeneous conditions. However, reaction parameters must be carefully controlled to prevent any further hydrogenation reaction while forming methyltetrahydrofuran (MTHF) and pentanoic acid, even though both products have a high added value [118, 119]. GVL can also be synthesized by catalytic transfer hydrogenation, using alcohols as the hydrogen donor. Several catalytic systems have recently been reported for the production of GVL via this process. Dumesic and Chia [86] compared different metal oxides in the H-transfer reduction of levulinate derivatives to GVL, finding ZrO_2 to be a highly active material for H-transfer in both batch reactions and a continuous-flow reactor. By using isopropanol/isobutanol as H-donors at 150 °C under a pressurized inert gas flow, the reaction reached a yield of 84% in 16 h. In the same way, Lin and coworkers [87] investigated an H-transfer reaction of ethyl levulinate (EL) over a ZrO_2 catalyst in a supercritical ethanol. By using ethanol as the H-donor, an integrated biomass utilization chain will be created, since ethanol can also be produced from biomass feedstock. Liu and coworkers [88] investigated the catalytic activity of hydrated zirconia ($ZrO(OH)_2 \cdot xH_2O$) for the H-transfer

Table 14.2 Heterogeneous catalysts used in H-transfer hydrogenation processes of biomass-derived oxygenated compounds. FA = formic acid

Substrate	Catalyst	H-donor	References
Levulinic acid	ZrO_2	Isopropanol/ isobutanol	[86]
	ZrO_2	Supercritical ethanol	[87]
	$Zr(OH)_x$	Supercritical ethanol, 2-propanol	[88]
	Zr-β	Secondary alcohols	[89, 90]
Ethyl levulinate	Zr-HBA	2-Propanol	[91]
	Ni-Raney	Propanol	[92]
	$Ru(OH)_x/TiO_2$	Propanol	[93]
	Supported Au	FA or butyl formate	[94, 95]
Hydroxymethylfurfural	Pd/C and H_2SO_4	FA	[96]
	Pd/C	FA	[97]
	Pd/C, Ru/C in ionic liquid	FA	[98]
	$Pd/ZrPO_4$	FA	[99]
	Cu–porous metal oxide	Supercritical methanol	[100]
	MgO	Methanol	[101]
	Ru/C	2-Propanol	[100, 102]
	Pd/C, Rh/C with H_2 pressure and $ZrCl_2$	Methanol	[103]
	Pd/Fe_2O_3	2-Propanol	[104]
	Ru/hydrotalcites	2-Propanol	[105]
Furfural	Pd/Fe_2O_3	2-Propanol	[104]
	MgO	Methanol	[101]
	Cu/SiO_2	Ethanol, propanol, and isopropanol	[106]
	Cu/Mg/Al/O		
	$Ru/RuO_2/C$	Primary and secondary alcohols	[107]
	Metal/carbon nitrile	FA	[108]
	Al/Mg hydrotalcites doped with La, Cu, Cr, Mn, Zr	Supercritical methanol	[109]
Lignin	Ni-Raney	2-Propanol	[110, 111]
Vanillin	Pd-supported catalyst	FA	[112]
Sorbitol/ mannitol	Ru/C	2-Propanol	[113]
Glycerol	Iron oxide and iron phosphate	C_3 alcohols	[114]
	$FeO_x–ZrO_2$	FA	[115, 116]
	Pd-supported catalysts	2-Propanol	[117]

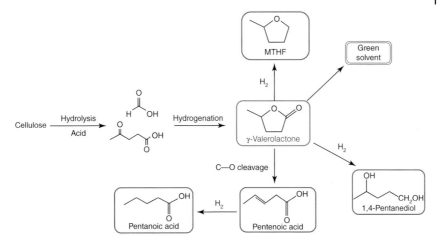

Figure 14.1 Reaction pathways for the conversion of γ-valerolactone (GVL) into fuels and chemicals.

reaction of EL in a supercritical ethanol. However, 93.6% EL conversion and 94.5% GVL selectivity were achieved when 2-propanol was used as a hydrogen donor at 200 °C with a reaction time of 1 h. Based on these results, a plausible mechanism for the H-transfer of levulinate esters to GVL via MPV reduction was suggested. The general description of the reaction pathway appears in Figure 14.2. Reduction steps typically followed the MPV reaction principle in which, at first, a six-membered ring transition state occurred, and then the intermediate ethyl 4-hydroxypentanoate (4-HPE) was formed. The newly formed 4-HPE was not stable and readily converted into GVL.

Microporous zirconosilicate molecular sieves (Zr-β) have recently been reported to be more active and stable catalysts compared to bulk zirconia. Chuah and coworkers [89] developed Zr-β zeolite as a robust, active catalyst for the hydrogen transfer reduction of LA to GVL by using 2-propanol as the H-donor. They also explained the high activity of Zr-β zeolite with the presence of both Lewis acidic sites with moderate strength and relatively few basic sites. The presence of a low amount of basic sites prevented catalyst poisoning due to the absorption of the acid reactant. In a batch reactor, GVL was formed with a selectivity higher than 96% and 100% LA conversion, whereas the quantitative

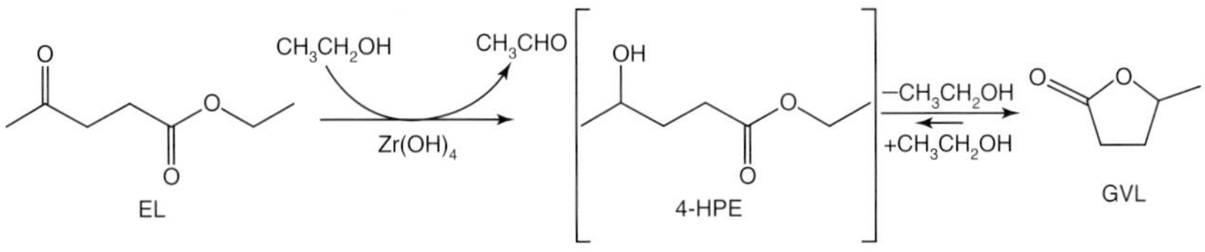

Figure 14.2 Reaction pathway for the conversion of ethyl levulinate (EL) toward γ-valerolactone (GVL) via MPV reduction.

conversion with >99% yield of GVL was obtained with a steady generation rate of $0.46 \, mol_{GVL} \, g_{Zr}^{-1} \, h^{-1}$.

Han and coworkers [91] reported a new porous Zr-containing catalyst with a phenate group in its structure, which was obtained by the coprecipitation of 4-hydroxybenzoic acid dipotassium salt and $ZrOCl_2$ (Zr-HBA) in water. In the presence of isopropanol, Zr-HBA was used as catalyst for the H-transfer of EL to GVL.

In addition to Zr-based oxide, some noble metal-supported catalysts were also investigated in LA transformation using the H-transfer process.

Fu and coworkers [92] reported a room-temperature H-transfer reaction using Ni-Raney preactivated in a 2-propanol solution; the catalyst, however, was sensitive to air, and for a long-time storage, an alcohol media was needed, which is a critical drawback in a practical large-scale synthesis of GVL.

Fujitani and coworkers [93] reported LA reduction via H-transfer over a cheap noble metal-supported catalyst: Ru hydroxide supported on anatase. Since Ru in the form of hydroxide had already been reported as an efficient catalyst in carbonyl compound hydrogenation, this study demonstrated that it worked for the reduction of biomass-derived carbonyl compounds as well. Further improvements were obtained by Yamashita and coworkers [120], by using a catalyst made up of uniformly dispersed Ru nanoparticles supported on Zr-containing spherical mesoporous silica. Under mild conditions, this system showed a superior activity for the hydrogenation of both LA and its ester; the Zr sites embedded within the spherical mesoporous silica were demonstrated (i) to have the ability to disperse and stabilize the Ru NPs, giving rise to an improved catalyst stability and reusability and (ii) to provide acidity, which greatly improved GLV selectivity. Cao and coworkers studied the $Au-ZrO_2$ catalyst to catalyze the hydrogenation of LA to GVL by using either FA or butyl formate [94, 95] as the H-donor. The superior performance of Au catalysts for formate-mediated transfer reduction was attributed to the nature of gold nanoparticles, which had been demonstrated to be an excellent catalyst in formate media decomposition. Indeed, they can directly convert a 1:1 aqueous mixture of LA and FA into GVL with high yields. This means that GVL can be synthesized from an integrated process, which starts from the solution of LA and FA derived from HMF synthesis and decomposition in an acidic aqueous medium.

The first example of LA/EL production starting from furfural using the H-transfer reaction was reported by Román-Leshkov and coworkers [90] (Figure 14.3) and received much attention because the catalytic system used in the process was commercial or hemicommercial zeolites. Microporous zirconosilicate molecular sieves (Zr-β) are more active and stable catalysts than bulk zirconia. In this case, furfural was chosen as a starting material because several well-established methods are available for the production of furfural from hemicelluloses. In this process, a combination of solid Lewis and Brønsted acids catalyzed the reaction in the presence of butanol as the H-donor, involving the steps of sequential H-transfer and ring opening. The highest yield of GVL approached 80% in 24 h at 120 °C.

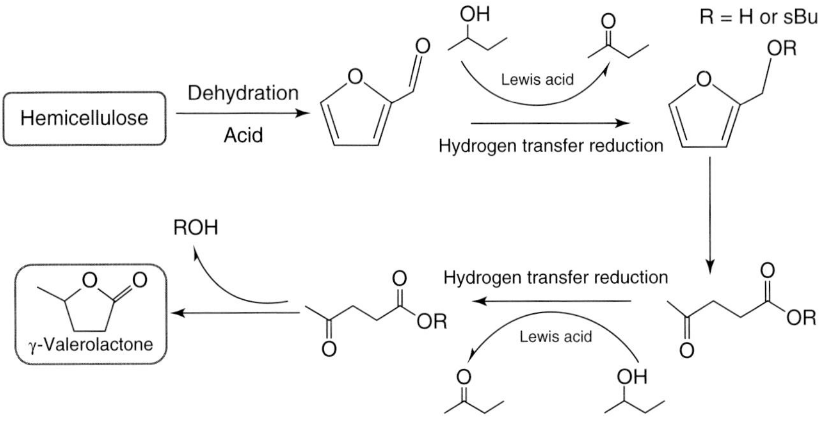

Figure 14.3 Production of GVL via MPV reduction starting from raw biomass.

14.4.2
Furan Derivative Hydrogenation

In recent years, methylfuran (MeF) and 2,5-dimethylfuran (DMF) – which are produced from the selective hydrogenation of biomass-derived furfural and HMF, respectively – have been considered as new-fashioned liquid biofuels for transportation, receiving a great deal of attention from many researchers worldwide. Furfural and HMF are considered the most important platform molecules obtained from the dehydration of C_5 and C_6 sugars, which are present in the hemicellulose and cellulose part of the biomass feedstock (Figure 14.4).

MeF and DMF are claimed to be more suitable compounds for the substitution of the second-generation bioethanol which is obtained from lignocellulose. In fact, there are several drawbacks in the use of ethanol as fuel in present-day

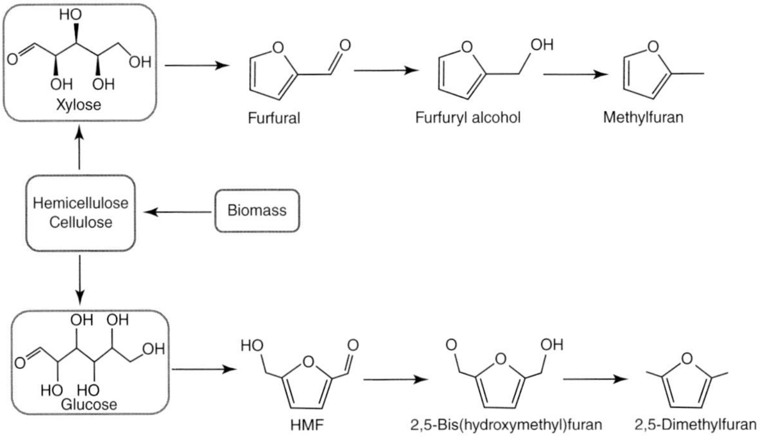

Figure 14.4 Reaction pathways for the conversion of biomass to form methylfuran (MeF) and 2,5-dimethylfuran (DMF).

Table 14.3 Comparison of the fuel properties of DMF versus ethanol and gasoline [103].

Property	DMF	Ethanol	Gasoline
Molecular formula	C_6H_8O	C_2H_6O	$C_{16}-C_{12}$
O/C	0.16	0.5	0
Gravimetric oxygen content (%)	16.7	34.8	0
Density at 20 °C (kg m^{-3})	889.7	790.9	744.6
Water miscibility at 25 °C (g l^{-1})	2.3	Miscible	Immiscible
Boiling point (°C)	93	78	32–200
Energy density (MJ l^{-1})	30	23.4	31
RON	119	110	95.8
Autoignition temperature (°C)	286	423	257

motor vehicles, namely, its higher oxygen content and lower energy density. On the contrary, MeF and DMF are characterized by higher energy density, higher boiling point, and higher octane number; moreover, they are immiscible in water. They even have a higher Research Octane Number (RON) value compared both to ethanol and to gasoline [103], as shown in Table 14.3.

Bioethanol is also slightly corrosive and must be used together with gasoline, increasing water adsorption in the fuel and leading to possible engine damages [121]. On the other hand, DMF can be used either pure – since it also has good lubricant properties – or blended with gasoline or diesel in today's direct-injection spark-ignition engines; blending DMF with gasoline or diesel, however, could increase the ignition time [122]. Considering the excellent properties of these new renewable fuels derived from lignocellulosic materials, many studies have been carried out with a view to optimizing their synthetic procedure.

Many hydrogenation processes using metal-supported catalysts and molecular hydrogen have been developed. At present, these processes are cost competitive with petrochemical technologies; thus the development of new approaches which decrease the number of reaction and purification processes is becoming fundamental. The hydrogen transfer approach could be considered as a significant way to produce biofuels while avoiding molecular hydrogen and, in some cases, also an expensive metal-based catalyst.

Up to now, research efforts have been focused on the H-transfer process for HMF reduction; the reported catalytic systems and the latest research achievements for the selective hydrogenation of HMF into DMF highlight the variety of the used hydrogen donors such as FA and alcohols while discussing the possible reaction mechanism. One of the key challenges for upgrading furans is the product selectivity; a mixture of side-chain ring-hydrogenated products and ring-opened products is often formed.

In 2010, Rauchfuss and Thananatthanachon [96] proposed a mild catalytic system, in which FA was first used as a hydrogen donor for the selective hydrogenation of HMF. When the reaction was carried out in tetrahydrofuran (THF) over the Pd/C catalyst, more than 95% DMF yield with 100% HMF conversion was observed at 70 °C after 15 h. Furthermore, a one-pot process for the synthesis of DMF from fructose was also investigated. In the presence of FA, H_2SO_4, Pd/C, and THF, fructose was initially dehydrated at 150 °C for 2 h, and the generated HMF was subsequently hydrogenated at 70 °C for 15 h, obtaining a 51% DMF yield. It is worth noting that when using FA as catalyst, it is possible to obtain three different reactions, thanks to its peculiar characteristics: it is an acid catalyst for the dehydration of fructose into HMF and a reagent for the deoxygenation of furanylmethanols, as well as a hydrogen donor for the hydrogenation of HMF into 2,5-dihydroxymethylfuran (DHMF).

The use of FA for these processes is very attractive because it can be derived from biomass and can be regenerated by the hydrogenation of formed CO_2. Furthermore, FA presence was reported to inhibit the decarbonylation of molecules, which can occur quite easily at 120 °C in the presence of Pd/C, even at a low reaction time (2–8 h) [97]. In 2012, with a similar catalytic system, De *et al.* [98] studied the conversion of fructose into DMF via HMF formation by using a microwave heating system and FA as the H-donor. The author described the first step, which involved HMF synthesis in the presence of FA and ionic liquid $[DMA]^+[CH_3SO_3]^-$ (DMA = *N,N*-dimethylacetamide). In the following steps, HMF was transformed into DMF by hydrogenation and hydrogenolysis reactions, using FA as the H-donor. When Ru/C was used, the maximum yield of DMF was achieved (32% from fructose and 27% from agar, respectively).

In addition to the reduction of aldehydic groups, some attempts were made to reduce and open the furan ring in order to produce linear diols such as 1,6-hexanediol, which is used extensively in the production of polyesters for polyurethane elastomers, coatings, adhesives, and polymeric plasticizers. Ebitani and coworkers [99] reported the direct synthesis of 1,6-hexanediol from HMF over $Pd/ZrPO_4$ by using FA as the H-donor. Their results indicated that the surface acidity, due the Brønsted acidic sites, is responsible for furan ring opening (C–O bond cleavage), while palladium catalyzes C=O and C=C hydrogenation.

From the standpoint of easy handling and large scale-up, in order to use FA as a hydrogen donor for DMF production, special corrosion-resistant equipment is needed, resulting in an increase of the corresponding costs and thus restricting to a large extent a wide range of applications of FA.

In addition to FA, some alcohols can also be used as the H-donor in HMF reduction. A new approach was reported by Riisager and coworkers [100] for the selective hydrogenation of HMF via hydrogen transfer, in which supercritical methanol was used both as a hydrogen donor and as reaction medium in the presence of a Cu-doped porous metal oxide. The author emphasized that production costs can be reduced, while operation safety can be improved to a certain extent, when methanol is used as a hydrogen donor instead of FA or H_2. However, in the reaction process, the critical temperature of methanol is very high (as high as 300 °C) and the selectivity of DMF is very low. Indeed, only 34% DMF yield can be obtained with 100% HMF conversion at 300 °C after 0.75 h of reaction.

The use of isopropyl alcohol, as a hydrogen donor as well as a reaction medium, was alternatively studied by Vlachos and coworkers [102] for HMF reduction.

When the reaction was conducted over the Ru/C-based catalyst, 100% conversion of HMF and a 81% of yield in DMF were achieved at 190 °C after 6 h. Unfortunately, when the recovered Ru/C was reused in the second cycle, HMF conversion and DMF yield had significantly decreased to 47% and 13%, respectively, showing a considerable deactivation of Ru/C even after its first use: a phenomenon which may be due to the formation of high molecular weight by-products on ruthenium surfaces. More recently, Pd and Rh supported on carbon were used for HMF hydrogenation in the presence of MeOH at 150 °C and 20 bar of H_2 pressure [103]. $ZrCl_2$ was used as cocatalyst since it is claimed to improve DMF selectivity, thanks to the presence of a strong synergistic effect between Pd and Zr; the addition of Zr salt to the reaction mixture has also been fundamental when Ru/C was used as the catalyst. DMF yield reached 39% in 2 h with a conversion of HMF around 75%. However, in the presence of methanol, HMF etherification occurred, forming 5-methoxymethylfurfural; this reaction was catalyzed by Lewis acidic sites which belonged to the used catalyst. Conversely, the use of THF as the solvent led to an 85% yield of DMF in 8 h with a complete conversion of HMF, thus demonstrating the inability of these systems to completely transform HMF into DMF in the presence of an alcoholic solution.

Another stable catalyst based on Pd supported on Fe_2O_3 was prepared and used by Hermans and coworkers [104]. The yield of DMF reached 72% when a continuous-flow reactor at 180 °C was used. Due to its similarity to HMF, furfural can be easily reduced via H-transfer using the same method already reported for this molecule. In fact, in the same paper it is reported that 2-propanol on Fe_2O_3-supported Cu, Ni, and Pd catalysts can carry out the sequential transfer hydrogenation/hydrogenolysis of furfural to MeF. An optimal yield of 57% of furfuryl alcohol and the formation of 10% of MeF were observed in a batch reactor at 180 °C after 7.5 h of reaction with Pd/Fe_2O_3. The remarkable activity of Pd/Fe_2O_3 in both transfer hydrogenation and hydrogenolysis is attributed to a strong metal–support interaction. Using a noble metal catalyst, the active site for H-production seems to be the same used for substrate hydrogenation.

Recently, Chilukuri and coworkers [105] reported a Ru-containing hydrotalcite catalyst for HMF conversion to DMF in 2-propanol, which was found to be both a good solvent and a good H-donor. Unfortunately, acetone was formed as by-product, and it must be separated from the final mixture, thus increasing the cost of the whole process.

An unconventional H-transfer reactant was used by Cavani and coworkers [101], who reported that HMF could be selectively reduced to 2,5-bis(hydroxymethyl)furan (BHMF – 100% yield) over MgO using methanol as both the H-donor and reaction solvent, under mild conditions (160 °C). In the same reaction conditions, furfural was hydrogenated and reached a furfuryl alcohol yield of 97%. MgO was demonstrated to be an excellent catalyst because it was able to activate methanol at low temperatures, which is the key step in this process; moreover, the only by-products formed were gaseous (CO, CO_2, CH_4), thus easily separable from the reaction mixture. The reaction mechanism was elucidated by means of reactivity tests and DFT calculations. The rate-determining

Figure 14.5 Sketch of the transition state for H-transfer in the presence of strong Lewis basic sites suggested in literature.

step of the process is the H-transfer from methanol to furfural carbonyl through a six-center transition state involving aldehyde and alcohol coordinated on a Mg$_3$C site. As a matter of fact, methanol and furfural absorb on the low-coordination defect site Mg$_3$C.

This mechanism is different from what is generally reported for strong Lewis basic sites (Figure 14.5), where the oxygen in the reducing alcohol absorbs on the metal ion and its hydrogen to the oxo-ion state. The carbonyl group forms a hydrogen bond with the surface hydroxyl group, thus leading to the formation of a seven-membered intermediate.

As far as furfural hydrogenation is concerned, a very recent paper has been published by Marchi and coworkers [106] on furfural hydrogenation by means of H-transfer with a copper-based catalyst (Cu/SiO$_2$ and Cu−Mg−Al) prepared by coprecipitation. The authors investigated the effect of the metal−support interaction and concluded that the smallest particles with a stronger interaction with the spinel-like matrix had a greater ability to transfer hydrogen than larger ones. Under optimal conditions (150 °C, 8 h), 100% conversion of furfural to furfuryl alcohol can be obtained. Moreover, ethanol, propanol, and isopropanol have been tested as H-donors, and the last one showed better performances. Vlachos and coworkers [107] studied the effect of the H-donor on Ru/RuO$_x$/C catalysts, using different primary and secondary alcohols. Secondary alcohols have shown greater activity in MeF production, and it has been discovered that alcohol polarity can affect the hydrogenolysis reaction. The optimum MeF yield of 76% was achieved using 2-butanol and 2-pentanol (180 °C, 10 h). Meanwhile, the author indicated that MeF yield rose when alcohol dehydrogenation activity was increased and alcohol polarity was decreased. Another paper has just been published by Vlachos and coworkers [108] with the aim of clarifying the reaction mechanism in MeF formation when Ru/RuO$_x$/C was used as the catalyst and 2-propanol as the H-donor. The reaction mechanism was studied using isotopically labeled chemicals and performing a mass fragmentation analysis. These mechanistic investigations highlighted the presence of an MPV Lewis acid-mediated pathway for furfural hydrogenation. The usual six-membered intermediate was formed, and the β-H of 2-propanol was transferred to the carbonyl carbon atom of the substrate in a concerted step, without transferring hydrogen to the surface of the catalyst, as is usually the case in metal-mediated hydrogenation.

Chen and coworkers [109] prepared a series of metal-supported carbon nitrile (M/CN, where M = Ag, Pt, Au, and Pd) catalysts which could induce the reduction of unsaturated compounds including furfural in water at room temperature, using FA as the hydrogen source. The author indicated that the reusability of such a hybrid catalyst is high due to the strong Mott–Schottky effect between metal nanoparticles and the carbon nitride support. The highest conversion of furfural was 64%, with 99% selectivity toward MeF.

14.4.3
Lignin and Sugar Hydrogenation

Lignin is considered one of the most abundant natural polymers. Among the components of the lignocellulosic biomass, lignin is the only one that contains aromatic structures. It is made up of numerous monomeric building blocks and is characterized by a very complex, three-dimensional structure, which is very difficult to transform and exploit for the production of building blocks. Up to now only hydrolysis, thermal cracking, and oxidative/reductive processes have been carried out, and lignin depolymerization under mild conditions is still the most difficult challenge. The high number of side reactions that can take place in lignin conversion, especially at high temperatures, leads to char formation, thus complicating the valorization process even more. The depolymerization and liquefaction of this material to obtain aromatics often require high temperature and high H_2 pressure. The development of an efficient fractionation procedure with an easy separation step for obtaining highly depolymerized bio-oil is still a topic of great interest. Lignin depolymerization leads to complex mixtures of phenols with high boiling points and high oxygen contents. Accordingly, the hydrodeoxygenation of the nonpyrolytic bio-oil-rendering arenes with low boiling points is needed. Warner *et al.* [110] developed a strategy for lignin depolymerization using supercritical methanol and Al/Mg hydrotalcite-like materials doped with La, Cr, Cu, Mn, and Zn. Copper-based catalysts promoted the formation of the highest content of soluble product, avoiding further recondensation reactions. This catalyst also made possible the formation of bio-oils with similar molecular weights and a significant content of monomeric and low-molecular weight oligomers. Rinaldi and Wang reported the chemical aspects of a hydrogen transfer reaction using Ni-Raney and 2-propanol in the transformation of bio-oils and lignin into arenes [111] and saturated compounds [112]. In order to investigate this mechanism, a model test starting from phenol was studied. The results revealed that phenol can be efficiently transformed into arenes over Ni-Raney using 2-propanol as the H-donor in an acidic medium. This work demonstrated that the H-transfer process using an alcohol as the donor can also be performed on lignin-derived compounds. In addition to lignin, Xiao and coworkers [113] reported on a study conducted on a model molecule: the reduction of vanillin using FA as the H-donor and Pd-supported catalysts. In some way, the field of use of H-transfer was enlarged by reducing 4-hydroxy-3-methoxybenzaldehyde, a common component of pyrolysis oil, into 2-methoxy-4-methylphenol (MMP), which is a potential future biofuel. Fukuoka and coworkers [114] reported on the catalytic hydrogen transfer of cellulose into sugar alcohols (sorbitol and mannitol) with a 47% yield, by using 2-propanol over a Ru/C catalyst. From this standpoint, together with some furan-based building block compounds, some macromolecules can also be reduced via the hydrogen transfer process.

14.4.4
Glycerol Dehydration/Hydrogenation

Glycerol – a biomass-derived C_3 triol available as a by-product of biodiesel fuel production – is considered to be a potential biorefinery feedstock. To date, various processes have been developed to obtain valuable chemicals from glycerol, among which some H-transfer reactions can be found.

The selective hydrogenation of acrolein to allyl alcohol is of great interest from both the industrial and the academic standpoints. However, since the hydrogenation of the C=C double bond is strongly favored over the hydrogenation of the C=O double bond, only in a few cases, allyl alcohol could be highly selectively produced from the hydrogenation of acrolein. Schüth and coworkers [115] reported a gas phase process for the conversion of glycerol to allyl alcohol over an iron oxide catalyst using propanol as the H-transfer donor via hydrogen transfer strategy (Figure 14.6).

Iron oxide was treated as a bifunctional catalyst which combines dehydration and hydrogenolysis. In addition to the pure iron oxide catalytic system, iron phosphate may offer another possibility for carrying out the H-transfer reduction of acrolein. Glycerol dehydration proceeds on the acidic sites of catalysts, while allyl alcohol production is assumed to be catalyzed by the nonacidic sites of catalysts through the hydrogen transfer mechanism. The selectivity in the second reaction step – the transfer hydrogenation to allyl alcohol – was close to 100%, yielding 20–25% of allyl alcohol at 320 °C; in these conditions, only a slight deactivation was observed after 72 h. Due to its low-price and easily synthetic approach, this catalyst has a high potential for scaling up. In an attempt to understand the mechanism involved in the hydrogen transfer reduction steps, several reactions between different alcohols and unsaturated aldehydes were carried out. The authors indicated that, indeed, iron oxide can process hydrogen transfer reaction, as was reported for MgO, ZrO_2, and $MgAlO_2$. Nevertheless, the detailed H-transfer mechanism was not clarified. It is possible that both the intermediates with hydroxyl groups and glycerol play the role of H-donors.

Doped iron oxides may be another possibility for carrying out the H-transfer reduction of acrolein. Tago and coworkers [116] reported that iron oxide-based catalytic systems including FeO_x/ZrO_2 mixed oxide- and alkali metal-supported FeO_x/ZrO_2 catalysts reduced acrolein to produce allyl alcohol. The highest yield of allylic alcohol was obtained over a $K/Al_2O_3 - ZrO_2 - FeO_x$ catalyst system.

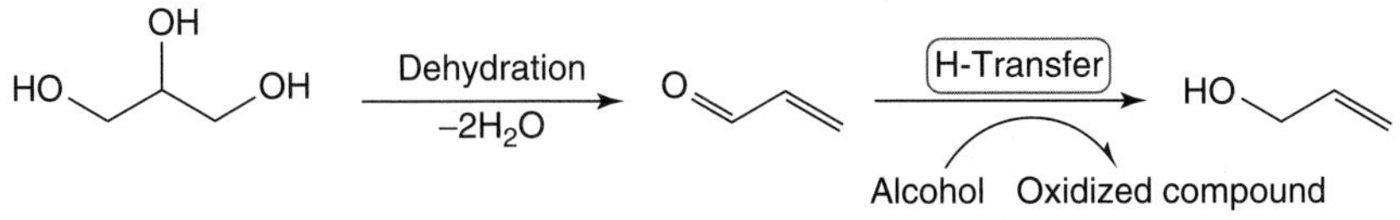

Figure 14.6 Possible reaction pathways from glycerol to allyl alcohol via H-transfer process.

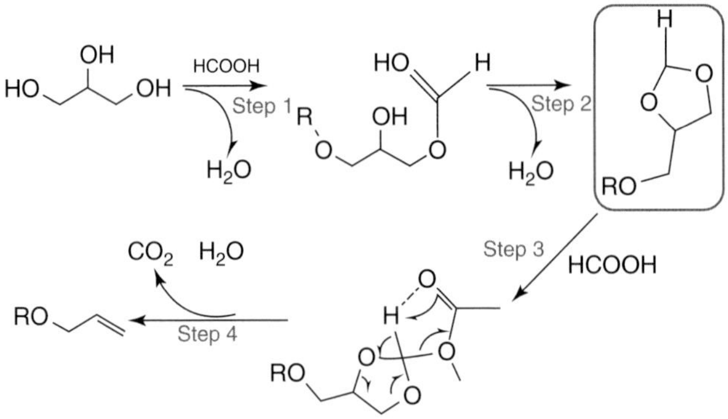

Figure 14.7 Mechanism of H-transfer using formic acid as H-donor and self-catalyst.

Further careful studies, using different premixed glycerol and H-donors (with acetol, acrolein, acetic acid, or FA) in an aqueous solution, were conducted. If acetol or FA were co-fed into the system, the yield of allyl alcohol increased. The authors concluded that the hydrogen transfer mechanism should take place between glycerol and either hydrogen atoms derived from FA formed during the reaction or active hydrogen species produced from the decomposition of H_2O by ZrO_2. Similarly, Ellman and coworkers [117] investigated the mechanisms involved in glycerol transfer toward allyl alcohol via H-transfer steps. With the aim of simplifying the system, FA was chosen as acid catalyst and H-donor to selectively produce allyl alcohol. The reaction mechanism is reported in Figure 14.7. During steps 1 and 2, FA acted as the acid catalyst for dehydration, whereas in steps 3 and 4 it acted as the H-donor. Thus this method substantially improves the yield and selectivity of the process.

Asakura and coworkers [123] reported the use of a palladium-supported catalyst to process glycerol hydrogenolysis via hydrogen transfer, where 2-propanol acted as both solvent and H-donor; moreover, both pure Pd- and bimetallic-supported catalysts (Pd–Co, Pd–Fe) were also tested. The relationship between the physical–chemical properties of palladium-based catalysts and their catalytic performance was elucidated. The electronic properties of Pd were modified when Co or Fe was introduced to form bimetallic catalysts. The author emphasized that the interaction between Pd and the other metal played an important role. As in other reported results, the conversion of glycerol involves glycerol dehydration to 1-hydroxyacetone and thus the subsequent hydrogenation of 1-hydroxyacetone to propylene glycol. H-transfer processes provide alternative H resources, but the main pathway toward the desired compounds was not influenced by the H-donor.

14.5
Industrial Applications of the MPV Reaction

Nowadays, the H-transfer reaction is not of significant preparative importance for the industry, and from the industrial standpoint, alcohol activation is the main concern. Nevertheless, if the catalytic system can provide alcohol activation with *in situ* H-transfer under mild conditions, these production processes may be applicable.

As for now, the MPV reduction is used to synthesize allyl alcohol from acrolein [124], and some industrial applications have been found in the fragrance and pharmaceutical industry. As an example, a patent assigned to Georgia Tech Research Corporation and American Pacific Corporation [125] describes the reduction of complex organic molecules, containing aldehydic or ketonic groups, to their corresponding alcohols. The described process uses $Al(OtBu)_3$ and $Al(Oi\text{-}Pr)_3$ as catalysts and makes possible the production of fine chemicals, basically used in the pharmaceutical industry.

Hoffmann-La Roche patented a process for the stereoselective reduction of carbonyl compounds via the MPV reaction [126]. The preferred carbonyl compounds were ketones with prochiral carbonyl carbon. The reaction was conducted under mild conditions (e.g., 50 °C or less), with aluminum isopropoxide, producing an excess of the desired optically active chiral alcohol. Mild conditions preserve the optical orientation of other asymmetric centers in the carbonyl compound starting materials.

The possible production of GVL from biomass derivatives by H-transfer reaction has also attracted the attention of many researchers. A first patent claims GVL production from fructose, HMF, or furfuryl alcohol over a solid base catalyst through the conversion of LA esters. GVL was finally obtained using Ni-Raney and a secondary alcohol (isopropyl propanol, 2-butanol, cyclohexanol) as hydrogen source, in inert gas [127]. Another patent for the production of GVL from EL was recently published [128]. Methanol, ethanol, isopropanol, and 1-butanol have been used as H-transfer reactants, while several different metal oxide catalysts were tested (Al_2O_3, MgO, ZrO_2 La_2O_3, and some Al_2O_3/MgO mixed oxides with different metal ratios).

On a small scale it may be strongly advantageous to perform transfer hydrogenation instead of performing hydrogenation with molecular hydrogen. As an example, the patent reported by Cheng *et al.* [129] refers to a method for the continuous catalytic reduction of methylallyl aldehyde through the recycling of aluminum isopropoxide. According to this method, the continuous reduction of methylallyl aldehyde is performed through the recycling of the catalyst and has the advantages of keeping the conversion rates of repeated catalytic reduction reactions above 85% and the selectivity above 90%. Reaction conditions were mild, the rate of the reduction reaction was high, and the aluminum isopropoxide catalyst can be recycled.

Very recently, a method for synthesizing sugar alcohols such as xylitol, mannitol, and sorbitol through the H-transfer process was patented by the Agency for Science, Technology and Research of Singapore [130]. According to the patent, the process uses diluted H_2SO_4, isopropanol as acidic solvent, and an alcoholic reducing agent to hydrolyze the biomass and produce sugar alcohols from cellulose and hemicellulose. The reported catalysts are mainly Ru-based materials supported on C or metal oxides (TiO_2, Al_2O_3, CeO_2).

14.6
Conclusions

Over the past decade, in the field of biomass conversion, great efforts have been devoted to the development of reductive processes capable of leading to the sustainable production of feedstock for the chemical industry.

Metal-catalyzed hydrogenation using molecular H_2 is one of the most used reactions for these transformations. Nevertheless, the hydrogenation of biomass-derived platform molecules using the H-transfer reaction could be an alternative approach for biomass valorization aiming for the development of more sustainable processes, which do not need high H_2 pressure and costly noble metal catalysts. Indeed, the H-transfer reaction may avoid any possible hazards connected with the presence of a gas with low molecular weight and high diffusivity in a chemical plant, since alcohols easily obtained from renewable resources may be used as a source of hydrogen. Various efforts have been undertaken in this sense in recent years. Indeed, using these processes, it has been possible to convert some of the main bio-based platform molecules such as LA, furfural, HMF, lignin, and glycerol into fuels, fuel additives, and chemicals.

By using H-transfer processes, LA can be converted into GVL, which is a green solvent and an intermediate for fuels (MTHF) and chemicals (e.g., pentenoic acid and pentanoic acid). MeF and DMF can be synthesized, respectively, from furfural and HMF, while the synthesis of allyl alcohol can be obtained from glycerol. Moreover, lignin depolymerization and deoxygenation using hydrogen transfer reaction were reported, and several milder processes for lignin valorization have been studied.

On the catalyst development front, it has been shown that heterogeneous Lewis acid, Lewis base, and metal-supported catalysts are effective for highly selective H-transfer reactions. Some different reaction mechanisms have been reported. The first one, accepted for most of the heterogeneous catalysts, especially for those which are characterized by Lewis acidic sites such as Al^{3+}, Zr^{4+}, Sn^{4+}, is similar to the mechanism observed with homogeneous alkoxide, where both the alcohol (donor) and the carbonylic compound (acceptor) coordinate to the same metal center, forming a six-membered ring intermediate. A different pathway can be observed for amphoteric oxides; in this case, both acidic and basic sites are often involved, leading to the heterolytic dissociation of alcohol, followed by the transfer of the α-hydrogen of alkoxide to the absorbed carbonylic compound. One drawback of the use of the H-transfer reaction is the reversibility of the process, leading to the control of selectivity driven by the thermodynamic properties of products and intermediates. Nevertheless, this disadvantage can be overcome by the use of a formate or methanol, since, in this case, the reaction is considered to be irreversible, with a loss of CO_2.

Acknowledgments

The University of Bologna is acknowledged for the financial support through the FARB Project "Catalytic transformation of biomass-derived materials into high added-value chemicals," 2014–2015. This work was cofunded through a SINCHEM Grant. SINCHEM is a Joint Doctorate program selected under the Erasmus Mundus Action 1 Programme (FPA 2013-0037).

References

1. Huber, G.W., Iborra, S., and Corma, A. (2006) *Chem. Rev.*, **106**, 4044–4098.
2. Azadi, P., Inderwildi, O.R., Farnood, R., and King, D.A. (2013) *Renewable Sustainable Energy Rev.*, **21**, 506–523.
3. Dusselier, M., Mascal, M., and Sels, B. (2014) in *Selective Catalysis for Renewable Feedstocks and Chemicals* (ed. K.M. Nicholas), Springer International Publishing, pp. 85–125.
4. Nakagawa, Y., Tamura, M., and Tomishige, K. (2013) *ACS Catal.*, **3**, 2655–2668.
5. Assary, R.S., Curtiss, L.A., and Dumesic, J.A. (2013) *ACS Catal.*, **3**, 2694–2704.
6. Williams, E.D., Krieger, K.A., and Day, A.R. (1953) *J. Am. Chem. Soc.*, **75**, 2404–2407.
7. de Graauw, C.F., Peters, J.A., van Bekkum, H., and Huskens, J. (1994) *Synthesis*, **10**, 1007–1017.
8. Ono, Y. and Hattori, H. (2011) *Solid Base Catalysis*, Springer, Berlin, Heidelberg, pp. 11–68.
9. Laue, T. and Plagens, A. (1994) *Namen- und Schlagwort-Reaktionen der Organischen Chemie*, Vieweg Teubner Verlag, pp. 221–223.
10. Ponndorf, W. (1926) *Angew. Chem.*, **39**, 138–143.
11. Palmer, M.J. and Wills, M. (1999) *Tetrahedron: Asymmetry*, **10**, 2045–2061.
12. Noyori, R., Yamakawa, M., and Hashiguchi, S. (2001) *J. Org. Chem.*, **66**, 7931–7944.
13. Sasson, Y. and Blum, J. (1975) *J. Org. Chem.*, **40**, 1887–1896.
14. Cohen, R., Graves, C.R., Nguyen, S.T., Martin, J.M.L., and Ratner, M.A. (2004) *J. Am. Chem. Soc.*, **126** (45), 14796–14803.
15. Ishii, Y., Nakano, T., Inada, A., Kishigami, Y., Sakurai, K., and Ogawa, M. (1986) *J. Org. Chem.*, **51**, 240–242.
16. Pasini, T., Solinas, G., Zanotti, V., Albonetti, S., Cavani, F., Vaccari, A., Mazzanti, A., Ranieri, S., and Mazzoni, R. (2014) *Dalton Trans.*, **43**, 10224–10234.
17. Enthaler, S., Junge, K., and Beller, M. (2008) *Angew. Chem. Int. Ed.*, **47**, 3317–3321.
18. Bolm, C., Legros, J., Le Paih, J., and Zani, L. (2004) *Chem. Rev.*, **104**, 6217–6254.
19. Correa, A., Garcia Mancheno, O., and Bolm, C. (2008) *Chem. Soc. Rev.*, **37**, 1108–1117.
20. Bullock, R.M. (2007) *Angew. Chem. Int. Ed.*, **46**, 7360–7363.
21. Ekström, J., Wettergren, J., and Adolfsson, H. (2007) *Adv. Synth. Catal.*, **349**, 1609–1613.
22. Sedelmeier, J., Ley, S.V., and Baxendale, I.R. (2009) *Green Chem.*, **11**, 683–685.
23. Polshettiwar, V. and Varma, R.S. (2009) *Green Chem.*, **11**, 1313–1316.
24. Geeta Pamar, M., Govender, P., Muthusamy, K., Krause, R.W.M., and Nanjundaswamy, H.M. (2013) *Orient. J. Chem.*, **29** (3), 969–974.
25. Urbano, F.J., Aramendía, M.A., Marinas, A., and Marinas, J.M. (2009) *J. Catal.*, **268**, 79–88.
26. Axpuac, S., Aramendía, M.A., Hidalgo-Carrillo, J., Marinas, A., Marinas, J.M., Montes-Jiménez, V., Urbano, F.J., and Borau, V. (2012) *Catal. Today*, **187**, 183–190.
27. Zhu, Y., Jaenicke, S., and Chuah, G.K. (2003) *J. Catal.*, **218**, 396–404.
28. Zhang, B., Tang, M., Yuan, J., and Wu, L. (2012) *Chin. J. Catal.*, **33**, 914–922.
29. Zhu, Y., Chuah, G., and Jaenicke, S. (2004) *J. Catal.*, **227**, 1–10.
30. Zhu, Y., Chuah, G., and Jaenicke, S. (2006) *J. Catal.*, **241**, 25–33.
31. Ramanathan, A., Castro Villalobos, M.C., Kwakernaak, C., Telalovic, S., and

Hanefeld, U. (2008) *Chem. Eur. J.*, **14**, 961–972.

32. Klomp, D., Maschmeyer, T., Hanefeld, U., and Peters, J.A. (2004) *Chem. Eur. J.*, **10**, 2088–2093.

33. Bortnovsky, O., Sobalík, Z., Wichterlová, B., and Bastl, Z. (2002) *J. Catal.*, **210**, 171–182.

34. Corma, A., Domine, M.E., and Valencia, S. (2003) *J. Catal.*, **215**, 294–304.

35. Carre, S., Gnep, N.S., Revel, R., and Magnoux, P. (2008) *Appl. Catal., A: Gen.*, **348**, 71–78.

36. Boronat, M., Corma, A., Renz, M., and Viruela, P.M. (2006) *Chem. Eur. J.*, **12**, 7067–7077.

37. Samuel, P.P., Shylesh, S., and Singh, A.P. (2007) *J. Mol. Catal. A: Chem.*, **266**, 11–20.

38. Van der Waal, C., Kunkeler, P.J., Tan, K., and van Bekkum, H. (1998) *J. Catal.*, **173**, 74–83.

39. Armedia, M.A., Borau, V., and Jimenez, C. (2001) *Appl. Catal., A*, **206**, 95–101.

40. Armedia, M.A., Borau, V., and Jimenez, C. (2001) *J. Mol. Catal. A*, **171**, 153–158.

41. Armedia, M.A., Borau, V., Jimenez, C., Marinas, J.M., Ruiz, J.R., and Urbano, F.J. (2003) *Appl. Catal., A*, **255**, 301–308.

42. Armedia, M.A., Borau, V., and Jimenez, C. (2003) *Appl. Catal., A*, **244**, 207–215.

43. Lopez, J., Valente, J.S., Clacens, J.M., and Figueras, F. (2002) *J. Catal.*, **208**, 30–37.

44. Armedia, M.A., Borau, V., Jimenez, C., Marinas, J.M., and Romero, F.J. (1999) *J. Catal.*, **183**, 119–127.

45. Ruiz, J.R., Jimenez-Sanchidrian, C., and Hidalgo, J.M. (2007) *Catal. Commun.*, **8**, 1036–1040.

46. Szollosi, G. and Bartok, M. (1999) *J. Mol. Catal. A: Chem.*, 265–273.

47. Szollosi, G. and Bartok, M. (1998) *Appl. Catal., A*, **169**, 263–269.

48. Szollosi, G. and Barrok, M. (1999) *J. Mol. Struct.*, **483**, 13–17.

49. Ivanov, V., Bachelier, J., Audry, F., and Lavalley, L. (1994) *J. Mol. Catal.*, **91**, 45–59.

50. Bartley, J.K., Xu, C., Lloyd, R., Enache, D.I., Knight, D.W., and Hutchings, G.J. (2012) *Appl. Catal., B*, **128**, 31–38.

51. Armedia, M.A., Borau, V., Jimenez, C., Marinas, J.M., Ruiz, J.R., and Urbano, F.J. (2001) *J. Colloid Interface Sci.*, **238**, 385–389.

52. Armedia, M.A., Borau, V., Jimenez, C., Marinas, J.M., Ruiz, J.R., and Urbano, F.J. (2003) *Appl. Catal., A*, **249**, 1–9.

53. Glinski, G. and Gadomska, A. (1998) *React. Kinet. Catal. Lett.*, **65**, 121–129.

54. Glinski, G. (2001) *React. Kinet. Catal. Lett.*, **72**, 133–137.

55. Glinski, G. (2009) *React. Kinet. Catal. Lett.*, **97**, 275–279.

56. Jyothi, T.M., Raja, T., and Rao, B.S. (2001) *J. Mol. Catal. A: Chem.*, **168**, 187–191.

57. Ruiz, J.R., Jimenez-Sanchidrian, C., Hidalgo, J.M., and Marinas, J.M. (2006) *J. Mol. Catal. A: Chem.*, **246**, 190–194.

58. Armedia, M.A., Borau, V., Jimenez, C., Marinas, J.M., Porras, A., and Urbano, F.J. (1996) *J. Catal.*, **161**, 829–838.

59. Kijenski, J., Glinski, M., and Czarnecki, J. (1991) *J. Chem. Soc., Perkin Trans. 2*, 1695–1698.

60. Kijenski, J., Glinski, M., and Quiroz, C.A. (1997) *Appl. Catal., A*, **150**, 77–84.

61. Jyothi, T.M., Raja, T., Sreekumar, K., Talawar, M.B., and Rao, B.S. (2000) *J. Mol. Catal. A: Chem.*, **157**, 193–198.

62. Ramos, J.J., Díez, V.K., Ferretti, C.A., Torresi, P.A., Apesteguía, C.R., and Di Cosimo, J.I. (2011) *Catal. Today*, **172**, 41–47.

63. Gliński, M. (2008) *Appl. Catal., A: Gen.*, **349**, 133–139.

64. Reddy, K.H.P., Anand, N., Venkateswarlu, V., Rao, K.S.R., and Burri, D.R. (2012) *J. Mol. Catal. A: Chem.*, **355**, 180–185.

65. Nagaraja, B.M., Padmasri, A.H., Seetharamulu, P., Hari Prasad Reddy, K., David Raju, B., and Rama Rao, K.S. (2007) *J. Mol. Catal. A: Chem.*, **278**, 29–37.

66. Stamatis, N., Goundani, K., Vakros, J., Bourikas, K., and Kordulis, C. (2007) *Appl. Catal., A: Gen.*, **325**, 322–327.

67. Alonso, F., Riente, P., Rodríguez-Reinoso, F., Ruiz-Martínez,

J., Sepúlveda-Escribano, A., and Yus, M. (2008) *J. Catal.*, **260**, 113–118.

68. Alonso, F., Riente, P., Rodríguez-Reinoso, F., Ruiz-Martínez, J., Sepúlveda-Escribano, A., and Yus, M. (2009) *ChemCatChem*, **1**, 75–77.

69. Su, F.-Z., He, L., Ni, J., Cao, Y., He, H.-Y., and Fan, K.-N. (2008) *Chem. Commun.*, 3531–3533.

70. He, L., Ni, J., Wang, L.-C., Yu, F.-J., Cao, Y., He, H.-Y., and Fan, K.-N. (2009) *Chem. Eur. J.*, **15**, 11833–11836.

71. Yu, J.-Q., Wu, H.-C., Ramarao, C., Spencer, J.B., and Ley, S.V. (2003) *Chem. Commun.*, 678–679.

72. Hammond, C., Schümperli, M.T., Conrad, S., and Hermans, I. (2013) *ChemCatChem*, **5**, 2983–2990.

73. Selvam, P., Sonavane, S.U., Mohapatra, S.K., and Jayaram, R.V. (2004) *Adv. Synth. Catal.*, **346**, 542–544.

74. Yoshida, K., Gonzalez-Arellano, C., Luque, R., and Gai, P.L. (2010) *Appl. Catal., A: Gen.*, **379**, 38–44.

75. Subramanian, T. and Pitchumani, K. (2012) *Catal. Sci. Technol.*, **2**, 296–300.

76. Neelakandeswari, N., Sangami, G., Emayavaramban, P., Ganesh Babu, S., Karvembu, R., and Dharmaraj, N. (2012) *J. Mol. Catal. A: Chem.*, **356**, 90–99.

77. Polshettiwar, V., Baruwati, B., and Varma, R.S. (2009) *Green Chem.*, **11**, 127–131.

78. Yamaguchi, K., Koike, T., Kim, J.W., Ogasawara, Y., and Mizuno, N. (2008) *Chem. Eur. J.*, **14**, 11480–11487.

79. Kim, J.W., Koike, T., Kotani, M., Yamaguchi, K., and Mizuno, N. (2008) *Chem. Eur. J.*, **14**, 4104–4109.

80. Gopiraman, M., Babu, S.G., Karvembu, R., and Kim, I.S. (2014) *Appl. Catal., A: Gen.*, **484**, 84–96.

81. Gao, Y., Jaenicke, S., and Chuah, G.-K. (2014) *Appl. Catal., A: Gen.*, **484**, 51–58.

82. Johnstone, R.A.W. and Wilby, A.H. (1985) *Chem. Rev.*, **85**, 120–170.

83. Radhakrishan, R., Do, D.M., Jaenicke, S., Sasson, Y., and Chuah, G.-K. (2011) *ACS Catal.*, **1**, 1631–1636.

84. Ono, Y. and Hattori, H. (2011) *Solid Base Catalysis*, Springer, Berlin, Heidelberg, pp. 308–315.

85. Boronat, M., Corma, A., and Renz, M. (2006) *J. Phys. Chem. B*, **110**, 21168–21174.

86. Chia, M. and Dumesic, J.A. (2011) *Chem. Commun.*, **47**, 12233–12235.

87. Tang, X., Hu, L., Sun, Y., Zhao, G., Hao, W., and Lin, L. (2013) *RSC Adv.*, **3**, 10277–10284.

88. Tang, X., Chen, H., Hu, L., Hao, W., Sun, Y., Zeng, X., Lin, L., and Liu, S. (2014) *Appl. Catal., B: Environ.*, **147**, 827–834.

89. Wang, J., Jaenicke, S., and Chuah, G.-K. (2014) *RSC Adv.*, **4**, 13481–13489.

90. Bui, L., Luo, H., Gunther, W.R., and Román-Leshkov, Y. (2013) *Angew. Chem. Int. Ed.*, **52**, 8022–8025.

91. Song, J., Wu, L., Zhou, B., Zhou, H., Fan, H., Yang, Y., Meng, Q., and Han, B. (2015) *Green Chem.*, **17**, 1626–1632.

92. Yang, Z., Huang, Y.-B., Guo, Q.-X., and Fu, Y. (2013) *Chem. Commun.*, **49**, 5328–5330.

93. Kuwahara, Y., Kaburagi, W., and Fujitani, T. (2014) *RSC Adv.*, **4**, 45848–45855.

94. Du, X.-L., He, L., Zhao, S., Liu, Y.-M., Cao, Y., He, H.-Y., and Fan, K.-N. (2011) *Angew. Chem. Int. Ed.*, **50**, 7815–7819.

95. Du, X.-L., Bi, Q.-Y., Liu, Y.-M., Cao, Y., and Fan, K.-N. (2011) *ChemSusChem*, **4**, 1838–1843.

96. Thananatthanachon, T. and Rauchfuss, T.B. (2010) *Angew. Chem. Int. Ed.*, **49**, 6616–6618.

97. Mitra, J., Zhou, X., and Rauchfuss, T. (2015) *Green Chem.*, **17**, 307–313.

98. De, S., Dutta, S., and Saha, B. (2012) *ChemSusChem*, **5**, 1826–1833.

99. Tuteja, J., Choudhary, H., Nishimura, S., and Ebitani, K. (2014) *ChemSusChem*, **7**, 96–100.

100. Hansen, T.S., Barta, K., Anastas, P.T., Ford, P.C., and Riisager, A. (2012) *Green Chem.*, **14**, 2457–2461.

101. Pasini, T., Lolli, A., Albonetti, S., Cavani, F., and Mella, M. (2014) *J. Catal.*, **317**, 206–219.

102. Jae, J., Zheng, W., Lobo, R.F., and Vlachos, D.G. (2013) *ChemSusChem*, **6**, 1158–1162.

103. Saha, B. and Abu-Omar, M.M. (2015) *ChemSusChem*, **8**, 1133–1142.

104. Scholz, D., Aellig, C., and Hermans, I. (2014) *ChemSusChem*, **7**, 268–275.

105. Nagpure, A.S., Venugopal, A.K., Lucas, N., Manikandan, M., Thirumalaiswamy, R., and Chilukuri, S. (2015) *Catal. Sci. Technol.*, **5**, 1463–1472.

106. Villaverde, M.M., Garetto, T.F., and Marchi, A.J. (2015) *Catal. Commun.*, **58**, 6–10.

107. Panagiotopoulou, P., Martin, N., and Vlachos, D.G. (2014) *J. Mol. Catal. A: Chem.*, **392**, 223–228.

108. Gilkey, M.J., Panagiotopoulou, P., Mironenko, A.V., Jenness, G.R., Vlachos, D.G., and Xu, B. (2015) *ACS Catal.* **5**(7), pp 3988–3994.

109. Gong, L.-H., Cai, Y.-Y., Li, X.-H., Zhang, Y.-N., Su, J., and Chen, J.-S. (2014) *Green Chem.*, **16**, 3746–3751.

110. Warner, G., Hansen, T.S., Riisager, A., Beach, E.S., Barta, K., and Anastas, P.T. (2014) *Bioresour. Technol.*, **161**, 78–83.

111. Wang, X. and Rinaldi, R. (2013) *Angew. Chem. Int. Ed.*, **52**, 11499–11503.

112. Wang, X. and Rinaldi, R. (2012) *Energy Environ. Sci.*, **5**, 8244–8260.

113. Wang, L., Zhang, B., Meng, X., Su, D.S., and Xiao, F.-S. (2014) *ChemSusChem*, **7**, 1537–1541.

114. Kobayashi, H., Matsuhashi, H., Komanoya, T., Hara, K., and Fukuoka, A. (2011) *Chem. Commun.*, **47**, 2366–2368.

115. Liu, Y., Tuysuz, H., Jia, C.-J., Schwickardi, M., Rinaldi, R., Lu, A.-H., Schmidt, W., and Schüth, F. (2010) *Chem. Commun.*, **46**, 1238–1240.

116. Konaka, A., Tago, T., Yoshikawa, T., Nakamura, A., and Masuda, T. (2014) *Appl. Catal., B: Environ.*, **146**, 267–273.

117. Arceo, E., Marsden, P., Bergman, R.G., and Ellman, J.A. (2009) *Chem. Commun.*, 3357–3359.

118. Alonso, D.M., Wettstein, S.G., and Dumesic, J.A. (2013) *Green Chem.*, **15**, 584–595.

119. Lange, J.-P., Price, R., Ayoub, P.M., Louis, J., Petrus, L., Clarke, L., and Gosselink, H. (2010) *Angew. Chem. Int. Ed.*, **49**, 4479–4483.

120. Kuwahara, Y., Magatani, Y., and Yamashita, H. (2015) *Catal. Today*, **258**, 262–269.

121. Climent, M.J., Corma, A., and Iborra, S. (2014) *Green Chem.*, **16**, 516–547.

122. Quian, Y., Zhu, L., Wang, Y., and Lu, X. (2015) *Renew. Sust. Energ. Rev.*, **41**, 633–646.

123. Mauriello, F., Ariga, H., Musolino, M.G., Pietropaolo, R., Takakusagi, S., and Asakura, K. (2015) *Appl. Catal., B: Environ.*, **166–167**, 121–131.

124. Sheldon, R.A., Arends, I., and Hanefeld, U. (2007) *Green Chemistry and Catalysis*, John Wiley & Sons, Inc..

125. Liotta, C.L., Pollet, P., Fisher, K.K., Dubay, W., and Stringer, J. (2013) Reduction of aldehydes and ketones to alcohols. US Patent US 2013/0096317 A1, assigned to Georgia Tech Research Corporation and American Pacific Corporation.

126. Brown, J.D., Cain, R.O., and Kopach, M.E. (1999) Stereoselective reduction of carbonyl compounds. EP Patent 0963972A2, assigned to F. Hoffmann-La Roche AG (CH).

127. Fu, R., Huang, Y., Yang, Z., and Guo, Q. (2013) Method for preparing gamma-valerolactone with high selectivity under mild condition. CN Patent 103012334, assigned to University of Science and Technology of China.

128. Song, Y., Lin, L., and Tang, X. (2013) Method for preparing gamma-valerolactone by transferring and hydrogenating levulinic acid and ester thereof. CN Patent 103497168 A, assigned to Xiamen University.

129. Cheng, X., Zhang, J., Li, F., Zhang, Q., and Qi, W. (2013) Method for continuous catalytic reduction of methylallyl aldehyde through recycling of aluminum isopropoxide. CN Patent 103664526, assigned to Shandong Yidali Chemical Industry Co. LTD.

130. Yi, G. and Zhang, Y. (2015) Methods for synthesizing a sugar alcohol. US Patent 20150057470, assigned to Technology and Research of Singapore Agency for Science (SG).

15
Selective Oxidation of Biomass Constitutive Polymers to Valuable Platform Molecules and Chemicals

Mihaela Florea and Vasile I. Parvulescu

15.1
Introduction

Different oxygen functionalities provide different reactivities and properties to organic molecules. Therefore their presence strongly correlates with the applications of complex oxygenate compounds in very diverse domains like commodity chemicals, pharmaceuticals, agrochemicals, polymers and other materials, and so on. Then, continuous growing in the population that is expected to be higher than 9 billion in 2050 [1] requires an increase of the production of complex molecules containing oxygen but also a diversification of their structures and properties.

For the moment, oil still represents the main raw material producing these compounds. However, its depletion and composition recommends more appropriate sources. For this reason, the identification of alternative sources and the development of new sustainable technologies based on the composition of these new resources have become a major challenge for the scientific community [2]. In this effort, the use of renewable resources, such as biomass, may represent an efficient alternative since these complex structures already have a high oxygen content that is associated to multiple functionalities.

However, this approach is not at all simple. It must firstly avoid any ethical conflict with agriculture/alimentation taking into account the predicted enormous growing in the population. Then, as it was shown by the new concept on metrics for the production of chemicals from renewable biomass, it should prove not only an economic advantage but also energetic and environmental efficiencies [3]. The selection of a particular biomass raw material is very important. Besides the previous economic criteria, its processing should correspond to significantly lower capital requirements thus enabling production at modest annual capacities [3].

Based on these, the biomass valorization becomes a hot investigation topic in the past decade, and there are many groups that dedicated a consistent effort in the identification of rational routes to capitalize this raw material into important platform molecules [4]. However, many of these studies were dedicated to the transformation of either cellulose/hemicellulose or of the glucose, as hydrolysis product, into chemical products and fuels and identification of proper catalysts

Chemicals and Fuels from Bio-Based Building Blocks, First Edition.
Edited by Fabrizio Cavani, Stefania Albonetti, Francesco Basile, and Alessandro Gandini.
© 2016 Wiley-VCH Verlag GmbH & Co. KGaA. Published 2016 by Wiley-VCH Verlag GmbH & Co. KGaA.

for such transformations [2e, 5]. As expected, some of these results were already analyzed in critical reviews and chapters [6].

On the other side, the direct selective oxidation of biomass into valuable chemicals and materials is also a very attractive subject. The oxidation of these structures is advantageous due to the presence of multiple reactive positions. In the current practice, there are many chemical products and commodities for which an intermediate selective oxidation process represents a critical step in the manufacture process [7]. The oxidation is the second largest process after polymerization and contributes with ~30% to total production in the chemical industry [8]. Moreover, at least half of the products obtained using catalytic processes are obtained by selective oxidation, and nearly all the monomers used in the production of fibers and plastics are obtained in this way [9].

Based on these, the present chapter proposes a critical analysis of the reports to date concerning the direct selective oxidation of the constitutive parts of the biomass, that is, cellulose/hemicellulose, starch, and lignin with capitalization in platform molecules and valuable products. However, since these processes should follow environmentally friendly versatile oxidation routes and methods, the chapter will analyze the role of the catalyst and the phase in which these processes are carried out. Very important for this process is also the E factor. The selected oxidation catalytic systems should stop the oxidation to the desired product and to avoid the total oxidation to CO/CO_2 under the applied reaction conditions.

15.2
Selective Oxidation of Cellulose

Cellulose is the main component of the plant cell walls and as a consequence the most abundant biomass source. Currently, it represents the basic building block for many textiles and papers [10]. Its structure corresponds to a natural polymer of linked sugar molecules (Figure 15.1).

Based on this, the selective transformation of cellulose into platform chemicals or directly in various chemicals, under mild conditions, is very appealing [11]. Furthermore, the possibility to achieve these transformations in only one step makes the applications even more attractive since it eliminates supplementary separations and purifications. The role of the catalyst and the polarity of the reaction medium in this approach is essential. Bi- or multifunctional catalysts may

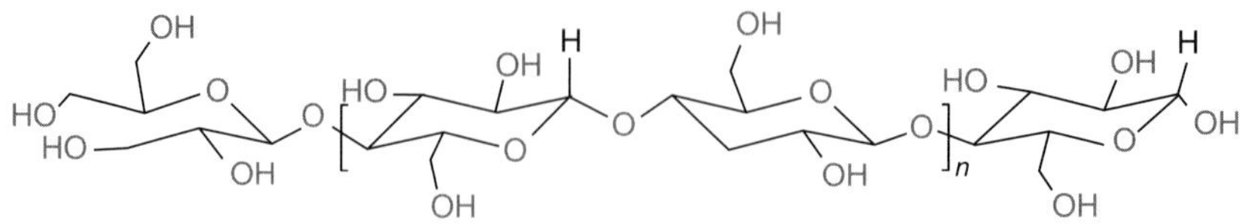

Figure 15.1 Cellulose structure.

accomplish several attributes, while an adequate polarity may control the reaction selectivity. Working in water or biphasic media is much preferable. In order to allow a good reactant accessibility to the catalyst active sites, the porous texture is also important when the catalysts are heterogeneous [5c, 12]. Scheme 15.1 depicts the general routes to obtain valuable chemical products via direct selective oxidation of cellulose. Such oxidations were achieved using both homogeneous and heterogeneous catalysts.

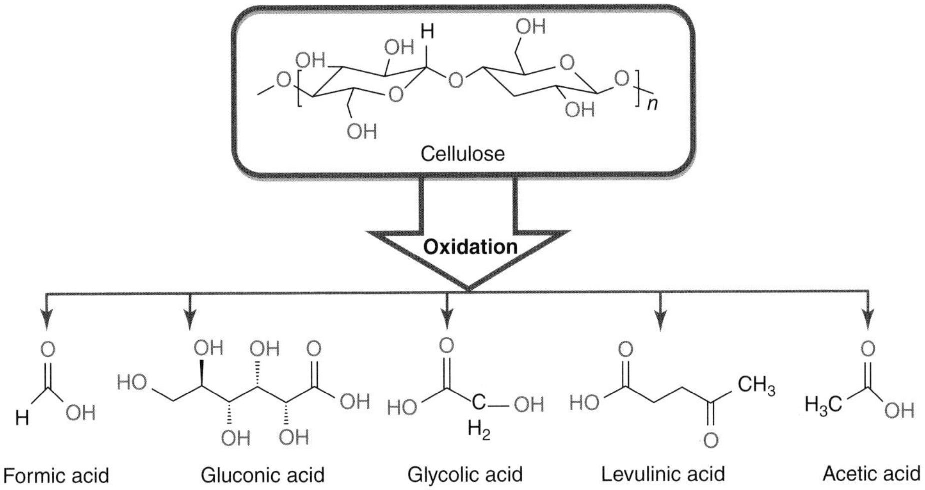

Scheme 15.1 General routes to valuable chemical products via direct selective oxidation of cellulose.

15.2.1
Platform Molecules Obtained via Selective Oxidation of Cellulose

15.2.1.1 Formic Acid

Formic acid (FA) is an important chemical with a high demand in several industries like agriculture, bulk chemistry, and pharmaceuticals [13]. The selective conversion of sugars into FA may be accomplished using stoichiometric quantities of periodic acid [14] or by thermal cracking [15]. However, such technologies have severe drawbacks both from the environmental and energetic points of view. The conversion of carbohydrate-based biomass to FA may also be realized in supercritical water conditions with hydrogen peroxide as oxidant that is a green procedure, but the reaction occurs with poor selectivities and is energetically unfavorable [15a, 16]. Table 15.1 summarizes the performances of different catalytic systems in the one-pot cellulose transformation to FA (Scheme 15.2).

Wasserscheid *et al.* [17] studied the oxidation of cellulose to FA in water, using oxygen, as the sole oxidant, and $H_5PV_2Mo_{10}O_{40}$, as homogeneous catalyst. Under mild temperatures (333–363 K) the Keggin-type heteropolyacid catalyst is able

Table 15.1 Catalytic systems for one-pot selective oxidation cellulose to formic acid.

Catalytic system	Temperature (K)/ time (h)	Conversion (%)/ FA yield (%)	References
$H_5PV_2Mo_{10}O_{40}$	353/26	0/0	[17]
$H_5PV_2Mo_{10}O_{40}$[a]	363/23	39/18.8	[18]
$H_8PV_5Mo_7O_{40}$[a]	363/24	76/28	[19]
$H_5PV_2Mo_{10}O_{40}$[b]	423/9	100/34	[20]
$H_4PVMo_{11}O_{40}$	453/3	100/67.8	[21]
$NaVO_3 + H_2SO_4$	433/120	100/64.9	[22b]
IL functionalized with $-SO_3H$ and $PMo_{11}VO_{40}^{4-}$	453/1	91.3/51.3	[23]

a) With additive *p*-toluenesulfonic acid (TSA).
b) With additive HCl.

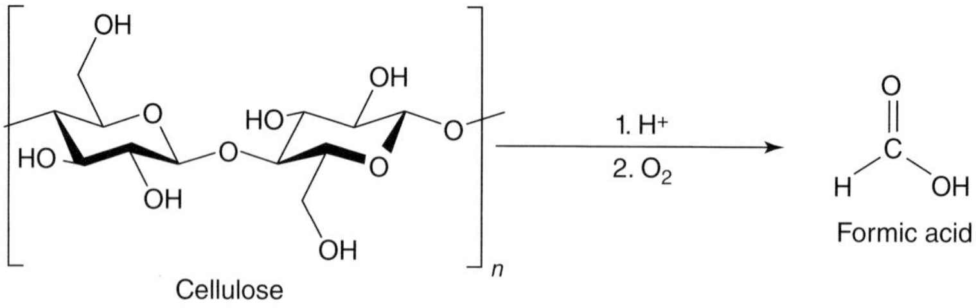

Scheme 15.2 Conversion of cellulose to formic acid.

to effectively convert water-soluble mono- and disaccharides to FA with yields of ~50% [17], but cellulose is hardly transformed even after long reaction times (26 h). However, in the presence of additives able to hydrolyze *in situ* cellulose (i.e., *p*-toluenesulfonic acid-TSA), the direct oxidation of water-insoluble cellulose with oxygen to FA occurred with a yield of 19% [18]. Part of the FA was also decomposed to CO_2. By increasing the content of the heteropolyacid in vanadium, $H_8PV_5Mo_7O_{40}$, the yield to FA increased to 28% [19]. These catalysts demonstrate in fact the role of pervanadyl species for such selective oxidations where higher vanadium-substituted catalysts can be completely reoxidized by oxygen at small pressures. Fu *et al.* also reported the use of such $H_5PV_2Mo_{10}O_{40}$ catalyst for the selective oxidation of biomass. Using a mineral acid as additive (HCl), the yields were superior to those with TSA (i.e., 34%) [20] but only at high reaction temperatures (i.e., 423 K). In the absence of the mineral acid, that is, in the absence of the hydrolysis, the conversion of cellulose occurred only to CO_2 and was enhanced by elevated temperatures. This confirms the capability of $H_5PV_2Mo_{10}O_{40}$ for the total oxidation of cellulose and its inability to directly produce FA under these reaction conditions. Han *et al.* found a positive effect of molybdenum in this process. Keggin-type vanadium-substituted phosphomolybdic acid catalyst with the $H_4PVMo_{11}O_{40}$ composition was found to convert various cellulosic biomass substrates to FA with yields as high as 68% in water at an oxygen pressure of 20 bar [21]. As an additional advantage, the homogeneous $H_4PVMo_{11}O_{40}$ catalyst can be recovered in solid form by distilling the products and solvent out and recycled at the same performances.

The presence of the mineral acids was also required by other oxidation catalysts showing good performances for the conversion of cellulose to FA. $NaVO_3$ in combination with H_2SO_4 allowed the one-pot conversion of cellulose to FA, when oxidizing by O_2 in aqueous solution (yield 65%) [22]. The hydrolyzed monosaccharides formed in the first step were readily oxidized to FA under the catalytic effect of VO_2^+. Very good yields (64%) with this catalytic system were also obtained starting from xylan (hemicellulose). When hydrolysis is deeper (i.e., from monosaccharides to levulinic acid (LA)), the catalytic oxidation occurs to acetic acid diminishing the selectivity in FA.

Coupling vanadium-polyoxometalate catalysts with ionic liquids (ILs) is also providing the one-pot transformation of cellobiose/cellulose to FA [24]. However, due to higher acidity, this system led to a significantly higher selectivity of LA (46.3%) compared to FA (26.1%). These catalysts are easily recyclable without any significant loss of activity. Heteropolyanion $PMo_{11}VO_{40}^{4-}$ in combination with SO_3H-functionalized ILs served also as bifunctional catalysts catalyzing the cellulose hydrolysis to glucose and further oxidation of glucose to FA with yields higher than 50% [23]. No typical heterogeneous catalysts have been reported to date as being active in this reaction.

15.2.1.2 Gluconic and Glycolic Acids

Gluconic and glycolic acids are important compounds widely used in organic synthesis, skincare products, pharmaceutical, and food industries [25]. The direct conversion of cellulose to glycolic acid was recently reported to occur with high yields (~50%) in the presence of phosphomolybdic acid ($H_3PMo_{12}O_{40}$) which acts as a bifunctional catalyst able to catalyze both the hydrolysis of cellulose and the subsequent oxidation reaction (Scheme 15.3). The reaction takes place in water using molecular oxygen as oxidant [26]. The catalytic behavior was merely attributed to the strong Brønsted acidity of this catalyst. Also, its moderate oxidative activity is essential for the selective oxidation of the aldehyde groups in the intermediates.

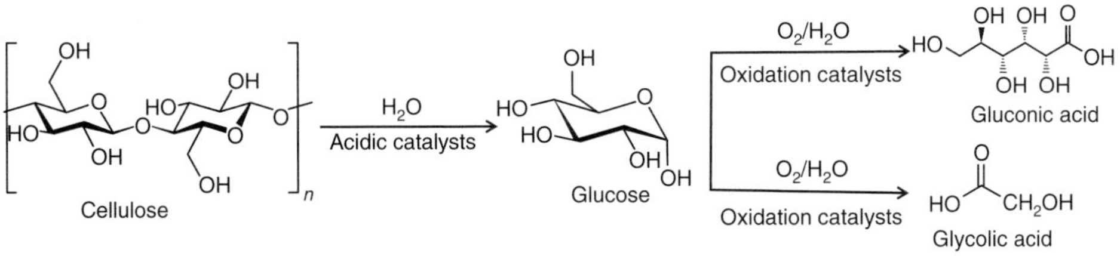

Scheme 15.3 Conversion of cellulose into gluconic and glycolic acids via hydrolysis and oxidation catalysis.

Gluconic acid is formed in high yields and high selectivities by oxidation of glucose in the presence of noble metal- (Au, Pt, Pd) supported catalysts [2e, 27]. However, the one-pot transformation of cellulose to gluconic acid and glycolic acid is a more desirable route for the synthesis of these molecules, and several studies have already investigated this reaction [28]. The deposition of Au nanoparticles onto Keggin-type polyoxometalate, $Au/Cs_xH_{3.0-x}PW_{12}O_{40}$, makes these catalysts also suitable for the oxidative conversion of cellobiose into gluconic acid in the presence of O_2. The reaction occurs with very good yields (60%) in water without any pH control. The role of the polyoxometalate was confirmed by comparison with typical metal oxide supports such as SiO_2, Al_2O_3, TiO_2, zeolites H-ZSM-5 and HY, or carbone nanotubes (CNTs) [29]. Like for the production of FA, this one-pot reaction requires two steps: (i) hydrolysis of cellulose to oligosaccharides and glucose that is promoted by $Cs_{1.2}H_{1.8}PW_{12}O_{40}$ and (ii) the oxidation of glucans by the small Au nanoparticles. Unfortunately, the $Cs_{1.2}H_{1.8}PW_{12}O_{40}$ system is unstable over long-term hydrothermal conditions, and $Au/Cs_{1.2}H_{1.8}PW_{12}O_{40}$ became deactivated for the conversion of cellulose during repeated use. Serious deactivation could be overcome by using $Au/Cs_{3.0}PW_{12}O_{40}$ in combination with $H_3PW_{12}O_{40}$. This combination afforded a gluconic acid yield of 85% for the conversion of the ball-milled cellulose and a gluconic acid yield of 39% for the conversion of microcrystalline cellulose. Also important, $H_3PW_{12}O_{40} - Au/Cs_{3.0}PW_{12}O_{40}$ could be recovered from the liquid phase and used repeatedly [29a], and further oxidation of gluconic acid by $Au/Cs_xH_{3.0-x}PW_{12}O_{40}$ is slow.

Onda *et al.* [30] investigated the catalytic performances of sulfonated carbon-supported Pt catalysts $(Pt/C-SO_3H)$ for the oxidative conversion of cellobiose. Working with cellobiose in water at 393 K and under an air atmosphere, these authors reported a yield of 46% in gluconic acid. The catalyst was highly water tolerant and showed the catalytic properties for the hydrolysis of polysaccharides and sequentially the air oxidation into gluconic acid in the one-pot process under hot water. The deposition of gold nanoparticles onto nitric acid-pretreated carbon nanotubes led to even more efficient catalysts for the direct selective oxidation of cellobiose by molecular oxygen to gluconic acid. In an aqueous medium, at 418 K, without any pH control the yield in gluconic acid was of 80% [28a]. The activity of these catalysts was associated with the presence of acidic groups on CNT surfaces that are able to catalyze the hydrolysis of cellobiose to glucose. Then, Au^0 nanoparticles catalyze the selective oxidation of glucose to gluconic acid by oxygen.

15.2.1.3 LA

LA is a suitable platform molecule for the production of hydrocarbon liquid fuels, fuel additives, and various chemicals [4]. LA is produced conventionally via an acid-catalyzed process using mineral acids as homogeneous catalysts [31]. To date, several approaches were reported for the liquid phase direct fragmentation of cellulose to LA following an oxidative route [32]. One possibility concerns the treatment of cellulose with lean air in water in the presence of a solid acid catalyst (zeolite β, zeolite Y, and ZSM-5) [33]. Less acidic oxides like ZrO_2, under the optimum reaction conditions, led also to good yields of LA (50%, at 513 K and 24 bar pressure of a 97.2%N_2 + 2.8%O_2 mixture). Redox, rather than acid/base, properties of the ZrO_2 catalyst played the central role on the oxidative deconstruction of cellulose. The proposed reaction pathway (Scheme 15.4) considered a gluconic acid intermediate mechanism [33]. Accordingly, a superoxide radical anion breaks the glycosidic bond of cellobiose leading to glucose and gluconic acid. Then, gluconic acid is deoxygenated through via Hofer–Moest decarboxylation [34] and suffers a series of consecutive dehydration/rehydration reactions. The mechanism was confirmed using gluconic acid as substrate. An important advantage of this process is its environmental friendliness, leading to a low humin formation.

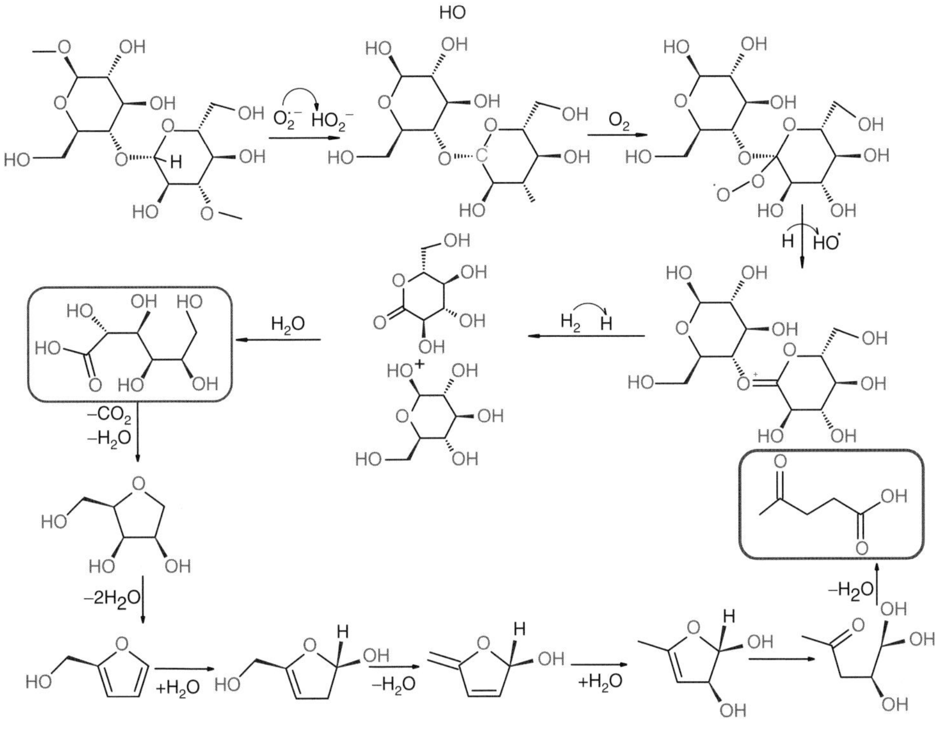

Scheme 15.4 Conversion of cellobiose to levulinic acid: reaction pathway of converting the cellobiose into glucose and gluconic acid by superoxide radical anions and reaction pathway of converting gluconic acid to levulinic acid via Hofer–Moest decarboxylation followed by consecutive dehydration/rehydration reactions. (Adapted from reference [33] with permission from The Royal Society of Chemistry.)

15.2.1.4 Succinic Acid

The use of a proper catalyst may allow the further oxidation of LA to succinic acid. In addition to the catalytic activity, its functionalization on silica-coated magnetic nanoparticles may ensure the easy separation from the unreacted solid biomass (Scheme 15.5). The recently reported Ru(III)/functionalized on silica-coated magnetic nanoparticles (MNP) satisfied these conditions providing selectivities in succinic acid higher than 95% [35].

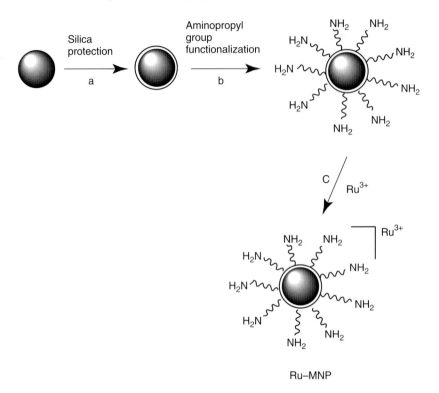

Scheme 15.5 Synthesis of a Ru–MNP catalyst for the direct oxidation to succinic acid. (From reference [35] with permission from The Royal Society of Chemistry.)

15.2.1.5 Acetic Acid

Acetic acid is another very important molecule with a very large annual production [36]. For the moment, petroleum is the most principal source in this production. The direct catalytic conversion of cellulose to acetic acid has been scarcely studied, especially due the low selectivity. So, today, there are only few reports on this subject. In the presence of oxygen, using a mixture of $NaVO_3 - H_2SO_4$ in aqueous solution, Marsh *et al.* [37] succeeded in the direct conversion of cellulose to acetic acid in a yield of 7.3%. These authors assumed the formation of acetic acid as a result of two parallel routes: the bond breakage of largely distributed acetyl groups in the hemicellulose structure and the oxidation of LA, as a product of a deep hydrolysis of polysaccharides (cellulose and hemicellulose). The increase of the H_2SO_4 concentration in the catalytic mixture increased the hydrolysis/transformation of polysaccharides to LA and its further oxidation to acetic acid [38].

15.3
Selective Oxidation of Lignin

Lignin is a heterogeneous, optically inactive polymer consisting of phenyl-propanoid interunits (coniferyl, sinapyl, and coumaryl alcohol polymers) which are linked by several covalent bonds (e.g., aryl–ether, aryl–aryl, carbon–carbon bonds) (Figure 15.2). The interunit linkages that connect the monomer units include the β-O-aryl ether (β-O-4'), resinol (β-β'), phenylcoumaran (β-5'), biphenyl (5-5'), diaryl ether (4-O-5'), and 1,2-diaryl propane (β-1') linkages [39]. Lignin, like cellulose and hemicellulose, is a major component of biomass and represents in fact the most abundant form of aromatic carbon in the biosphere. Because of this specificity the processing of lignin is more complicated and requires more specific and effective catalysts.

In spite of these, the catalytic selective oxidation of lignin appears to be very interesting for the production of more complex aromatic compounds with additional functionality. It occurs with water as the only by-product of oxidation and may take place without adding any other stoichiometric reagent [40].

Literature reported several examples of catalytic oxidative cleavage of lignin and of selective oxidation of its phenolic or nonphenolic model compounds (veratryl, vanillyl, syringyl alcohols) in which the scope was to promote the reactivity of the most abundant linkages in lignin, that is, the β-O-4'-glycerolaryl ether and β-O-4'-ethanoaryl linkages [5b, 41]. Most of these approaches resulted in the formation of benzoic acids, phenolic aldehydes, or different quinone derivatives as products. Such products may serve as valuable starting materials in fine chemical industries.

The target in selectivity depends on the further valorization. Thus, the production of vanillin requires a cleavage of both C–C and C–O bonds, while other applications demand the conservation of the alkyl chain [42]. However, working with such structural complex molecules is not simple, and many reports replaced lignin by model compounds, like those presented in Scheme 15.6. Working with these allows a better understanding of the chemical reactivity and of the parameters affecting the process.

15.3.1
Selective Oxidation of Lignin in the Presence of Homogenous Catalysts

Selective oxidation of lignin followed different routes using various catalytic approaches like biocatalysis or organometallic catalysis (including biomimetic catalysis) [43]. Most of the experiments were carried out in the presence of environmentally friendly oxidants such as oxygen (O_2) or hydrogen peroxide (H_2O_2) [40b, 44] suggesting that, indeed, homogeneous catalysts may represent a tool to lignin oxidation [40].

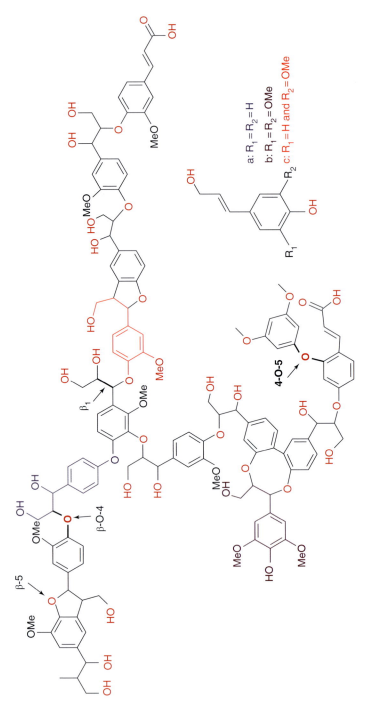

Figure 15.2 Structure of lignin and primary precursors. (a) *trans-p*-Coumaryl alcohol, (b) *trans*-sinapyl alcohol, and (c) *trans*-coniferyl alcohol.

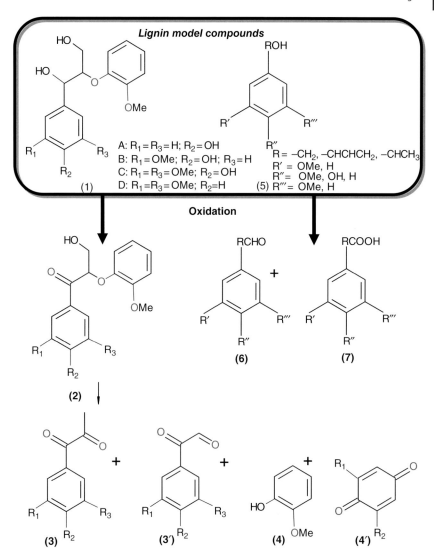

Scheme 15.6 Typical lignin model compounds (**1**, **5**), oxidation products (**2**, **6**, **7**), and C–O bond cleavage products **3**, **3′** and **4**, **4′**.

Among these, homogeneous vanadium complexes, as those reported by Toste *et al.* (Figure 15.3), convert selectively dimeric lignin model nonphenolic compounds containing a β-O-4 linkage (i.e., (**1D**)) [41d]. For example, complexes (**a–c**) yielded benzylic alcohol oxidation with the formation of (**2D**), as major product, and small amounts of C–O bond cleavage products (**3′**) and (**4**). Ligands with larger angles (complexes (**d) and (e**)) led to a higher activity (>95%) and selectivity for the C–O bond cleavage. This demonstrated that changes in the ligand structure can divert the reactivity of vanadium–oxo complexes from the typical alcohol

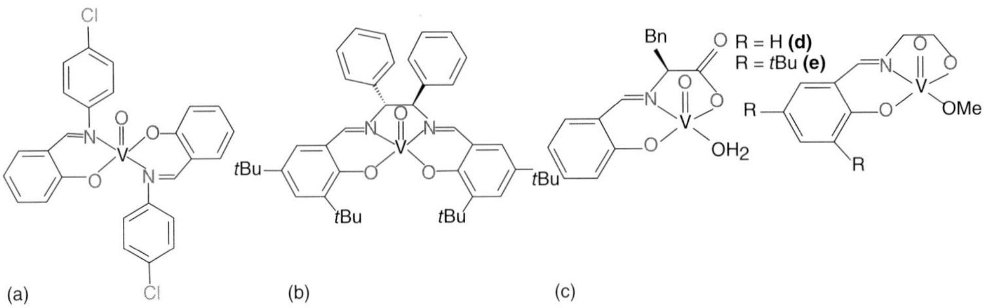

Figure 15.3 Homogenous vanadium-based complexes. (From reference [41d] with permission from The Royal Society of Chemistry.)

oxidation to an unprecedented C–O bond cleavage reaction. This last transformation was suggested to proceed through a ketyl radical generated by a hydrogen atom transfer to a vanadium (V) complex. Oxygen is not essential for the reaction, although it increases the reaction rate.

As an alternative, Hanson *et al.* [45] investigated other vanadium complexes as catalysts for the oxidative C–C bond cleavage of lignin model compounds. Among these, the 8-quinolinate complex (Figure 15.4a) was found to catalyze selectively the oxidation of the phenolic lignin model compound **1C**. This mediates a new type of reaction in which the C–C bond between the aryl ring and the adjacent alcohol group is broken to give 2,6-dimethoxybenzoquinone (**4′**), acrolein derivative (**3**), and ketone (**2**) [46]. In contrast, the catalyst presented in Figure 15.4b favorize the C–O bond cleavage, with the formation of products (**3′**) and (**3**), and ketone (**2**).

Aqueous polyoxometalates in the presence of alcohols catalyze the conversion of Kraft lignin from spruce wood into chemicals of industrial interest directly in the presence of oxygen [47]. Vanillin, methyl vanillate, ethyl vanillate, and

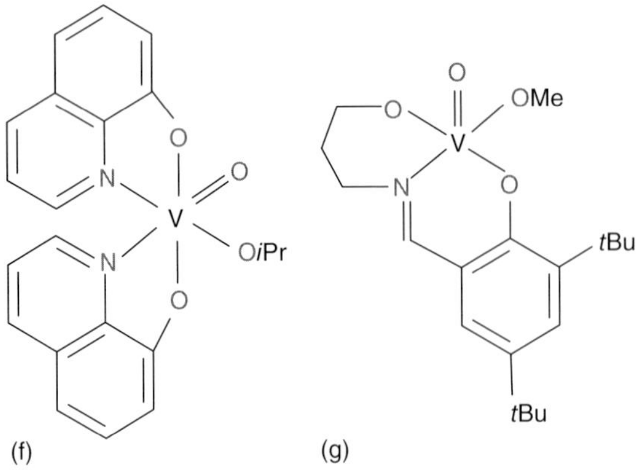

Figure 15.4 Vanadium catalysts for the oxidation of lignin models.

other monomeric products are thus produced by direct conversion of lignin with $H_3PMo_{12}O_{40}$ in either methanol or ethanol. The addition of alcohols enhanced the yield of monomeric products by preventing the recondensation of the lignin fragments.

In an effort to increase the yields of monomeric products, a mixture of catalysts (**f**) (5 mol%) and (**g**) (5 mol%), with NEt_3 (10 mol%) was evaluated for the oxidation of (**1C**) in ethyl acetate. Complete consumption of (**1C**) was observed after heating the mixture at 353 K for 48 h, where a mixture of C–O (**3′** (24%)) and C–C bond cleavage products (**4′** (36%), **3** (30%)), and ketone **2** (14%) was obtained. FA (4%) was also detected by ^1H and ^{13}C NMR spectroscopy [45b, 46]. Although the yields of monomeric products were slightly increased by using a combination of both catalysts, the mass balance (about 74%) and the presence of a brown precipitate suggested that some undesired side reactions still occurred.

The selectivity in the oxidation of lignin can be also controlled by the choice of the transition metal. Thus, the Cu^I/TEMPO (TEMPO = (2,2,6,6-tetramethylpiperidin-1-yl)oxy) catalyst system showed good selectivity for alcohol oxidation, with the formation of veratryl aldehyde (compound (**6**)) as the major product. However, the formation of unidentified by-products still limits its utility [48]. Working with pyridine (CuCl/pyridine/TEMPO catalyst) allowed the oxidation of the nonphenolic lignin model compound (**1D**) leading to a direct C–C bond cleavage that afforded the substituted benzaldehyde (compound (**6**)) [45b]. However, the selectivity for the oxidative C–C bond cleavage exhibited by soluble copper catalysts with simple lignin models is hampered by more complex β-O-4 models such as those presented by (**1C**) and (**1D**) [49]. In these cases, the catalyst activity was inhibited by the phenol by-product. Further reduction of the catalyst loading led primarily to the ketone derived from the oxidation of the secondary benzylic alcohol.

Changing the ligand is also influencing the performances in this reaction. Stahl and coworkers succeeded the aerobic oxidation of a complex β-O-4 lignin model compound (**1C**) using a Cu(OTf)/bpy/TEMPO/*N*-methylimidazole catalyst and reported the selective synthesis of veratryl aldehyde (compound **6** (Z = H, R = CH_2O, X = Y = OCH_3)) as the major product [50]. The cleavage of the C–C bond was suggested to occur via the oxidation of the primary alcohol followed by a retroaldol reaction (Scheme 15.6 compounds (**6**) and (**4**)).

Oxidative processes of lignin were also reported in the presence of a simple mixed Cu–Fe homogeneous catalyst system affording 14 wt% aldehydes (compound (**6**)) [51] and for $CuSO_4$ in an IL yielding up to 30% aldehyde [52]. Also, Ambrose *et al.* [53] reported the study of an IL tagged Co(salen), on which veratryl alcohol (compound (**5**)) was selectively oxidized to veratraldehyde (compound (**6**)) using air or pure oxygen as the source of oxygen. The benzyl functionality in veratryl alcohol was also selectively oxidized to form veratraldehyde, with excellent turnover frequencies (1440 and 1300 h^{-1} using $CoCl_2 \cdot 6H_2O$ and Co(salen), respectively) [54]. However, phenolic functional groups contained in guaiacol, syringol, and vanillyl alcohol remained intact, although the benzyl alcohol group in the latter was oxidized to form vanillin.

Phenolic and nonphenolic monomeric model compounds, as vanillyl alcohol or veratryl alcohol (compound (**5**) with $Y = -OCH_3$, $-OH$ or H), were treated with a methyltrioxo rhenium (MTO)/hydrogen peroxide mixture in acetic acid. Complex mixtures of products were obtained, including aldehyde and carboxylic acid derivatives that must originate from the oxidation of the side-chains. Benzoquinones and muconolactones derived from oxidative ring cleavage of the aryl groups were detected also in significant yields [55]. Similar results were observed in the case of more complex β-O-4 dimeric compounds, such as (**1B**). In this case, products of alkyl side-chain oxidation and successive cleavage at the α-position or at the β-position, namely, compounds (**3**) and (**4**), but also products of aromatic ring oxidation were recovered.

Recently, metal-free catalytic systems were also reported for the oxidation of lignin model compounds and extracted lignin. Stahl *et al.* [50] identified mixtures of TEMPO derivatives and mineral acids as highly effective catalytic systems for the chemoselective aerobic oxidation of benzylic alcohols (with isolated yields of compound (**2**) up to 94%). Good results were also obtained with 4-acetamido-TEMPO, where AcNH-TEMPO acts in combination with HNO_3 or HCl as cocatalysts (1 bar O_2, 318 K, CH_3CN/H_2O as the solvent 24 h). Other solvents such as acetic acid, ethylacetate, and dioxane were less effective. However, besides some advantages these homogeneous catalysts can lead to a secondary pollution and high recycling costs that restricts their industrial utilization.

15.3.2
Selective Oxidation of Lignin in the Presence of Heterogeneous Catalysts

The use of the heterogeneous catalysts may eliminate the disadvantages of homogenous catalysis but suffers in terms of activity. A way to convert lignin into aromatic aldehydes with increased yields, using heterogeneous catalysis, is the so-called wet aerobic oxidation (WAO) process, which often uses oxygen as oxidizing agent. Noble metals have been used as catalysts in the lignin WAO process, in order to obtain intermediary oxidation products, such as vanillin, syringaldehyde, and p-hydroxybenzaldehyde [56], but they are expensive, which is an impediment for their commercial applications. Therefore, the use of an earth-abundant metal as a catalyst could provide additional benefits in terms of availability and costs. Some inexpensive metal ions such as iron, copper, and cobalt have also shown activity in the lignin oxidation process [51, 57], but the severe operating conditions (403–603 K; 5–20 bar) limited the application of this technology. Perovskite catalysts such as $LaFe_{1-x}Cu_xO_3$ ($x = 0$, 0.1, 0.2) prepared by the sol–gel method exhibited a reasonable activity in the catalyzed WAO of lignin. Compared to the noncatalyzed processes, the lignin conversion and yields to aromatic aldehydes were significantly improved under these conditions [58]. $LaMnO_3$ was also recently reported as an efficient and recyclable heterogeneous catalyst for WAO of lignin (57.0% conversion after 3 h) [59]. p-Hydroxybenzaldehyde was obtained on this catalyst in a yield 31.3% higher than that of the noncatalytic process.

The nanostructured Co_3O_4 catalyst was also evaluated for the liquid phase oxidation of different lignin-derived substructured compounds (see (**5**) from Scheme 15.6: veratryl, vanillyl, sinapyl, coniferyl, *p*-coumaryl alcohols) to the corresponding aldehydes. The reactions were carried out under base-free conditions in water indicated a substrate dependency activity. Thus, the oxidation of veratryl alcohol occurred in the presence of oxygen leading, under optimized conditions, to conversions higher than 85% and selectivities higher than 96% to the corresponding aldehyde [60].

Inspired by the catalytic activity demonstrated by the homogeneous cobalt complexes and nanostructures, Weckhuysen *et al.* investigated the ability of Co-ZIF-9 as catalyst for the oxidation of aromatic molecules expected to be separated from lignin depolymerization. Oxidation of molecules containing benzyl alcohols, phenols, methoxy groups, and benzyl ethers (i.e., the same substructure compounds (**5**) from Scheme 15.6) occurred indeed on this catalyst but with inferior performances than on cobalt-based homogeneous catalysts. For veratryl alcohol, using a wide range of solvents, the maximum yield in veratraldehyde was of 46%. Stability tests were carried as well with this catalyst [61].

Co(salen) showed high activity in the oxidation of phenolic and nonphenolic β-O-4 aryl ether lignin model compounds (**5**) [62]. However, its deposition onto SBA-15 led to even higher activities for the oxidation of a lignin model phenolic dimmer **1B** [63]. These results were attributed to an effect of the cleavage of the $C_β$–O bond, providing 2-methoxyphenol (compound (**4**)) selectively and with relatively high turnover numbers.

$Pd/γ$-Al_2O_3 was investigated with the scope to produce aldehyde-type compounds (compound (**6**)) from lignin. Experiments carried out in both batch slurry and continuous fluidized-bed reactors, under an oxygen partial pressure between 2 and 10 bar at temperatures between 373 and 413 K [56b], showed that the degradation of lignin led to aromatic aldehydes following a complex series of parallel reaction network. The influence of additives in the selective oxidation of (**1D**) to (**2D**) was also studied with Pd/C catalysts [64]. In the absence of any additive the yield of oxidation was of 68%, but the presence of additives enhanced these performances [65]. FA alone promotes a disproportionation reaction of the benzyl alcohol for both benzyl alcohols and lignin-fragmented compound models. The use of formate instead of FA gave C–O bond cleavage in different β-O-4′-ethanolaryl ether positions of these lignin compound models. The reaction starts with an initial dehydrogenation. The role of the additive is to maintain the palladium activity only in the dehydrogenation reaction. In this way, the aryl ethyl ketone can be prepared selectively under mild reaction conditions. In the presence of formate, a reductive cleavage of the weaker C–O ether bond [66] also occurs in a fast subsequent reaction step. However, the more challenging β-O-4′-glycerolaryl ether models were reactive only at the temperature higher than 393 K and under an overstoichiometric amount of a hydrogen donor.

The use of ILs as solvents for both homogenous and heterogeneous conversions of lignin and lignin model compounds into value-added aromatic chemicals is also of interest [67]. Their use in oxidation reactions may enhance the selectivity [68].

The use of nanoparticle IL-stabilized catalysts is another alternative for the oxidation of lignin providing a bridge between homogeneous and heterogeneous catalysis. Nanopalladium(0) catalysts associated to a pyridinium salt of iron bis(dicarbollide) as cocatalyst demonstrated efficiency for the oxidation of benzyl alcohol and substituted derivatives to produce aromatic aldehydes with yields between 77 and 93% using O_2 as oxidant and $[C_4mim]-[PF_6]/[C_4mim][MeSO_4]$ mixture as solvent [69]. The solvating properties of ILs are not the only properties generating advantage in these reactions. These solvents can also stabilize the catalysts or even the reactive intermediates, playing an important role in the mechanisms and leading sometimes to different results than those obtained in organic solvents. Zakzeski *et al.* [54] used $CoCl_2 \cdot 6H_2O$ and Co(salen) catalysts in combination with $[C_{2mim}][Et_2PO_4]$ suggesting that IL favored the coordination of the substrate to cobalt. Stärk *et al.* investigated the oxidation of syringaldehyde, as a model compound, using $Mn(NO_3)_2$ as catalyst (20 wt%), O_2 as oxidant (0.7 bar), and $[C_2mim][CF_3SO_3]$ as solvent. The reaction led to the formation of 2,6-dimethoxybenzoquinone, a product that was not obtained with molecular solvents [70]. The hydrolytic cleavage of β-O-4 ether bonds in lignin model compounds was also studied using $AlCl_3$, $FeCl_3$, and $CuCl_2$ and C_4mim as IL [71]. These studies indicated $AlCl_3$ as more effective in cleaving the β-O-4 bond than $FeCl_3$ or $CuCl_2$. Thus, in the presence of $FeCl_3$ and $CuCl_2$, after 120 min, at 423 K, the conversion reached 100%, where about 70% of the β-O-4 bonds were hydrolyzed, while $AlCl_3$ led, for the same level of conversion, an about 80% linkage of these bonds. In fact, this catalytic activity was associated with the stability of these chlorides. $AlCl_3$ hydrolyzes much easier to hydrochloric acid than $FeCl_3$ and $CuCl_2$, leading thus to hydrochloric acid that is the real acid catalyst in this reaction.

Nitrogen-containing graphene materials were also used as catalysts for the oxidative decoupling of α-O-4- and β-O-4-type lignin model compounds using *tert*-butyl hydroperoxide as oxidant in water. Under optimized conditions the monomeric acidic products were obtained in yields of 45% in compound (7) [72]. The reaction follows a free-radical mechanism, and the structures of key intermediates were identified by free-radical trapping experiments.

Based on these reports one may conclude that in spite of the numerous achievements, the selective oxidative valorization of biomass is still a challenging direction. Further catalyst development is needed to increase the control over the chemical and mass distribution of aldehyde and acid products from oxidative fragmentation of the lignin extracts. The stability of the catalyst also represents a key parameter that has to be taken into account.

15.4
Selective Oxidation of Starch

Starch is a glucose polysaccharide that is formed from a large number of glucose units connected by glycosidic bonds. It consists of two types of molecules, that

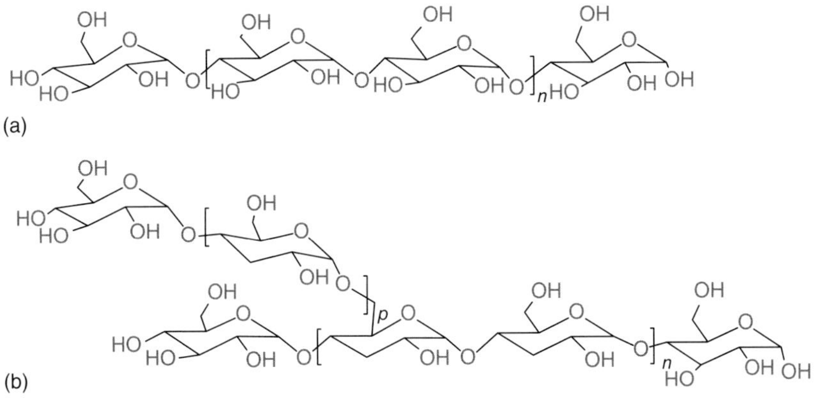

(a)

(b)

Figure 15.5 The structures of starch consisting of (a) amylose and (b) amylopectin.

is, the linear and helical amylose that is composed from R-1,4-glycoside linkages and the branched amylopectin. While amylose is the water-soluble component of starch, amylopectin is water insoluble (Figure 15.5). The R-linkages make the polymer amorphous [73]. For instance, the most important applications of starch are in the paper, textile, and domestic industries [74].

The selective nondegrading oxidation of starch focuses on hydroxyl groups, primarily at C-2, C-3, and C-6 positions, that are replaced in this process by carboxyl and carbonyl groups, while the granular structure remains intact (Scheme 15.7). Therefore, in such transformations, the carboxyl and carbonyl contents of oxidized starch are generally used to indicate the level of oxidation. Industrial processes for nondegrading starch oxidation used transition metal salts (Cu^{2+}, Fe^{2+}) as homogeneous catalysts and mineral oxidizing agents, such as NaOCl [75], N_2O_4 [76], or $NaIO_4$ [75c, 77]. The use of these oxidizing agents also correlates to the selectivity of the process. N_2O_4 is merely selective for oxidation to carboxyl groups,

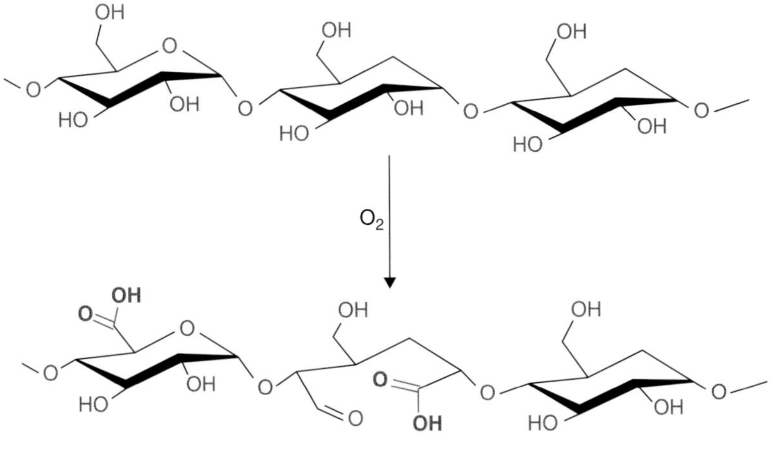

Scheme 15.7 Schematic representation of starch selective oxidation.

while $NaIO_4$ stops the reaction to aldehydes. Although efficiently, these processes result also in an overoxidation that generates environmental issues. Therefore several catalytic approaches have been proposed to improve the nondegrading starch oxidation. Thus, both bio- and chemical catalyses have been used to carry out such processes. The new approaches considered as oxidants alkaline peroxides [74], nitrogen oxides [78], or peracetic acid [79]. However, these oxidants are also raising environmental problems since large amounts of wastes were still produced.

Replacing these oxidants with hydrogen peroxide improved somehow the efficiency of these processes. Working with transition metal catalysts, like iron [80], copper [81], or tungsten salts [75a, 80] in rather high concentrations of metal ions (0.01–0.1% based on dry starch), have been found to both activate in a suited form hydrogen peroxide and to solve the environmental problems. $FeSO_4$ and $CuSO_4$ are efficient catalysts to transform starch into polyhydroxycarboxylic acids in high yields, and their productivity of 0.61 mol CO_2H per 100 g [82] is among the best results reported to date. However, these catalysts have also a drawback since due to complexation properties of the oxidized starch, the metals are retained by carboxyl functions. For example, the use of copper-based catalysts generates the product coloration and induces toxicity in modified products and plant effluents [83]. Therefore, even in this case additional treatments such as washing with chelating agents are necessary.

Starch was selectively oxidized at the C6 position in the presence of a NaOCl/NaBr mixture combined with the TEMPO radical [84]. The primary hydroxyl groups in potato starch were also oxidized by the TEMPO mediator in the presence of laccase enzyme and oxygen as a primary oxidant [85] or by using greener oxidants such as O_2 and H_2O_2 [86] or with H_2O_2 under alkalinic and acidic reaction conditions in the presence of copper, iron, and tungstate catalysts. All these processes yielded carboxyl and carbonyl groups [80].

An efficient catalytic system for the starch oxidation to carboxylic acids is also the $CH_3ReO_3/H_2O_2/Br^-$ system dissolved in a water–acetic acid mixture. The oxidation product may serve as a material acting as a water superadsorbent. However, this transformation requires high amounts of CH_3ReO_3 which have to be regenerated after the use because of deactivation [87]. The mechanistic considerations shown in Scheme 15.8 took into consideration the formation of hypobromite in the presence of an excess of hydrogen peroxide [87]. This constitutes a reasonable explanation for the absence of aldehyde in the reaction product and the very high selectivity in the carboxylic acid [85].

High yields in the nondegrading oxidation of starch were also obtained with H_2O_2 as oxidant in the presence of a water-soluble iron tetrasulfophthalocyanine complex leading to more hydrophilic starch without any major chain breaking. This corresponded in fact to the oxidation of C2, C3, and C6 positions into carbonyl and carboxyl functions, yielding 1.5 carboxyl and 5.6 carbonyl functions per 100 glucose units [72, 83, 88]. The oxidation was best achieved by incipient wetness method whereby catalysts dissolved in small volumes of water were impregnated in the starch powder under continuous mixing, followed by addition of hydrogen peroxide to the impregnated solid. Typically, the substrate/catalyst ratio was

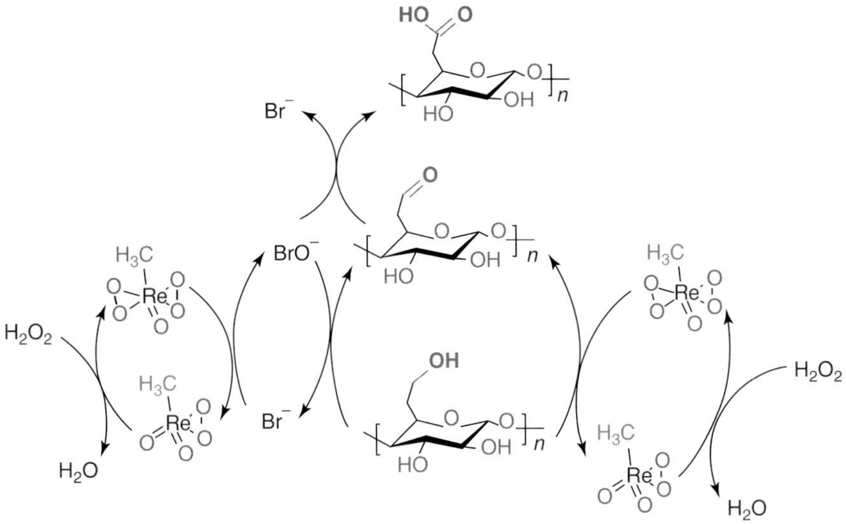

Scheme 15.8 Proposed simplified mechanism of the starch oxidation by $CH_3ReO_3/H_2O_2/Br^-$ system. (Adapted from reference [87]).

of only 25.800/1. The process was verified working with starches obtained from different crops (potato, wheat, rice, corn), and the resulting materials were successfully tested in the formulation of coating agents and cosmetics. The ultrasound treatment of starch prior to oxidation [89] led to an increased oxidation degree.

The performances of strong Brønsted acid polyoxometalates in starch nondegrading selective oxidation were recently reported. Among different compositions, $Cs_3H_2PMo_{10}V_2O_{40}$ afforded the higher degrees of oxidation (0.59 mol per 100 g starch) at a low catalyst loading [90] in the presence of H_2O_2 as oxidizing agent. These studies demonstrated that the strong Brønsted acidity could also promote the nondegrading oxidation of starch. High surface areas were also beneficial for this process.

The combination of POMs with ILs (as the case of choline chloride) also enhanced the catalytic oxidation of starch [91]. The polyoxometalate-immobilized IL $[(CH_3)_3NCH_2CH_2OH]_5PV_2Mo_{10}O_{40}$ synthesized via a precipitation and ion exchange combined method using choline chloride and $H_5PMo_{10}V_2O_{40}$ exhibited high efficiency in catalytic oxidation of starch with H_2O_2. The reaction occurred with a high oxidation degree (0.59 mol CO_2H per 100 g) under mild conditions (at 70 °C, for 10 h) that constitutes a comparable behavior to those of traditional homogeneous catalysts [82].

More complex catalytic systems like $Na_4Co(H_2O)_6V_{10}O_{28}\cdot18H_2O$ [92] and $Ag_{3.5}(NH_4)_{1.5}PMo_{10}V_2O_{40}$ [93] also acted as catalysts for the nondegrading oxidation of starch with oxygen or H_2O_2 under atmospheric pressure at low temperatures. On $Na_4Co(H_2O)_6V_{10}O_{28}\cdot18H_2O$ system, only negligible breakages of C–O–C linkages were detected under investigated conditions, and the catalyst was recycled at least six times without the obvious loss of activity. This behavior was assigned to a synergistic effect between the $[V_{10}O_{28}]^{6-}$ anion and the Co cation, resulting in 1.35 carboxyl and 2.1 carbonyl functions per 100 glucose units. $Ag_{3.5}(NH_4)_{1.5}PMo_{10}V_2O_{40}$ is another heterogeneous catalyst, which allows an easy recycle and reuse and exhibits high activity leading to a high oxidative degree.

15.5
Conclusions

Direct selective catalytic oxidation of the constitutive parts of the biomass such as cellulose/hemicellulose, starch, and lignin with capitalization in either platform molecules or materials is very challenging. Since biomass already contains substantial amounts of oxygen, its use represents the most proper source for the production of the compounds requiring its presence. From this perspective, its transformation should only tailor the introduction of new functionalities and the change of the size. In this respect, the direct transformation will avoid successive transformation steps, separation costs, and formation of wastes and will make the synthesis of these products more profitable.

Reports to date indicated that such transformations are possible in the presence of a catalyst. Good yields were already reported with some homogeneous catalytic systems. However, to be sustainable, the catalysts should be recovered and recycled. In this stage this objective is still challenging.

Also, the role of the solvent in these processes is far to be trivial. Solvents participate in the activation of the biomass and allow a better behavior of the catalysts. ILs already demonstrated their unique efficiency in the solvation of biomass polymers and in selective oxidations. ILs also allow both solubilization and activation of oxygen.

In conclusion, the research on selective catalytic oxidation of the constitutive parts of the biomass is still in a pioneer stage looking for more efficient solutions. However, the achievements obtained to date recommends this route as one of the most promising routes for the valorization of biomass to valuable platform molecules and chemicals.

References

1. United Nations, W. P. P. T. R. see: http://www.un.org/esa/population/publications/wpp2008/wpp2008_highlights.pdf.
2. (a) Fornasiero, P. and Graziani, M. (2011) *Renewable Resources and Renewable Energy: A Global Challenge*, 2nd edn, CRC Press; (b) Dusselier, M., Mascal, M., and Sels, B. (2014) in *Selective Catalysis for Renewable Feedstocks and Chemicals*, vol. 353 (ed K.M. Nicholas), Springer International Publishing, pp. 1–40; (c) Gallezot, P. (2008) *ChemSusChem*, **1**, 734–737; (d) Sheldon, R.A. (2014) *Green Chem.*, **16**, 950–963; (e) Alonso, D.M., Wettstein, S.G., and Dumesic, J.A. (2012) *Chem. Soc. Rev.*, **41**, 8075–8098.
3. Sheldon, R.A. and Sanders, J.P.M. (2015) *Catal. Today*, **239**, 3–6.
4. Bozell, J.J. and Petersen, G.R. (2010) *Green Chem.*, **12**, 539–554.

5. (a) Luque, R. and Balu, A.M. (2013) *Producing Fuels and Fine Chemicals from Biomass Using Nanomaterials*, Taylor & Francis; (b) Corma, A., Iborra, S., and Velty, A. (2007) *Chem. Rev.*, **107**, 2411–2502; (c) Climent, M.J., Corma, A., and Iborra, S. (2011) *Green Chem.*, **13**, 520–540; (d) Gallezot, P. (2007) *Green Chem.*, **9**, 295–302; (e) Gallezot, P. (2011) *Catal. Today*, **167**, 31–36; (f) Gallezot, P. (2012) *Chem. Soc. Rev.*, **41**, 1538–1558.

6. (a) Luterbacher, J.S., Rand, J.M., Alonso, D.M., Han, J., Youngquist, J.T., Maravelias, C.T., Pfleger, B.F., and Dumesic, J.A. (2014) *Science*, **343**, 277–280; (b) Simakova, O.A., Davis, R.J., and Murzin, D.Y. (2013) *Biomass Processing over Gold Catalysts*, Springer; (c) Corma, A., de la Torre, O., and Renz, M. (2012) *Energy Environ. Sci.*, **5**, 6328–6344; (d) de Jong, E. and Gosselink, R.J.A. (2014) in *Bioenergy Research: Advances and Applications* (eds V.K. Gupta, M.G.T.P. Kubicek, and J.S. Xu), Elsevier, Amsterdam, pp. 277–313; (e) Gallezot, P. (2013) in *New and Future Developments in Catalysis* (ed S.L. Suib), Elsevier, Amsterdam, pp. 1–27; (f) Jin, L., Kuo, C.-h., and Suib, S.L. (2013) in *New and Future Developments in Catalysis* (ed S.L. Suib), Elsevier, Amsterdam, pp. 253–270; (g) Osmundsen, C.M., Egeblad, K., and Taarning, E. (2013) in *New and Future Developments in Catalysis* (ed S.L. Suib), Elsevier, Amsterdam, pp. 73–89.

7. (a) Cavani, F. and Teles, J.H. (2009) *ChemSusChem*, **2**, 508–534; (b) Cavani, F. (2010) *J. Chem. Technol. Biotechnol.*, **85**, 1175–1183; (c) Cavani, F. and Ballarini, N. (2009) *Modern Heterogeneous Oxidation Catalysis*, Wiley-VCH Verlag GmbH & Co. KGaA, pp. 289–331; (d) Cavani, F. (2010) *Catal. Today*, **157**, 8–15.

8. (a) Thayer, A.M. (1992) *Chem. Eng. News Arch.*, **70**, 27–49; (b) Thomas, J.M. and Raja, R. (2006) *Catal. Today*, **117**, 22–31.

9. (a) Cavani, F., Centi, G., Perathoner, S., and Trifirò, F. (2009) *Sustainable Industrial Chemistry: Principles, Tools and Industrial Examples*, Wiley-VCH Verlag GmbH; (b) Arpentinier, P., Cavani, F., and Trifirò, F. (2001) *The Technology of Catalytic Oxidations: Chemical, Catalytic & Engineering Aspects*, Technip; (c) Centi, G., Cavani, F., and Trifirò, F. (2001) *Selective Oxidation by Heterogeneous Catalysis*, Springer.

10. Klemm, D., Heublein, B., Fink, H.-P., and Bohn, A. (2005) *Angew. Chem. Int. Ed.*, **44**, 3358–3393.

11. Centi, G. and Santen, R.A.v. (2007) *Catalysis for Renewables*, Wiley-VCH Verlag GmbH & Co. KGaA, pp. 413–423.

12. Deng, W., Zhang, Q., and Wang, Y. (2014) *Catal. Today*, **234**, 31–41.

13. Reutemann, W. and Kieczka, H. (2000) *Ullmann's Encyclopedia of Industrial Chemistry*, Wiley-VCH Verlag GmbH & Co. KGaA.

14. Schöpf, C. and Wild, H. (1954) *Chem. Ber.*, **87**, 1571–1575.

15. (a) Jin, F., Yun, J., Li, G., Kishita, A., Tohji, K., and Enomoto, H. (2008) *Green Chem.*, **10**, 612–615; (b) Taccardi, N., Assenbaum, D., Berger, M.E.M., Bosmann, A., Enzenberger, F., Wolfel, R., Neuendorf, S., Goeke, V., Schodel, N., Maass, H.J., Kistenmacher, H., and Wasserscheid, P. (2010) *Green Chem.*, **12**, 1150–1156.

16. (a) Yu, D., Aihara, M., and Antal, M.J. (1993) *Energy Fuel*, **7**, 574–577; (b) Calvo, L. and Vallejo, D. (2002) *Ind. Eng. Chem. Res.*, **41**, 6503–6509.

17. Wolfel, R., Taccardi, N., Bosmann, A., and Wasserscheid, P. (2011) *Green Chem.*, **13**, 2759–2763.

18. Albert, J., Wolfel, R., Bosmann, A., and Wasserscheid, P. (2012) *Energy Environ. Sci.*, **5**, 7956–7962.

19. Albert, J., Luders, D., Bosmann, A., Guldi, D.M., and Wasserscheid, P. (2014) *Green Chem.*, **16**, 226–237.

20. Li, J., Ding, D.-J., Deng, L., Guo, Q.-X., and Fu, Y. (2012) *ChemSusChem*, **5**, 1313–1318.

21. Zhang, J., Sun, M., Liu, X., and Han, Y. (2014) *Catal. Today*, **233**, 77–82.

22. (a) Niu, M., Hou, Y., Ren, S., Wang, W., Zheng, Q., and Wu, W. (2015) *Green Chem.*, **17**, 335–342; (b) Wang, W., Niu, M., Hou, Y., Wu, W., Liu, Z., Liu, Q.,

Ren, S., and Marsh, K.N. (2014) *Green Chem.*, **16**, 2614–2618.

23. Xu, J., Zhang, H., Zhao, Y., Yang, Z., Yu, B., Xu, H., and Liu, Z. (2014) *Green Chem.*, **16**, 4931–4935.

24. Li, K., Bai, L., Amaniampong, P.N., Jia, X., Lee, J.-M., and Yang, Y. (2014) *ChemSusChem*, **7**, 2670–2677.

25. Ramachandran, S., Fontanille, P., Pandey, A., and Larroche, C. (2006) *Food Technol. Biotechnol.*, **44**, 185–195.

26. Zhang, J., Liu, X., Sun, M., Ma, X., and Han, Y. (2012) *ACS Catal.*, **2**, 1698–1702.

27. (a) Besson, M. and Gallezot, P. (2000) *Catal. Today*, **57**, 127–141; (b) Besson, M., Lahmer, F., Gallezot, P., Fuertes, P., and Fleche, G. (1995) *J. Catal.*, **152**, 116–121; (c) Ishida, T., Kinoshita, N., Okatsu, H., Akita, T., Takei, T., and Haruta, M. (2008) *Angew. Chem. Int. Ed.*, **47**, 9265–9268.

28. (a) Tan, X., Deng, W., Liu, M., Zhang, Q., and Wang, Y. (2009) *Chem. Commun.*, **46**, 7179–7181; (b) Zhang, J., Liu, X., Hedhili, M.N., Zhu, Y., and Han, Y. (2011) *ChemCatChem*, **3**, 1294–1298.

29. (a) An, D., Ye, A., Deng, W., Zhang, Q., and Wang, Y. (2012) *Chem. Eur. J.*, **18**, 2938–2947; (b) Amaniampong, P.N., Li, K., Jia, X., Wang, B., Borgna, A., and Yang, Y. (2014) *ChemCatChem*, **6**, 2105–2114.

30. Onda, A., Ochi, T., and Yanagisawa, K. (2011) *Catal. Commun.*, **12**, 421–425.

31. Serrano-Ruiz, J.C., Luque, R., and Sepulveda-Escribano, A. (2011) *Chem. Soc. Rev.*, **40**, 5266–5281.

32. Lin, H. (2013) Catalytic Process of Conversion Biomass into Hydrocarbon Fuels, U.S. Patent 20130079566 A1.

33. Lin, H., Strull, J., Liu, Y., Karmiol, Z., Plank, K., Miller, G., Guo, Z., and Yang, L. (2012) *Energy Environ. Sci.*, **5**, 9773–9777.

34. Stapley, J.A. and BeMiller, J.N. (2007) *Carbohydr. Res.*, **342**, 407–418.

35. Podolean, I., Kuncser, V., Gheorghe, N., Macovei, D., Parvulescu, V.I., and Coman, S.M. (2013) *Green Chem.*, **15**, 3077–3082.

36. Moore, J., Stanitski, C., and Jurs, P. (2009) *Principles of Chemistry: The Molecular Science*, Cengage Learning.

37. Niu, M., Hou, Y., Ren, S., Wu, W., and Marsh, K.N. (2015) *Green Chem.*, **17**, 453–459.

38. Chang, C., Cen, P., and Ma, X. (2007) *Bioresour. Technol.*, **98**, 1448–1453.

39. (a) Brunow, G. (2005) *Biopolymers Online*, Wiley-VCH Verlag GmbH & Co. KGaA; (b) Chakar, F.S. and Ragauskas, A.J. (2004) *Ind. Crops Prod.*, **20**, 131–141.

40. (a) Partenheimer, W. (2009) *Adv. Synth. Catal.*, **351**, 456–466; (b) Zakzeski, J., Bruijnincx, P.C.A., Jongerius, A.L., and Weckhuysen, B.M. (2010) *Chem. Rev.*, **110**, 3552–3599.

41. (a) Huber, G.W., Iborra, S., and Corma, A. (2006) *Chem. Rev.*, **106**, 4044–4098; (b) Crestini, C., Crucianelli, M., Orlandi, M., and Saladino, R. (2010) *Catal. Today*, **156**, 8–22; (c) Lange, H., Decina, S., and Crestini, C. (2013) *Eur. Polym. J.*, **49**, 1151–1173; (d) Son, S. and Toste, F.D. (2010) *Angew. Chem. Int. Ed.*, **49**, 3791–3794; (e) Crestini, C., Caponi, M.C., Argyropoulos, D.S., and Saladino, R. (2006) *Bioorg. Med. Chem.*, **14**, 5292–5302; (f) Herrmann, W.A., Weskamp, T., Zoller, J.P., and Fischer, R.W. (2000) *J. Mol. Catal. A: Chem.*, **153**, 49–52; (g) Rahimi, A., Ulbrich, A., Coon, J.J., and Stahl, S.S. (2014) *Nature*, **515**, 249–252; (h) Pilla, S. (2011) *Handbook of Bioplastics and Biocomposites Engineering Applications*, John Wiley & Sons Inc.

42. Collinson, S.R. and Thielemans, W. (2010) *Coord. Chem. Rev.*, **254**, 1854–1870.

43. (a) Kuwahara, M., Glenn, J.K., Morgan, M.A., and Gold, M.H. (1984) *FEBS Lett.*, **169**, 247–250; (b) Tien, M. and Kirk, T.K. (1983) *Science*, **221**, 661–663.

44. (a) Argyropoulos, D.S. (2001) *Oxidative Delignification Chemistry: Fundamentals and Catalysis*, American Chemical Society; (b) Van Dyk, J.S. and Pletschke, B.I. (2012) *Biotechnol. Adv.*, **30**, 1458–1480.

45. (a) Hanson, S.K., Baker, R.T., Gordon, J.C., Scott, B.L., and Thorn, D.L. (2010) *Inorg. Chem.*, **49**, 5611–5618; (b) Sedai, B., Diaz-Urrutia, C., Baker, R.T., Wu, R., Silks, L.A.P., and Hanson, S.K. (2011) *ACS Catal.*, **1**, 794–804.

46. Hanson, S.K., Wu, R., and Silks, L.A.P. (2012) *Angew. Chem. Int. Ed.*, **51**, 3410–3413.

47. Voitl, T. and Rudolf von Rohr, P. (2008) *ChemSusChem*, **1**, 763–769.

48. Hoover, J.M. and Stahl, S.S. (2011) *J. Am. Chem. Soc.*, **133**, 16901–16910.

49. Sedai, B. and Baker, R.T. (2014) *Adv. Synth. Catal.*, **356**, 3563–3574.

50. Rahimi, A., Azarpira, A., Kim, H., Ralph, J., and Stahl, S.S. (2013) *J. Am. Chem. Soc.*, **135**, 6415–6418.

51. Wu, G.X., Heitz, M., and Chornet, E. (1994) *Ind. Eng. Chem. Res.*, **33**, 718–723.

52. Liu, S., Shi, Z., Li, L., Yu, S., Xie, C., and Song, Z. (2013) *RSC Adv.*, **3**, 5789–5793.

53. Ambrose, K., Hurisso, B.B., and Singer, R.D. (2013) *Can. J. Chem.-Rev. Can. Chim.*, **91**, 1258–1261.

54. Zakzeski, J., Jongerius, A.L., and Weckhuysen, B.M. (2010) *Green Chem.*, **12**, 1225–1236.

55. Crestini, C., Pro, P., Neri, V., and Saladino, R. (2005) *Bioorg. Med. Chem.*, **13**, 2569–2578.

56. (a) Sales, F.G., Maranhao, L.C.A., Lima Filho, N.M., and Abreu, C.A.M. (2006) *Ind. Eng. Chem. Res.*, **45**, 6627–6631; (b) Sales, F.G., Maranhao, L.C.A., Lima, N.M., and Abreu, C.A.M. (2007) *Chem. Eng. Sci.*, **62**, 5386–5391.

57. Tejuca, L.G. and Fierro, J.L.G. (2000) *Properties and Applications of Perovskite-Type Oxides*, Taylor & Francis.

58. Zhang, J., Deng, H., and Lin, L. (2009) *Molecules*, **14**, 2747–2757.

59. Deng, H., Lin, L., Sun, Y., Pang, C., Zhuang, J., Ouyang, P., Li, Z., and Liu, S. (2008) *Catal. Lett.*, **126**, 106–111.

60. Mate, V.R., Jha, A., Joshi, U.D., Patil, K.R., Shirai, M., and Rode, C.V. (2014) *Appl. Catal., A-Gen.*, **487**, 130–138.

61. Zakzeski, J., Debczak, A., Bruijnincx, P.C.A., and Weckhuysen, B.M. (2011) *Appl. Catal., A-Gen.*, **394**, 79–85.

62. (a) Rajagopalan, B., Cai, H., Busch, D., and Subramaniam, B. (2008) *Catal. Lett.*, **123**, 46–50; (b) Cedeno, D. and Bozell, J.J. (2012) *Tetrahedron Lett.*, **53**, 2380–2383.

63. Badamali, S.K., Luque, R., Clark, J.H., and Breeden, S.W. (2011) *Catal. Commun.*, **12**, 993–995.

64. Dawange, M., Galkin, M.V., and Samec, J.S.M. (2015) *ChemCatChem*, **7**, 401–404.

65. (a) Sawadjoon, S., Lundstedt, A., and Samec, J.S.M. (2013) *ACS Catal.*, **3**, 635–642; (b) Galkin, M.V. and Samec, J.S.M. (2014) *ChemSusChem*, **7**, 2154–2158.

66. Younker, J.M., Beste, A., and Buchanan, A.C. III, (2011) *ChemPhysChem*, **12**, 3556–3565.

67. Chatel, G. and Rogers, R.D. (2014) *ACS Sustainable Chem. Eng.*, **2**, 322–339.

68. (a) Cimpeanu, V., Parvulescu, A.N., Parvulescu, V.I., On, D.T., Kaliaguine, S., Thompson, J.M., and Hardacre, C. (2005) *J. Catal.*, **232**, 60–67; (b) Hardacre, C. and Parvulescu, V. (2014) *Catalysis in Ionic Liquids: From Catalyst Synthesis to Application*, Royal Society of Chemistry; (c) Parvulescu, V.I. and Hardacre, C. (2007) *Chem. Rev.*, **107**, 2615–2665.

69. Zhu, Y., Li, C., Sudarmadji, M., Min, N.H., Biying, A.O., Maguire, J.A., and Hosmane, N.S. (2012) *Chemistryopen*, **1**, 67–70.

70. Stärk, K., Taccardi, N., Bösmann, A., and Wasserscheid, P. (2010) *ChemSusChem*, **3**, 719–723.

71. Jia, S., Cox, B.J., Guo, X., Zhang, Z.C., and Ekerdt, J.G. (2011) *Ind. Eng. Chem. Res.*, **50**, 849–855.

72. Gao, Y., Zhang, J., Chen, X., Ma, D., and Yan, N. (2014) *ChemPlusChem*, **79**, 825–834.

73. Tester, R.F., Karkalas, J., and Qi, X. (2004) *J. Cereal Sci.*, **39**, 151–165.

74. Whistler, R.L. and Schweiger, R. (1959) *J. Am. Chem. Soc.*, **81**, 3136–3139.

75. (a) Floor, M., Schenk, K.M., Kieboom, A.P.G., and van Bekkum, H. (1989) *Starch – Stärke*, **41**, 303–309; (b) Teleman, A., Kruus, K., Ämmälahti, E., Buchert, J., and Nurmi, K. (1999) *Carbohydr. Res.*, **315**, 286–292; (c) Nieuwenhuizen, M.S., Kieboom, A.P.G., and van Bekkum, H. (1985) *Starch – Stärke*, **37**, 192–200; (d) Kuakpetoon, D. and Wang, Y.-J. (2001) *Starch – Stärke*, **53**, 211–218; (e) Floor, M., Kieboom, A.P.G., and van Bekkum, H. (1989) *Starch – Stärke*, **41**, 348–354.

76. Kochkar, H., Morawietz, M., and Hölderich, W.F. (2001) *Appl. Catal., A-Gen.*, **210**, 325–328.

77. Veelaert, S., de Wit, D., Gotlieb, K.F., and Verhé, R. (1997) *Carbohydr. Polym.*, **33**, 153–162.

78. Painter, T.J., Cesàro, A., Delben, F., and Paoletti, S. (1985) *Carbohydr. Res.*, **140**, 61–68.

79. Chong, W.T., Uthumporn, U., Karim, A.A., and Cheng, L.H. (2013) *LWT Food Sci. Technol.*, **50**, 439–443.

80. Parovuori, P., Hamunen, A., Forssell, P., Autio, K., and Poutanen, K. (1995) *Starch – Stärke*, **47**, 19–23.

81. Manelius, R., Buleon, A., Nurmi, K., and Bertoft, E. (2000) *Carbohydr. Res.*, **329**, 621–633.

82. Aouf, C., Harakat, D., Muzart, J., Estrine, B., Marinkovic, S., Ernenwein, C., and Le Bras, J. (2010) *ChemSusChem*, **3**, 1200–1203.

83. Sorokin, A., Kachkarova-Sorokina, S., Donzé, C., Pinel, C., and Gallezot, P. (2004) *Top. Catal.*, **27**, 67–76.

84. (a) de Nooy, A.E.J., Besemer, A.C., and van Bekkum, H. (1994) *Recl. Trav. Chim. Pays-Bas*, **113**, 165–166; (b) de Nooy, A.E.J., Besemer, A.C., and van Bekkum, H. (1995) *Carbohydr. Res.*, **269**, 89–98; (c) Bragd, P.L., van Bekkum, H., and Besemer, A.C. (2004) *Top. Catal.*, **27**, 49–66.

85. Mathew, S. and Adlercreutz, P. (2009) *Bioresour. Technol.*, **100**, 3576–3584.

86. Sheldon, R.A. and Arends, I.W.C.E. (2006) *J. Mol. Catal. A: Chem.*, **251**, 200–214.

87. Herrmann, W.A., Zoller, J.P., and Fischer, R.W. (1999) *J. Organomet. Chem.*, **579**, 404–407.

88. (a) Kachkarova-Sorokina, S.L., Gallezot, P., and Sorokin, A.B. (2004) *Chem. Commun.*, 2844–2845; (b) Pierre, G. and Alexander, B.S. (2008) *Catalysis of Organic Reactions*, CRC Press, pp. 263–270.

89. Tolvanen, P., Sorokin, A., Maki-Arvela, P., Leveneur, S., Murzin, D.Y., and Salmi, T. (2011) *Ind. Eng. Chem. Res.*, **50**, 749–757.

90. Chen, X., Liu, Y., Wang, H., Yuan, M., Wang, X., and Chen, Y. (2014) *RSC Adv.*, **4**, 11232–11239.

91. Chen, X., Souvanhthong, B., Wang, H., Zheng, H., Wang, X., and Huo, M. (2013) *Appl. Catal., B-Environ.*, **138**, 161–166.

92. Chen, X., Yan, S., Wang, H., Hu, Z., Wang, X., and Huo, M. (2015) *Carbohydr. Polym.*, **117**, 673–680.

93. Chen, X., Wang, H., Xu, J., Huo, M., Jiang, Z., and Wang, X. (2014) *Catal. Today*, **234**, 264–270.

16
Deoxygenation of Liquid and Liquefied Biomass

Thuan Minh Huynh, Udo Armbruster, and Andreas Martin

16.1
Introduction

The basement of the present energy and chemical industry is largely depending on the use of coal, crude oil, and natural gas (e.g., [1]). Eighty-seven percent of the energy consumed worldwide in 2010 and 87% and 92% of energy consumed in the US and China, respectively, had fossil sources [2]. Only a minor portion consists of other resources applied in the energy sector such as nuclear and hydroelectric power, wind and geothermal sources, solar energy, and biomass. For example, biomass covered 8% of the German electric power consumption, being equal to 47.9 TWh in 2013 (for comparison 33.7 TWh in 2010) [3, 4]. The European Union power consumption based on biomass came to 123.3 TWh in 2010 [4]. These numbers mainly include solid fuels and biogas; liquid fuels only play a minor role. The picture in the chemical industries looks similar; biomass-related products are rather scarce [5–7]. Only 8–10% of the feedstock of the European chemical industry (6–7 Mton compared to about 75 Mton of fossil raw materials) is bio-based and used for the production of various chemicals. However, it is very clear that the previously mentioned fossil energy sources are finite; simultaneously they are suspected for the threat of climate change. The maximum in crude oil production (so-called "peak oil") is expected soon and, thus, our world is presently facing a feedstock change with respect to energy production and chemical industry [8, 9]. Therefore, sustainable and green resources are needed to serve as alternatives [5, 10–12].

The recent two decades revealed a significant increase in use of renewables for energy supply and production of chemicals. In particular, biofuel industry basing on plant oils, sugar cane, and corn dramatically increased their production of biodiesel (FAME – fatty acid methyl ester) and bioethanol. For the most part, these oxygenates are added to conventional fuels in amounts up to 10 vol%. However, also flexible fuel vehicles (FFVs) can be met accepting much higher oxygenate portions (up to 85 vol% of ethanol or methanol in general). However, there are also several efforts to upgrade oxygenates such as FAME to hydrocarbons being very similar to conventional, fossil sources-based fuels. One such example is the

Chemicals and Fuels from Bio-Based Building Blocks, First Edition.
Edited by Fabrizio Cavani, Stefania Albonetti, Francesco Basile, and Alessandro Gandini.
© 2016 Wiley-VCH Verlag GmbH & Co. KGaA. Published 2016 by Wiley-VCH Verlag GmbH & Co. KGaA.

NExBTL process of the Finnish company Neste Oil Oyj, by which FAME is hydrogenated completely to hydrocarbons, for example, heptadecane [13]. In addition, huge knowledge has been accumulated to convert different biomass sources into valuable chemicals using various biorefinery concepts (e.g., [14–16]). These concepts work well if having expensive and rather small-scale products on the agenda (e.g., [15, 17–21]). Larger processes like the dehydration of bioethanol to ethylene seem to be not economic at present. However, the resources for such kind of biofuels and even large-scale chemicals are limited; they need fertile land and water, they partly destroy nature and environment, and they are often in competition to nutrition demands. Thus, the world is looking for a loophole to bridge the gap between increasing demand for liquid fuels and various platform chemicals, respectively, and future depletion of fossil sources as well as inexpensive, abundantly available, and sustainable raw materials (e.g., [22]).

For such purpose, lignocellulosic biomass might have a great potential for the production of fuels, chemicals, and other carbon-based materials [23]. Three main classes of organic materials are present: cellulose, hemicellulose, and lignin in portions of approximately 35–45%, 25–35%, and 15–25%, respectively, depending on the respective biomass. Unfortunately, one serious drawback for manufacture of motor fuels and diverse chemicals from such raw materials is their oxygen content varying in a range of 35–45 wt% in dry matter. Thus, biomass reveals a general sum formula of $CH_{1.4}O_{0.6}$ in contrast to hydrocarbons or liquid fuels, showing a sum formula close to CH_2. A useful comparison of the energy content of biomass and fossil fuels is given by their O:C and H:C ratios, also known as the *van Krevelen diagram* (e.g., [24, 25]). The lower the respective ratios the greater the energy content of the material. Thus, wheat straw reveals about 17 MJ kg^{-1} (wood chips or lignite reveal similar low values) in contrast to diesel fuel that offers 43 MJ kg^{-1} (e.g., [26]).

In addition, its complex nature, characterized by a high polymerization degree, makes it very difficult to get defined products by chemical transformations. However, there are several options to use lignocellulosic biomass for the production of liquid fuels and/or chemicals including thermochemical and biochemical conversions [25]. The first step of the thermochemical-induced conversion exhibits bond breaks leading to smaller molecules and thus to liquids and/or gases (Figure 16.1) [25]. This can be carried out by pyrolysis [27], hydrothermal liquefaction [28–30], or catalytic liquefaction [31–33]. Downstream removal of the remaining oxygen is needed, thus deoxygenation in the absence of additionally fed hydrogen or hydrodeoxygenation (HDO) have to follow [34, 35].

Pyrolysis runs in a temperature window of 200–900 °C. Low-temperature or mild pyrolysis is also known as *torrefaction* (e.g., [27, 36–38]). Biomass is heated to 200–350 °C in a low oxygen-containing atmosphere. Thus, all moisture is removed as well as a part of the volatile matter to produce a high-quality solid biofuel that can be used for combustion and gasification. Slow pyrolysis at around 400 °C mainly leads to charcoal; however, lower carboxylic acids, aldehydes, alcohols, and tar compounds are typical by-products [27]. Higher temperatures mainly lead to an increased amount of liquid products, but also useful combustible pyrolysis gas and solid pyrolysis coke are obtained (e.g.,

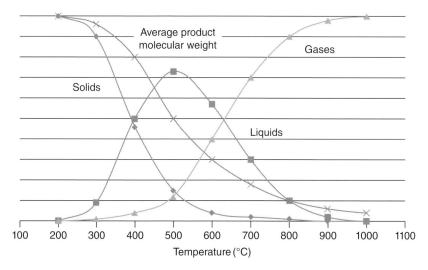

Figure 16.1 Simplified scheme of product groups obtained by pyrolytic cracking at different temperatures consequently leading to a drop in molecular weight of product molecules.

[27]). Beneficially, liquefied biomass has a high energy density and is easier to transport compared to solid biomass like straw, agricultural residues, or wood. In particular, fast or flash pyrolysis is widely used to manufacture so-called biocrude or bio-oil [27, 39–41]. Outstanding examples are the bioliq® process developed at KIT Karlsruhe/Germany [27, 42, 43], the BtO® process developed by Pytec GmbH Lüneburg/Germany together with the Thünen-Institute for Wood Chemistry at Hamburg/Germany, based on the ablative pyrolysis [44], or the rotating cone principle developed by the Dutch BTG company [45]. Figure 16.2 shows the composition of a typical product from BtO® process ($T = 500\,°C$, $t_{reaction} = 1\,s$). Higher temperatures lead to an increased proportion of gases. In the end, chemical bonds can be completely cleaved by the gasification route to get CO and H_2 (syngas); subsequently, methanol or Fischer–Tropsch (FT) syntheses might be used for production of oxygenates, olefins, or hydrocarbons that can be used as liquid fuel or chemicals [27]. Much is known on this route from coal

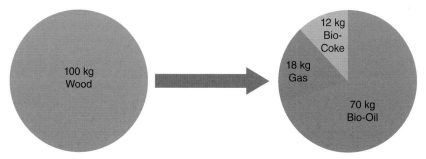

Figure 16.2 Products obtained by fast pyrolysis (BtO® process) at about 500 °C and a reaction time of 1 s.

and wood gasification from past 150 years, although process flexibility and plant design may cause difficulties for operation in large scale.

Upgrading of liquefied biomass is necessary for further use as a drop-in fuel or cofeed in conventional refineries because of high oxygenate and water content [34]. This can be carried out as already mentioned by HDO but also by a deoxygenation in absence of hydrogen; an outstanding example for the latter is the so-called aqueous phase reforming (APR) process leading from biomass to chemicals (e.g., [46, 47]). Among the available upgrading strategies, HDO supported by catalysts is considered as most effective technology [48–50]. Catalysts for HDO are traditional hydrodesulfurization (HDS) catalysts, such as Ni- or Co-containing MoS_2/Al_2O_3, or transition or noble metal catalysts, as, for example, Pd/C. However, carbon deposition often limits catalysts lifetime.

The following paragraphs might give a current overview on various deoxygenation strategies including the use of model compounds and real biocrude feeds such as triglycerides, pyrolysis oils, and black liquor, a waste stream from pulp and paper industry.

16.2
General Remarks on Deoxygenation

There are several pathways that have been proposed for partial or total deoxygenation of liquid or liquefied biomass such as hydrotreating, catalytic cracking, APR, steam reforming, esterification, gasification, and so on, (e.g., [35, 40, 41, 51–54]), in which two main routes have been extensively investigated such as catalytic hydrotreating or HDO and catalytic cracking or simple deoxygenation. In the latter case, the oxygen is removed as carbon dioxide, carbon monoxide, water and short-chain oxygenates via several reactions like cracking, decarboxylation, decarbonylation, dehydration and water gas shift reaction. In general, this deoxygenation is carried out in presence of cracking catalysts (e.g., zeolites, silica-alumina, etc.) at atmospheric or lower pressure [48, 53]. The resulting oil has a lower heating value and H:C ratio because the reaction runs without external hydrogen source. In contrast, HDO is a route which is processed in presence of hydrogen and catalysts, that is, oxygen can be mainly eliminated as water, leading to a high liquid product yield compared to those of catalytic cracking route [48]. HDO normally plays a minor role in standard refineries because of the very low concentration of oxygen in the crude oil feed. However, driven by the demand for alternative and renewable liquid fuels, especially from biomass resources, HDO has evolved quickly into a major technology platform [55, 56].

The analysis and evaluation of catalytic HDO of bio-oil are very difficult due to the complexity of the feeds, requiring more elaborated analysis techniques and still remaining a challenge up to now. As a result, studies on various individual oxygenates as model compounds could provide more insight into the chemistry before transferring to real feeds. Fundamentally, the oxygenated compounds can be classified in several groups: acids, alcohols, phenols, sugars,

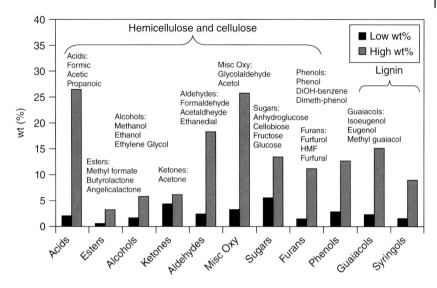

Figure 16.3 Typical detected chemical composition of bio-oils. Reproduced with permission of American Chemical Society, 2006, ref. [54].

aldehydes, ethers, esters, and so on (Figure 16.3). From a thermochemical point of view, among these groups, the oxygen elimination from phenols and aromatic ether groups will be difficult compared to that from alcohol and aliphatic ethers [56]. On the other hand, the reactivity of the different functional groups under hydroprocessing conditions follows the order: aldehydes, ketones < aliphatic ether < aliphatic alcohols < carboxylic acids < phenolic ethers < phenols < diphenyl ether < dibenzofuran, as, for example, reported by Elliott [34]. Consequently, the hydroprocessing of aldehydes and ketones run at low temperature (<200 °C), whereas phenolic ethers and phenols may need 300 °C or more. For these reasons, monomer phenolic compounds (phenol, catechol, anisole, guaiacol) or dimers have attracted more attention for investigation compared to other groups (acid, aldehydes, ketones, furans).

16.3
Deoxygenation of Model Compounds

16.3.1
Phenol and Alkylphenols

Considerable amounts of phenol and substituted phenols have been found in bio-oil, especially those derived from lignin as one of the main component in lignocellulosic biomass. However, the conversion of phenol to oxygen-free compounds is difficult. Therefore, phenol seems to be an excellent probe compound for understanding the fundamental chemistry of catalytic HDO reaction and measuring catalyst activity. Several reaction networks have been reported for phenol deoxygenation (e.g., to cyclohexane, benzene, methylcyclopentane), which are dependent mainly on types of catalyst and the reaction conditions. There are two general main pathways to eliminate oxygen from phenol, either the direct deoxygenation (DDO) of phenol to benzene or the hydrogenation (HYD) of the aromatic ring followed by dehydration for oxygen removal [57–59].

The latter pathway involves several steps such as phenol hydrogenation to cyclohexenol, an unstable intermediate that rapidly tautomerizes to cyclohexanone, which is further hydrogenated to cyclohexanol. Subsequently, cyclohexanol may dehydrate to form cyclohexene and finally be hydrogenated to cyclohexane. Scheme 16.1 represents our proposed reaction network of HDO of phenol in the presence of high amount of water as a solvent on monometallic and bimetallic Ni-based catalyst (Ni, Ni–Co, Ni–Cu) supported on different acidic materials (H-ZSM-5, H-Beta, H-Y, and ZrO_2) at 250 °C, 50 bar H_2 at room temperature (RT), based on several separate tests with feed and intermediates [60, 61]. In contrast, Zhao *et al.* proposed that hydrogenolysis of phenol to benzene is inhibited, and hydrogenolysis of cyclohexanol to cyclohexane is also suppressed when bifunctional catalysts (Pt/C, Ru/C, Pd/C combined with a mineral acid, Ni/H-ZSM-5, and Ni/Al_2O_3-HZSM-5) are used at the given reaction conditions (200 °C, 50 bar H_2 at RT) [62, 63]. However, hydrogenation of phenol is thermodynamically favorable at low temperature; high temperature promotes the formation of benzene [64].

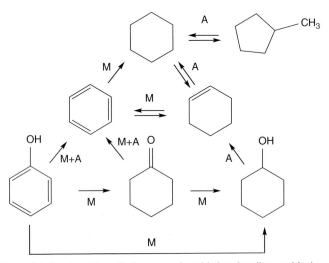

M: metal sites (metal or alloy)   A: acid sites (zeolites, oxides)

Scheme 16.1 Proposed reaction pathways of phenol HDO over supported Ni-based catalysts.

In addition, acid–base properties of supports, solvent properties, catalyst types, and reaction conditions influence reaction mechanisms (e.g., [58, 65–68]). Romero *et al.* [65] proposed a mechanism for HDO of 2-ethylphenol over conventional HDS catalysts (Ni- or Co-promoted MoS_2) in gas phase (340 °C, 70 bar). Sulfur vacancy sites are generated due to the weak bond between molybdenum and sulfur. Thus, oxygen atoms from the fed 2-ethylphenol can be observed on these active sites of a MoS_2 slab edge and then the direct C–O bond cleavage with the formation of ethylbenzene is initiated. The vacancy is subsequently recovered by elimination of water.

Another HDO reaction mechanism has been proposed for supported metal catalysts [66, 69]. The combination of transition metals and different supports basing on oxophilic metals (MoO_3, Cr_2O_3, WO_3, ZrO_2, Al_2O_3) or zeolites (H-ZSM-5, H-Beta, H-Y, etc.) enables the discussed pathways. Furthermore, the adsorption mode also affects the reaction network and the mechanism. For example, *p*-cresol can be adsorbed on the catalyst surface via either the hydroxyl group (terminal position), leading to the removal of oxygen to form toluene, or the coplanar position, undergoing aromatic ring saturation to form 4-methylcyclohexanol [66].

16.3.2
Guaiacol and Substituted Guaiacols

These compounds possess two oxygen-containing functional groups (hydroxyl and methoxyl attached to aromatic ring) and have been investigated extensively as model compounds (e.g., [70, 71]). Various pathways have been reported for guaiacol conversion toward a variety of products such as phenol, catechol, benzene, cyclohexane, methyl-substituted phenols (e.g., [70]) as indicated in Scheme 16.2. The possible routes are as follows: (i) migration of methyl groups

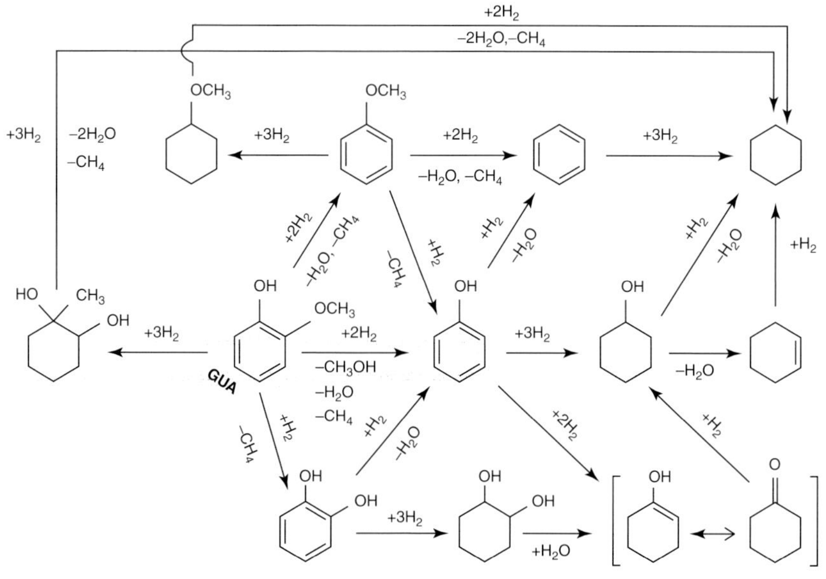

Scheme 16.2 Possible reaction pathways of HDO of guaiacol over Ni-based catalysts. Reproduced with permission of Royal Society of Chemistry, 2010, ref. [107].

to the aromatic ring followed by hydrogenation to form 1-methylcyclohexane-1,2-diol, (ii) direct demethoxylation to form phenol and subsequent formation of benzene or cyclohexane, (iii) direct demethylation into catechol and its further hydrogenation to 1,1-cyclohexanediol, and (iv) direct hydrogenolysis of the hydroxyl group to form anisole. Additionally, methyl-substituted aromatics were also observed, formed possibly by demethylation/methylation reactions catalyzed by acidic supports.

16.3.3
Lignin-Derived Molecules

Apart from the aforementioned monomers, phenolic dimers have been involved in HDO studies owing to their large amount in lignin-derived bio-oil. Zhao and Lercher [72, 73] reported the aqueous phase HDO of phenolic dimers on bifunctional catalysts (Pd/C and H-ZSM-5 or Ni/H-ZSM-5); the result is presented in Scheme 16.3. Several consecutive steps are involved: (i) hydrogenolysis of the ether to phenol and arenes on metal sites and (ii) hydrogenation/dehydration by both metal and acid sites. The observed products are mainly hydrocarbons of C_6–C_{16} range due to selective cleavage of substituted hydroxyl and ketone groups, whereas the C–C bond in the 5-5, β-1, and β-β structure of dimers is retained. Recently, Strassberger *et al.* [74] studied the HDO of β-O-4 lignin-type dimers (supported Cu-containing catalysts, 150 °C, 25 bar), revealing that mainly both β-O-4 cleavage products (phenol) and HDO products (ethylbenzene) are formed.

16.3.4
Short-Chain Carboxylic Acids

Larger quantities of carboxylic acids (mainly C_1–C_3 acids) might be present in bio-oils and cause strong corrosion impact. They have to be removed by various ways: (i) C–O bond cleavage (hydrogenolysis) followed by hydrogenation/dehydration/hydrogenation steps to finally form alkanes, (ii) decarboxylation or decarbonylation might form methane, and (iii) hydrogenation to form alcohols [75, 76]. Olcay *et al.* [76] reported the aqueous phase hydrogenation of acetic acid over a range of supported metal catalysts (110–290 °C, 51.7 bar). Ru/C was found to be most effective in both acetic acid conversion and ethanol selectivity at $T < 160$ °C. Elliott and Hart [77] found that acetic acid is hydrogenated to ethanol at lower temperatures but decomposes mainly to CH_4 and CO_2 at higher temperature over Ru/C (150–300 °C, 138 bar). In the case of propionic acid, Alotaibi *et al.* [78] reported that hydrogenolysis of C–C bonds plays an important role in the gas-phase reaction over Pd/SiO$_2$ and Pt/SiO$_2$ catalysts under atmospheric pressure ($T = 250$–400 °C) with 100% selectivity to ethane, whereas propanal and 1-propanol are dominant by hydrogenation over Cu/SiO$_2$. In the aqueous phase HDO at high pressure (64 bar), C–C bonds were cracked to form methane and ethane as the dominant products on Ru/ZrO$_2$ catalyst at lower temperature (190 °C) [79].

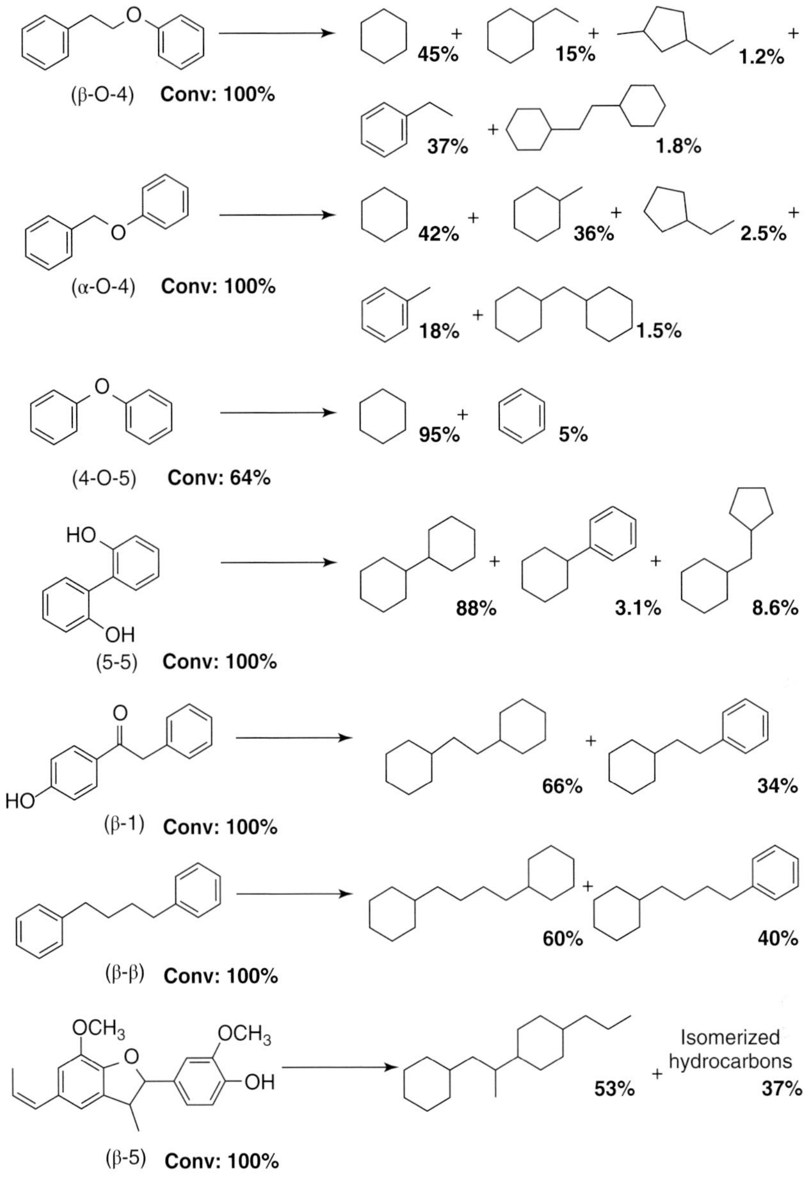

Scheme 16.3 HDO of phenolic dimers on Ni/HZSM-5 catalyst. Adapted with permission from ref. [72]. Copyright 2012 Wiley.

16.3.5
Furans, Furfurals, and Benzofurans

HDO of furans and furfurals (furan, methylfuran, dimethylfuran, furfural, benzofuran, or dibenzofuran) has been reported in literature (e.g., [56, 80, 81]). Two different routes (HYD and DDO) have been described for HDO of furan, methylfuran, and dimethylfuran, mainly leading to hydrocarbons (isomers of alkanes, alkenes, and alkadienes) [56]. Sitthisa and Resasco [81] studied the HDO of furfural in gas phase on SiO_2-supported Pd, Cu, and Pd–Cu catalysts under atmospheric pressure of hydrogen at $210-290\,^\circ$C; hydrogenation to furfuryl alcohol or decarbonylation to furan are the main routes. Furfuryl alcohol is the main product by using Cu/SiO_2 as catalyst, whereas decarbonylation is the dominant reaction over Pd/SiO_2. Zhao *et al.* [62, 63] reported that the HDO of furfural revealed two parallel reactions resulting in 64% pentane (by HDO) and 36% tetrahydropyran (by dehydration), respectively.

Benzofuran is also a common probe compound to evaluate HDO catalyst performance with regard to hydrocarbon formation (ethylbenzene, ethylcyclohexane, or methylcyclohexane). A variety of catalysts like sulfided or reduced Mo-containing materials [82, 83], NiMoP/Al_2O_3 [84], or $SiO_2-Al_2O_3$ supported Pt, Pd, and bimetallic Pt–Pd [85] are used.

16.4
Deoxygenation of Liquid and Liquefied Biomass

16.4.1
Triglycerides

Triglyceride-rich biomass has been emerging as potential feedstock for renewable liquid fuel since two decades. Biodiesel production is based on transesterification of triglycerides with methanol to FAME [86]. Beside conventional sources, the next-generation feedstock like nonedible oil, waste cooking oil, or algae oils is close to come. To overcome some drawbacks of biodiesel (product stability, gumming tendency, and lower energy content) and to make it suitable as a component for diesel fuel, residual oxygen and unsaturation should be removed prior to further use. Deoxygenation of free fatty acids can be performed via decarboxylation (forming alkane), decarbonylation (forming alkene), or HDO to get linear alkane without carbon loss [87]. Decarboxylation seems to be more favorable than decarbonylation, as Gibb's free energy is lower and required heat of reaction is much less [88]. Very comprehensive research work on deoxygenation of fatty acids was published during the last decade by the group of Murzin [89–92]. Very recently, Gosselink *et al.* [93] reported on a novel catalyst based on tungsten for selective deoxygenation of biomass-derived glyceride feedstock, in which supported tungsten oxide is highly active toward HDO products, whereas tungsten oxide catalyzed the decarboxylation/decarbonylation pathways.

Catalytic hydrotreating of waste cooking oil to white diesel was systematically studied over $NiMoS/Al_2O_3$ by Bezergianni *et al.* [94, 95]. The authors reported that the hydrotreating leads to an increase in the H/C ratio from 0.15 to 0.17 and a decrease of density from 0.858 to 0.757 g cm^{-3} with a negligible concentration of O, N, and S in the product oil. HDO of the next generation of triglycerides biomass (microalgae oil) to alkanes has been proposed intensively, for example, by the group of Lercher [96, 97]. They received 78 wt% yield of liquid alkanes (including 60 wt% of C_{18} alkanes) at complete conversion over 10 wt% Ni/H-Beta (260 °C, 40 bar H_2, 8 h) which is close to the theoretical yield (84 wt%) [96]. The progress on the deoxygenation of triglyceride biomass (vegetable, nonedible, microalgae, and waste cooking oil) was recently reviewed by Mohammad *et al.* [98] and Zhao *et al.* [99].

16.4.2
Bio-Oils from Hydrothermal Liquefaction or Fast Pyrolysis

The upgrading of hydrothermal liquefaction oil (LO) began in 1979 as the first pilot scale plant for liquefaction of wood started operation. The HDO of such LOs has been conducted extensively using the same catalysts and conditions as the conventional petroleum hydrotreating processes. The oxygen content can be reduced to near zero and the resulting upgraded oils showed excellent fuel quality when being analyzed with the standard petroleum products tests [100]. The preliminary tests showed that sulfided CoMo catalysts are highly active compared to the corresponding oxides. Ni-based catalysts showed similar activity compared to the sulfided CoMo catalyst, but a larger gas fraction and a loss of activity are observed during long-term tests [34]. The development of HDO of LOs in the past decades has been comprehensively described in some reviews [34, 101, 102].

Compared to the LO feed, the upgrading of fast pyrolysis oil (FPO) feed is a difficult task due to the reactivity of the condensed bio-oil recovered [34]. The transfer of hydrotreating conditions from HDO of LOs to FPOs seems to be unsuitable because of the high levels of char/coke production causing reactor plugging. Thus, a two-stage hydrotreatment concept was developed, in which FPO was first stabilized below 300 °C with either sulfided NiMo or CoMo catalysts [103]. Alternatively, nonisothermal hydrotreatment and low-severity HDO over the conventional HDS catalysts were also proposed [104, 105]. However, the conventional sulfided hydrotreating catalysts seem to be insufficient and unstable for FPO hydroprocessing due to low sulfur but high oxygenates and water contents in the feed [34]. In addition, the conventional support (Al_2O_3) is not stable, thus boosting the search for stable and active nonsulfided catalysts and advanced processing concepts [34].

Various supported noble metal catalysts (e.g., Ru/C, Ru/Al_2O_3, Ru/TiO_2, Pd/C, Pt/C) and conventional hydrotreating catalysts ($NiMo/Al_2O_3$, $CoMo/Al_2O_3$ as the benchmarks) were intensively studied for this process in batch reactor, for example, by Wildschut *et al.* [106]. In order to understand the effect of the different catalysts and the reaction conditions, a van Krevelen plot seems to be a common

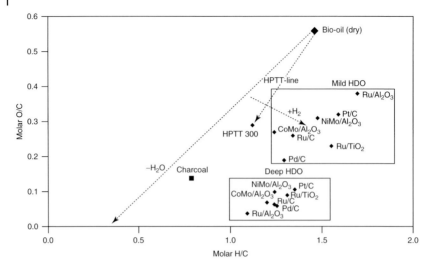

Figure 16.4 van Krevelen plot based on the elemental compositions (dry basis) of the mild and deep HDO over various catalysts. Reproduced with permission of American Chemical Society, 2009, ref. [106].

tool for evaluation as shown in Figure 16.4 [106]. A high degree of deoxygenation could be achieved with a suitable catalyst under severe conditions (350 °C, 200 bar). Ru/C seems to be a promising catalyst for HDO of FPO, but the loss of activity is evident during catalytic cycles, probably due to significant coke deposition accompanied by a decrease in pore volume and metal dispersion. It is likely that various parallel and/or consecutive reactions take place during HDO or high pressure thermal treatment (HPTT) [107–109] (Figure 16.5).

Elliott *et al.* [110] carried out HDO of mixed wood-derived FPO in a continuous flow reactor using Pd/C catalyst. The oxygen content of the hydrotreated bio-oil was reduced from 34 to 20 wt% at liquid hourly space velocity (LHSV) = 0.7 h⁻¹ and to 10 wt% at LHSV = 0.25 h⁻¹ with an oil yield of 62% based on dry matter (340 °C, 138 bar). However, the plugging in the front end of the catalyst bed and a pressure drop were observed during 10–100 h onstream. The tests with several bio-oils from other feedstock (corn stover, oak, poplar) indicated that the effect of feed nature is negligible in terms of oil yield and hydrogen consumption.

It can be realized from these studies that oxygenates containing C=O and/or C=C double bonds, particularly carbonyl compounds (aldehydes, ketones) and olefins, are highly reactive and responsible for the chemical instability of FPO. This means that stabilization via hydrogenation of the reactive compounds plays an important role. Therefore, two-stage HDO processes have been widely investigated [108, 110, 111]. Venderbosch *et al.* [108] reported the mild hydrotreating of FPO in presence of hydrogen and Ru/C catalyst up to 250 °C as a stabilization step. Elliott *et al.* [112] described a two-stage hydrotreating for upgrading of FPO obtained from pine sawdust fast pyrolysis. The first stage was

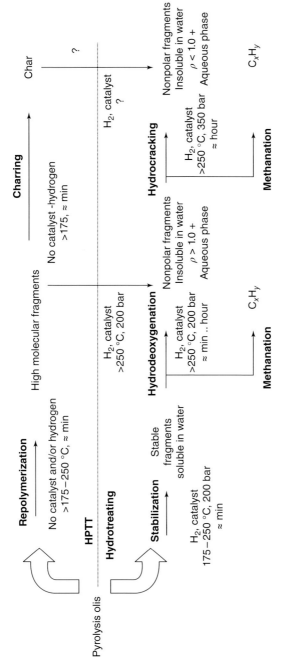

Figure 16.5 Proposed reaction pathway of HPTT and hydrotreating of pyrolysis oils. Reproduced with permission of Royal Society of Chemistry, 2010, ref. [107].

catalyzed by Ru/C at 175 or 250 °C, the second by sulfided CoMo/C at 400 °C. Consequently, product oils were obtained with 35–45 wt% yield based on dry matter ($\rho = 0.82–0.92\,\mathrm{g\,ml}^{-1}$, 0.2–2.7 wt% O content). However, pressure drop and plugging of the catalyst bed were noticed by char formation.

Notably, bio-oils are acidic and possess high water content (15–30 wt%), thus an aqueous phase processing has been recognized as an alternative approach [113]. In this regard, Vispute et al. [113, 114] proposed the HDO of wood-derived FPO or its aqueous fraction via an integrated catalytic processing. The feed is first hydrogenated using Ru/C (125 °C, 100 bar), followed by upgrading toward olefins, aromatics, and light alkanes in presence of ZSM-5 zeolite. The yield of named products was further increased by a subsequent HDO step using Pt/C (250 °C, 100 bar) without reactor plugging during 5 days on operation [114].

Hydrogen consumption in HDO reaction is an important aspect because it is in line with the degree of deoxygenation, the higher heating value (HHV), and finally it rules the operating cost of the process. The hydrogen consumption increases nonlinearly with the degree of HDO [108]. Roughly 300 Nm^3 of hydrogen might be consumed per ton of FPO at the degree of oxygen removal of 70%, which is significantly lower than the complete deoxygenation requires (approximately 600 $\mathrm{Nm}^3\,\mathrm{ton}^{-1}$ in extrapolation). This can be explained by the reactivity of the various oxygen functionalities in bio-oil: the highly reactive components (ketones, aldehydes) are easily converted with low hydrogen consumption, whereas the more complex molecules (phenols, guaiacols, dimers) are recalcitrant to deoxygenation and have to be submitted to an initial hydrogenation/saturation and therefore the hydrogen consumption exceeds the stoichiometric prediction [56]. It is likely that the degree of deoxygenation correlates with the oil yield, that means full removal of oxygen would lead to rather low product yields; for instance, the oil yield drops from 55% to 30% when the degree of deoxygenation increases from 78% to 100% [115]. Mercader et al. found that the hydrogen consumption correlates with HHV of product oil at about 22 $\mathrm{Nl(H_2)/MJ}$ [116]. One strategic question for researchers is therefore to which extent the deoxygenation should be pushed?

16.4.3
Black Liquor

Apart from the previously mentioned liquefied biomass, utilization of industrial lignin-derived feeds (e.g., black liquor) from pulp and paper industry and future lignocellulosic biorefineries might increase the value of existing processes as well as the resource efficiencies. There are several routes for usage of black liquor such as gasification, thermal treatment, and hydrotreating [117]. Few studies on HDO of black liquor have been carried out owing to the complexity of the feed. Elliott and Oasmaa [118] reported that black liquor can be converted into a mixture of hydrocarbons and phenols by two-stage HDO at 280 °C (40 min) and 380 °C (65 min) over conventional HDS catalysts (sulfided CoMo/Al_2O_3, NiMo/Al_2O_3, and NiMo/zeolite). The degree of deoxygenation reached 75%

and further increased to 85% with acid-washed black liquor (to remove sodium before reaction). The authors also stated that addition of water may result in high yields of distillates and higher product quality than without water addition. Oasmaa *et al.* [119] conducted HDO runs of different technical lignins from the corresponding black liquors using several commercial metal oxide catalysts (NiMo/Al_2O_3–SiO_2, NiMo/zeolite, and Cr_2O_3/Al_2O_3) at 300–400 °C for 40 min. Oil yields of 49–71 wt% were received. The analyses showed low molecular weight components in the oil products and the presence of various alkyl benzenes, phenols, and polycyclic aromatics.

16.5
Deoxygenation in Absence of Hydrogen

Simple pyrolysis of biomass can already lead to partial removal of oxygen, most preferred via decarboxylation, decarbonylation, or dehydration. The more oxygen is removed via these reactions, the less external hydrogen is required. In principle, pyrolysis and/or hydrothermal liquefaction of any biomass toward bio-oil can be seen also as a first step of the deoxygenation pathway. There exist two main routes, thermal or catalytic cracking, following well-known basic mechanisms, for example, the cleavage of C–C bonds in carboxylic acids or esters to form CO_2, CO, and value-added fuel-like hydrocarbons.

16.5.1
Pyrolysis Oils

A further effective upgrading of FPOs requires the use of catalysts and/or external hydrogen. Most work on HDO has been discussed in the previous sections. In addition, some studies on the catalytic cracking of pyrolysis oil are known (e.g., [120, 121]). However, the major challenges are noticed (e.g., nozzle plugging, irreversible catalyst deactivation) owing to significant formation of coke, tar, and char [122]. On the other side, the degree of deoxygenation of some FPOs was the starting point for investigating the feasibility of direct coprocessing together with crude oil in standard refineries, as related infrastructure and mature unit operations are available. Prior to the coprocessing, the upgrading of FPOs must be performed to make them miscible with fossil feeds. Several attempts have been addressed on mild HDO and subsequent coprocessing using lab-scale facility Fluid Catalytic Cracking (FCC) or HDS test (e.g., [115, 116, 123]). Mercader *et al.* [116] demonstrated that the coprocessing of 20 wt% HDO oil and 80 wt% standard feedstock (long residue) is successful even if the oxygen content of HDO oil (17–28 wt% on dry basis) is high. Similar product yields, that is, gasoline (44–46 wt%) and light cycle oil (23–25 wt%) are retained compared to the base feed. The authors also reported that a competition between HDS and HDO is observed during coprocessing in HDS unit and the degree of HDS is reduced. Moreover, *in situ* catalytic fast pyrolysis is considered as a developing process for stabilized bio-oil production; however, the development of stable and robust is needed to overcome the fast deactivation [124].

16.5.2
Triglycerides

In general, the utilization of triglycerides for deoxygenation routes competes with other outlets like food and pharmaceutical markets and transesterification to biodiesel. It has been stated that pyrolysis processes have some advantages over the biodiesel production, being the most popular alternative nowadays, namely, better product qualities [125] and lower processing costs [126]. The reaction pathway and the final product compositions depend on the degree of saturation, as it might affect the sequence of deoxygenation and C–C bond cleavage. An intrinsic drawback of decarboxylation/decarbonylation of triglycerides is the loss of a part of biomass carbon as CO_x, in contrast to hydrogen-based processes, where oxygen is removed as water.

It is well-known that triglycerides start to decompose at around 200 °C. Their pyrolysis via direct thermal or catalytic cracking has been investigated for more than one century and is extensively reviewed in [127]. Most frequently used catalyst types include transition metal catalysts, molecular sieve type (zeolite) catalysts (as used in conventional crackers), activated alumina, and sodium carbonate [126]. A large variety of catalysts has been reported, but depending on feedstock, catalysts and reactions conditions, a broad range of conversion, hydrocarbon yields, and oxygen contents in the product fractions are obtained. It seems that zeolite catalysts increase the fraction of gaseous by-products and promote the formation of aromatic products. Mesoporous materials lower the formation of gas fraction and increase the yield for diesel-like hydrocarbons rather than gasoline. Transition metals give the opportunity to tune acidity and redox properties at the same time, and high conversions and yields can be obtained. Sodium carbonate produces less aromatic compounds; however, 70% yield for alkanes can be obtained together with approximately 20% of aromatics. Typically, studies were carried out in both batchwise and continuous modes from 300 to 500 °C at atmospheric pressure, but temperatures up to 1000 °C have been reported. During such runs, triglycerides are converted into aliphatic, unsaturated, and aromatic hydrocarbons, accompanied by oxygenated compounds like ketones and carboxylic acids with overall yields around 80% and more [89, 126]. As an example, for conversion of canola oil in a fixed-bed reactor using different heterogeneous catalysts, for example, zeolites and alumina, conversions of 54–100% have been reported [128]. In general, the obtained hydrocarbon fractions show lower viscosities and higher cetane numbers than the parent triglycerides.

Own studies with sun flower oil [129] revealed that the deoxygenation efficiency was enhanced by adding KOH as reaction partner. NMR spectroscopy proved to be an effective tool to quantify the various hydrocarbon fractions; it was evidenced that the fraction of aromatics increased steadily with temperature to 15 wt% at 400 °C. At the same time, accelerated C–C bond cleavage at the cost of deoxygenation was observed.

16.5.3
Long-Chain Carboxylic Acids

A special case of triglyceride conversion toward fuels is the deoxygenation of natural long-chain carboxylic acids. The economic potential for the conversion of free fatty acids toward fuels may be rather low; however, many researchers use them as model compounds to study reactivity of intermediates and mechanisms in deoxygenation by thermal [130] or catalytic processes in batch or continuous mode. Extensive studies on decarboxylation of different fatty acids (C_{17}–C_{20} and C_{22}) and derivatives (e.g., ethyl esters) have been carried out applying noble metal and Ni- or Mo-containing catalysts using several carriers (C, MgO, SiO_2, Al_2O_3) in semibatch reactor around 300 °C, also using dodecane as solvent [88–92, 131–133]. Complete conversion of stearic acid was achieved over a 5% Pd/C (Sibunit) catalyst with a $C_{17}H_{36}$ selectivity of 99%. Apart from supported Pd catalysts, Ni exhibited good hydrocarbon selectivity on SiO_2 and Cr_2O_3 supports. Often also reference data for treatment in presence of small partial pressures of hydrogen are reported. Mäki-Arvela *et al.* described the linearly increasing conversion of stearic acid with temperature but inversely affected selectivity to alkanes over 5% Pd/C. The yield of aromatics increased by ninefold with rising of temperature from 300 to 360 °C [92]. Deoxygenation of biodiesel (FAME) to improve its quality has been studied with model compounds like methyl stearate over Pt/Al_2O_3 (in liquid phase and in gas phase [134]) and ethyl stearate over Pd/C in liquid phase [135]. Formation of hydrocarbons with one C atom less was observed, together with some oxygenated by-products in gas phase.

Own studies on the deoxygenation of various fatty acids with different chain lengths (caprylic acid, lauric acid, and stearic acid) over Pd-containing catalysts revealed that the chain length of the fatty acid had a strong impact on the conversion at comparable reaction conditions (batch autoclave, 300–350 °C, 6 h, reaction pressure 75 bar, N_2 atmosphere) [136]. The best performing catalyst was 10% Pd/C with 63% undecane yield at 327 °C using lauric acid as feed.

Such studies might be of particular interest for feedstock that contain a large fraction of acids, the most prominent being tall oil from pulping (world capacity >1 Mton per annum) that contains fatty acids and resin acids up to 55 and 60 wt%, respectively, besides other oxygenated compounds. Pyrolytic conversion of fatty acid salts at 750 °C yields a high amount of unsaturated hydrocarbons together with minor amounts of monoaromatics, whereas conversion of resin acid salts mainly results in aromatic hydrocarbons [137]. In addition, the upgrading of tall oil fatty acids was also studied over Pd/C catalysts at 300–350 °C in batch autoclaves under He pressure and obtained high yields for heptadecane and heptadecene [138].

16.5.4
Biomass and Liquefied Biomass Deoxygenation by Hydrothermal Processes

As an alternative to using external hydrogen, thermal or catalytic cracking, hydrothermal processes have attracted high attention in the last decades [139].

Several special features of water make it a plausible option: (i) many liquefied biomass derivatives contain already large fractions of water; (ii) water is cheap; (iii) water can act as reactant, for example, via hydrolysis of C–O bonds; (iv) water may serve as a source of reactive hydrogen; and (v) water properties (pH, polarity, viscosity) can be tuned with pressure and temperature.

To take advantage of these features at reasonable reaction rates, a minimum temperature of 250 °C is necessary, which generates a vapor pressure of already 40 bar. This is the typical range for APR, where some reactive but high-boiling biomass derivatives (e.g., polyols) can be easily gasified to hydrogen and CO_x. Other biomass derivatives are more recalcitrant, in particular lignin, and need higher temperatures.

For processes operating in the lower temperature range, the use of catalysts is crucial to obtain high reaction rates. Surprisingly, it was found that alkali salts can act as homogeneous catalysts in biomass liquefaction under hydrothermal conditions from 200 to 400 °C and outperform most other catalyst types [140]. The basic effect is the enhancement of the rate of the water–gas shift reaction toward CO_2 by one order of magnitude compared to alkali free systems. This is an interesting aspect, as biomass may contain up to 0.5 wt% of potassium already. A special feature of hydrothermal processing is the crucial role of formate $HCOO^-$, which can form via a series of elementary steps and serves as a highly active hydrogen donor. Other important species in this catalytic system are CO_3^{2-}, HCO_3^-, and OH^-:

$$K_2CO_3 + H_2O \rightarrow KHCO_3 + KOH \tag{16.1}$$

$$KOH + CO \rightarrow HCOOK \tag{16.2}$$

$$HCOOK + H_2O \rightarrow KHCO_3 + H_2 \tag{16.3}$$

$$2KHCO_3 \rightarrow H_2O + K_2CO_3 + CO_2 \tag{16.4}$$

$$H_2O + CO \leftrightarrows HCOOH \leftrightarrows H_2 + CO_2 \tag{16.5}$$

The temperature window from 280 to 370 °C is usually assigned to hydrothermal upgrading processes. Up to these temperatures, the use of catalysts is beneficial, but finding stable catalysts is a demanding task. Beyond the critical point of water (374 °C, 221 bar), reaction rates are very high even in absence of catalysts, but cracking and gasification get dominant.

16.5.4.1 APR

Hydrocarbon reforming with water to produce hydrogen from fossil fuels is done at technical scale with steam at 800–900 °C and 20–30 bar; meanwhile also pilot plants for biomass gasification with supercritical water (SCW) have been erected [141]. APR – with most of the water being in liquid state – provides milder reaction conditions and has been intensively investigated by the group of Dumesic for conversion of renewables into hydrogen [142–144]. They tested heterogeneous, noble metal-containing catalysts (Ru, Pt) and concluded that hydrogen selectivity in APR mainly depends on nature of active metal, support, and pH value. It was found that acidic solutions favor alkane formation whereas neutral/basic solutions lead to high hydrogen selectivities [145]. The reaction conditions allow the transformation of alcohols, polyols, and sugars toward hydrogen, which at the end stems from the organic material as well as from water (for example, see equations 16.6–16.8 in the following). Due to the thermodynamic equilibrium at low temperature and the impact of water–gas shift reaction, CO_2 is released instead of CO (as in steam reforming).

In case of glycerol reforming the reaction network starts with the decomposition of glycerol:

$$C_3H_8O_3 \leftrightarrows 3CO + 4H_2 \qquad \Delta H(250\,^\circ C) = 349 \ \text{kJ mol}^{-1} \qquad (16.6)$$

The reaction is coupled with the water–gas shift reaction:

$$CO + H_2O \leftrightarrows CO_2 + H_2 \qquad \Delta H(250\,^\circ C) = -41 \ \text{kJ mol}^{-1} \qquad (16.7)$$

The overall reaction equation is formulated as

$$C_3H_8O_3 + 3H_2O \leftrightarrows 3CO_2 + 7H_2 \qquad \Delta H(250\,^\circ C) = 227 \ \text{kJ mol}^{-1} \quad (16.8)$$

One big advantage of APR compared to steam or supercritical reforming is related to energetics: the high heat capacity of water necessitates a high amount of external heat to reach the process temperatures of the latter [146]. Thus, the specific energy demand for hydrogen production strongly depends on the process temperature. To compensate for the lower reaction rates, effective catalysts are required.

Effective heterogeneous APR catalysts have to be active in C–C bond cleavage and water–gas shift reaction but must not facilitate C–O bond cleavage and subsequent hydrogenation of CO or CO_2 [143]; this requires a balance between redox and acid–base properties. Furthermore, they must be stable at hydrothermal conditions. Most favored systems use either noble metals (Pt, Ru) or cheaper Ni as active compound on support with large surface area. A recent review by our group gives a more detailed overview on catalyst systems, focusing on glycerol APR [147]. It also discusses the efficiency of *in situ* formed hydrogen in simultaneous glycerol hydrogenolysis toward propanediols. Instead of producing molecular hydrogen via APR in competition with existing steam reforming technology, the *in situ* generation of reactive hydrogen could be coupled with another hydrogen-consuming process like deoxygenation of recalcitrant liquefied biomass [148]. Prerequisite are cheap hydrogen donors and the choice of a suited feedstock to be upgraded. If the latter selectively reacts via the deoxygenation pathway instead of cracking or gasification at given APR conditions, many of the discussed advantages can be realized in a single process. Initial reactions like hydrolysis or dehydration can further promote the efficiency of the deoxygenation process, whereas coking is negligible.

Following this concept, aiming at the deoxygenation of larger molecules from renewable resources, already Dumesic and coworkers [46] reported the formation of renewable alkanes ($C_1 - C_6$) from APR of polyols (sugars) like sorbitol using a $Pt/SiO_2 - Al_2O_3$ catalyst at 250 °C and 40 bar with 64% alkane selectivity. Though APR is mostly seen as a technology for hydrogen generation from water-soluble biomass, some technical development to demonstration plant scale has been made to treat polyols, bio-oil, or its water-soluble fraction at APR conditions to produce biogasoline [149].

Similarly, other reactions that provide hydrogen *in situ* may play a role as well. An interesting concept is catalytic transfer hydrogenation, for example, using well-known hydrogen donors like 2-propanol or formic acid in presence of suitable catalysts for hydrogenolysis reactions [150–152]. The best hydrogen donors include simple molecules like cyclohexene, hydrazine, formic acid, and formates [153]. Some of these donor molecules can be recycled externally, as known for 2-propanol. Interestingly, catalytic transfer hydrogenation can be carried out effectively over heterogeneous catalysts in the same temperature range as APR processes [147].

16.5.4.2 Hydrothermal Upgrading

Several processes have been developed during the last three decades that aim at the conversion of biomass at even more severe hydrothermal conditions (>300 °C, >100 bar), preferably in continuous operation mode. By doing this, some of the aforementioned features of APR might be utilized for the valorization of more recalcitrant feedstock.

HTU® (Hydrothermal Upgrading) process was developed by Shell in 1982 and then realized in a demonstration scale with a capacity of $100 \, kg \, h^{-1}$. This continuous process operates at 300–350 °C, 120–180 bar, and contact times of 5–20 min. The process was claimed to be able to convert almost all types of biomass. Due to thermal impact and *in situ* generated hydrogen, deoxygenation is possible even in absence of catalyst, and a remarkable fraction of the oxygen is found in CO_2. The process ends up with a biocrude, representing approximately 50 wt% of the feed, together with 25% of gas (90% CO_2), 20% of water, and 10% of water-soluble oxygenates. This biocrude has already a lower heating value of $30-35 \, MJ \, kg^{-1}$ but contains still about 10 wt% of oxygen, and therefore it needs further treatment steps (hydrogenation, cracking) to match the quality standards of conventional diesel fuels. At current, the commercialization of the process is stopped.

Continuing from this point, other groups have developed processes that operate similar or at even higher temperatures and claim that the production of high-quality fuels is then possible within one step. New Oil Resources L.L.C. claims hydrothermal biomass treatment up to 370 °C within 5 min reaction time [154, 155], and 70–80% of the feed heating value is fixed in the liquid products. Changing World Technologies company operated a demonstration plant with a daily throughput of 250 ton of waste biomass, which was converted at 200–300 °C and 40 bar into 60 000 l of oil (mostly linear hydrocarbons) [28, 156]. This plant is not in operation anymore.

A more sophisticated technology based on hydrothermal upgrading is claimed by several researchers and mostly known as *CatLiq® process* [157]. The continuous reactor operates at 280–350 °C and 220–250 bar; an internal recycle is used to establish an extremely rapid heating of fresh feed which suppresses coking in upstream parts. The process also takes advantage of admixing K_2CO_3 to the aqueous biomass feed, giving better control of the pH value and enhancing the reaction rate in water–gas shift. Finally, heterogeneous zirconia catalysts are used to promote deoxygenation. A demonstration plant with a capacity to convert 480 l per day of wet biomass and organic waste into bio-oil has been erected in Denmark by SCF Technologies A/S in 2007. The bio-oil yield amounts to 30–35% [28].

According to known process data, all these processes are definitely able to remove oxygen from liquid biomass streams to a large extent. However, they yet have not proved to produce high-quality fuels at an economic scale. In most cases, additional unit operations are necessary to remove the residual oxygen.

16.5.4.3 Supercritical Water Processing

SCW has received much attention because of its properties being completely different from ambient water (e.g., [158]). It is completely miscible with gases, has high solvent power for nonpolar hydrocarbons due to its low dielectric constant, and can take part in elementary reactions. The typical temperatures and pressures above 400 °C and 250 bar on the one side offer ways to innovative chemical processes and allow high reaction rates, but on the other side corrosion and plugging by inorganic matter make SCW processing a very demanding technology. Nevertheless, pilot plants have been erected, mostly aiming at the total oxidation or gasification of organic matter (including biomass) up to 700 °C without catalysts. The required amount of heat to reach these temperatures is extraordinary high and cracking reactions get more and more predominant, and this sets the limits for conversion of biomass to fuels in SCW.

Several groups have investigated the potential to valorize biomass under these conditions. Many natural and industrial polymers can be decomposed within few minutes; for example, cellulose is completely converted into glucose and by-products [159]. Wet phytomass decomposition at 330–410 °C and 300–500 bar leads within 15 min to phenols, furfurals, $C_2–C_5$ acids, and aldehydes [160]; conversion of wood biomass in presence of Na_2CO_3 gave maximum yields of heavy oil at 380 °C, and similar products were identified [161]. The reaction pathways strongly depend on the ion product of water, promoting either ionic or radical steps. Pulp from softwood (47% oxygen content) was converted at 360–380 °C in the presence of NaOH, but coke was the main product in all experiments [162].

16.6
Conclusions and Outlook

Liquefied biomass has potential for replacing partly fossil fuels due to its high energy density and heating value compared to parent biomass. However, the direct use is mostly impossible because of the immiscibility with conventional fuels, the high oxygen content, and the considerable amount of water and thus posttreatment by deoxygenation is necessary.

Studies on model compounds provide insights in the reaction pathways and mechanisms, the influence of reaction conditions, and the role of catalysts to maximize the yield of oxygen-free products. However, the transfer of reaction conditions and processing concepts to the real feeds and upscaling remain big challenges, as well as the lifetime of catalysts.

The deoxygenation of triglyceride biomass (e.g., waste cooking oil, microalgae oil) runs toward fuel-like alkanes at mild reaction conditions compared to any hydrothermal process; the specific energy input will be less as no water ballast has to be heated.

The treatment of bio-oils derived from lignocellulosic biomass is much more complicated. The bio-oil obtained from any liquefaction process can be converted easily to nearly oxygen-free hydrocarbons using the catalysts and reaction conditions of conventional hydroprocessing in standard refineries. The limiter is always the initial liquefaction process. Hydrothermal treatment needs high pressure (roughly 200 bar), which means high investment and running costs (to heat the large water fraction). In contrast, fast pyrolysis is carried out at ambient pressure and around 500 °C, but the product contains high oxygenate and water content. As a result, the conditions suited for hydroprocessing of liquefaction oil seem to be inefficient for pyrolysis oils and catalysts deactivate rapidly.

The process design for biofuels from liquefied biomass must fulfill the specifications for commercial standard gasoline and diesel. To take this into account, coprocessing of LOs or FPOs with conventional streams in standard refinery units (e.g., FCC) seems to be a feasible short-term concept to bring biomass utilization to the market by taking advantage of existing refinery infrastructure. In the long-term period, research should focus on (i) fast pyrolysis without subsequent stabilization step and (ii) improvement of liquid-phase upgrading processes by highly active and stable catalysts. To reach both targets, bifunctional catalysts offering balanced redox (noble metals, Ni, or bimetallic systems) and acid–base properties seem to be a suitable choice as they can catalyze all the elementary steps in the preferred deoxygenation pathways. With regard to catalyst lifetime, improved hydrothermal stability and resistance against coking are crucial, not to forget the resistance against heteroatoms in biomass.

One of the most challenging aspects for hydrotreating of liquefied biomass (as well as of fossil fuels in standard refineries) is the continuously increasing specific hydrogen consumption (driven by lower feedstock quality and fuel specifications). Hydrogen supply is a major driver in operating costs of the deoxygenation processes but is today produced mostly from fossil hydrocarbons. This is of course contradictory to the overall concept of converting "green" feedstock into fuels. Thus, any technology that lowers the specific hydrogen consumption would be attractive. This has been addressed by the development of several hydrothermal processes which might generate a part of required hydrogen *in situ*; however, it seems that the economic breakeven point has not been reached yet. The chance to produce fuels from biomass via deoxygenation technologies at current depends on the linkage to existing technologies (hydrogen production, refineries), which can serve as jump start until competitive biorefinery concepts will be available. It is rather unlikely that standalone processes will prevail.

The further improvement of any deoxygenation technology should focus on the reduction of investment (number of unit operations, materials of construction) and operating costs (e.g., hydrogen source). However, this only makes sense if at the same time effective supply chains can provide biomass with a minimum energy loss caused by harvesting, pretreatment, and transportation. All these factors determine the optimum plant capacity. Finally, the specific product costs will still have to compete with fossil fuels for several decades. To evaluate the potential of deoxygenation, the commercial NExBTL process of the Finnish company Neste Oil Oyj plays a leading role.

References

1. Chu, S. and Majumdar, A. (2012) *Nature*, **488**, 294–303.
2. Wang, H., Male, J., and Wang, Y. (2013) *ACS Catal.*, **3**, 1047–1070.
3. Agentur für Erneuerbare Energien http://www.unendlich-viel-energie .de/erneuerbare-energie/strom-aus-biomasse (accessed 27 September 2015).
4. FNR http://fnr.de/marktanalyse/ marktanalyse.pdf (accessed 27 September 2015).
5. Nicholas, K.M. (ed) (2014) *Selective Catalysis for Renewable Feedstocks and Catalysis*, Topics in Current Chemistry, Springer, Cham, Heidelberg, New York, Dordrecht, London.
6. Vennestrøm, P.N.R., Osmundsen, C.M., Christensen, C.H., and Taarning, E. (2011) *Angew. Chem. Int. Ed.*, **50**, 10502–10509.
7. Sakakura, T., Chol, J.-C., and Yasuda, H. (2007) *Chem. Rev.*, **107**, 2365–2387.
8. Aleklett, K., Höök, M., Jakobsson, K., Lardelli, M., Snowden, S., and Söderbergh, B. (2010) *Energy Policy*, **38**, 1398–1414.
9. Kuparinen, K., Heinimö, J., and Vakkilainen, E. (2014) *Biofuels, Bioprod. Biorefin.*, **8**, 747–754.
10. Dincer, I. (2000) *Renewable Sustainable Energy Rev.*, **4**, 157–175.
11. Carpenter, D., Westover, T.L., Czernik, S., and Jablonski, W. (2014) *Green Chem.*, **16**, 384–406.
12. Saxena, R.C., Adhikari, D.K., and Goyal, H.B. (2009) *Renewable Sustainable Energy Rev.*, **13**, 167–178.
13. Neste Oil Oyj (2004) Process for producing a hydrocarbon component of biological origin. EP Patent 1396531.
14. Chheda, J.N., Huber, G.W., and Dumesic, J.A. (2007) *Angew. Chem. Int. Ed.*, **46**, 7164–7183.
15. Kamm, B., Gruber, P.R., and Kamm, M. (eds) (2010) *Biorefineries – Industrial Processes and Products*, Wiley-VCH Verlag GmbH, Weinheim.
16. Cherubini, F. (2010) *Energy Convers. Manage.*, **51**, 1412–1421.
17. Mehdi, H., Fábos, V., Tuba, R., Bodor, A., Mika, L.T., and Horváth, I.T. (2008) *Top. Catal.*, **48**, 49–54.
18. Prüße, U., Jarzombek, P., and Vorlop, K.-D. (2012) *Top. Catal.*, **55**, 453–459.
19. Kulik, A., Martin, A., Pohl, M., Fischer, C., and Köckritz, A. (2014) *Green Chem.*, **16**, 1799–1806.
20. Walther, G., Knöpke, L.R., Rabeah, J., Checinski, M.P., Jiao, H., Bentrup, U., Brückner, A., Martin, A., and Köckritz, A. (2013) *J. Catal.*, **297**, 44–55.
21. Martin, A. and Richter, M. (2011) *Eur. J. Lipid Sci. Technol.*, **113**, 100–117.
22. Vertès, A.A., Qureshi, N., Blaschek, H.P., and Yukawa, H. (eds) (2010) *Biomass to Biofuels – Strategies for Global Industries*, John Wiley & Sons, Ltd., Chichester.

23. Regalbuto, J.R. (2009) *Science*, **325**, 822–824.

24. McKendry, P. (2002) *Bioresour. Technol.*, **83**, 37–46.

25. McKendry, P. (2002) *Bioresour. Technol.*, **83**, 47–54.

26. Biomass Energy Centre http://www .biomassenergycentre.org.uk/portal/ page?_pageid=75,20041&_dad=portal&_ schema=PORTAL (accessed 27 September 2015).

27. Dahmen, N., Henrich, E., Kruse, A., and Raffelt, K. (2010) in *Biomass to Biofuels – Strategies for Global Industries* (eds A.A. Vertès, N. Qureshi, H.P. Blaschek, and H. Yukawa), John Wiley & Sons, Ltd., Chichester, pp. 91–122.

28. Toor, S.S., Rosendahl, L., and Rudolf, A. (2011) *Energy*, **36**, 2328–2342.

29. Zhang, Y. (2010) in *Biofuels from Agricultural Wastes and Byproducts* (eds H.P. Blaschek, T.C. Ezeji, and J. Scheffran), Wiley-Blackwell, Ames, IA, pp. 201–232.

30. Tekin, K. and Karagöz, S. (2013) *Res. Chem. Intermed.*, **39**, 485–498.

31. Behrendt, F., Neubauer, Y., Oevermann, M., Wilmes, B., and Zobel, N. (2008) *Chem. Eng. Technol.*, **31**, 667–677.

32. Patil, P.T., Armbruster, U., Richter, M., and Martin, A. (2011) *Energy Fuels*, **25**, 4713–4722.

33. Patil, P.T., Armbruster, U., and Martin, A. (2014) *J. Supercrit. Fluids*, **93**, 121–129.

34. Elliott, D.C. (2007) *Energy Fuels*, **21**, 1792–1815.

35. Zacher, A.H., Olarte, M.V., Santosa, D.M., Elliott, D.C., and Jones, S.B. (2014) *Green Chem.*, **16**, 491–515.

36. IEA Bioenergy http://www.ieabcc.nl/ publications/IEA_Bioenergy_T32_ Torrefaction_review.pdf (accessed 27 September 2015).

37. van der Stelt, M.J.C., Gerhauser, H., Kiel, J.H.A., and Ptasinski, K.J. (2011) *Biomass Bioenergy*, **35**, 3748–3762.

38. Nhuchhen, D.R., Basu, P., and Acharya, B. (2014) A Comprehensive Review on Biomass Torrefaction, *Int. J. Renewable Energy Biofuels*, **2014** (2014). doi: 10.5171/2014.506376, Article ID 506376.

39. Meier, D. and Faix, O. (1999) *Bioresour. Technol.*, **68**, 71–77.

40. Butler, E., Devlin, G., Meier, D., and McDonnell, K. (2011) *Renewable Sustainable Energy Rev.*, **15**, 4171–4186.

41. Bridgwater, A.V. (2012) *Biomass Bioenergy*, **38**, 68–94.

42. KIT http://www.ieatask33.org/app/ webroot/files/file/2013/Workshop_ Gothenburg/19/Kolb.pdf (accessed 27 September 2015).

43. Bioliq http://www.bioliq.de/english/ index.php (accessed 27 September 2015).

44. PyNe http://www.pyne.co.uk/ Resources/user/PYNE%20Newsletters/ PyNews%2017.pdf (accessed 27 September 2015).

45. BTG Biomass Technology Group http://www.btgworld.com/en/rtd/ technologies/fast-pyrolysis (accessed 27 September 2015).

46. Huber, G.W., Cortright, R.D., and Dumesic, J.A. (2004) *Angew. Chem. Int. Ed.*, **43**, 1549–1551.

47. Huber, G.W., Chheda, J.N., Barrett, C.J., and Dumesic, J.A. (2005) *Science*, **308**, 1446–1450.

48. Mortensen, P.M., Grunwaldt, J.-D., Jensen, P.A., Knudsen, K.G., and Jensen, A.D. (2011) *Appl. Catal., A*, **407**, 1–19.

49. He, Z. and Wang, X. (2013) *Catal. Sustainable Energy*, **1**, 28–52. doi: 10.2478/cse-2012-0004

50. Ruddy, D.A., Schaidle, J.A., Ferrell, J.R. III,, Wang, J., Moens, L., and Hensley, J.E. (2014) *Green Chem.*, **16**, 454–490.

51. Czernik, S. and Bridgwater, A.V. (2004) *Energy Fuels*, **18**, 590–598.

52. Bridgwater, A.V. (2012) *Environ. Prog. Sustainable Energy*, **31**, 261–268.

53. Graça, I., Lopes, J.M., Cerqueira, H.S., and Ribeiro, M.F. (2013) *Ind. Eng. Chem. Res.*, **52**, 275–287.

54. Huber, G.W., Iborra, S., and Corma, A. (2006) *Chem. Rev.*, **106**, 4044–4098.

55. Furimsky, E. (1983) *Catal. Rev. Sci. Eng.*, **25**, 421–458.

56. Furimsky, E. (2000) *Appl. Catal., A*, **199**, 147–190.

57. Hong, D.-Y., Miller, S.J., Agrawal, P.K., and Jones, C.W. (2010) *Chem. Commun.*, **46**, 1038–1040.

58. Massoth, F.E., Politzer, P., Concha, M.C., Murray, J.S., Jakowski, J., and Simons, J. (2006) *J. Phys. Chem. B*, **110**, 14283–14291.

59. Deepa, A.K. and Dhepe, P.L. (2014) *ChemPlusChem*, **79**, 1573–1583.

60. Huynh, T.M., Armbruster, U., Pohl, M.-M., Schneider, M., Radnik, J., Hoang, D.-L., Phan, B.M.Q., Nguyen, D.A., and Martin, A. (2014) *ChemCatChem*, **6**, 1940–1951.

61. Huynh, T.M., Armbruster, U., Phan, B.M.Q., Nguyen, D.A., and Martin, A. (2014) *Chim. Oggi*, **32**, 40–44.

62. Zhao, C., Kou, Y., Lemonidou, A.A., Li, X., and Lercher, J.A. (2009) *Angew. Chem. Int. Ed.*, **48**, 3987–3990.

63. Zhao, C., Kasakov, S., He, J., and Lercher, J.A. (2012) *J. Catal.*, **296**, 12–23.

64. Gandarias, I., Barrio, V.L., Requies, J., Arias, P.L., Cambra, J.F., and Güemez, M.B. (2008) *Int. J. Hydrogen Energy*, **33**, 3485–3488.

65. Romero, Y., Richard, F., and Brunet, S. (2010) *Appl. Catal., B*, **98**, 213–223.

66. Wan, H., Chaudhari, R., and Subramaniam, B. (2012) *Top. Catal.*, **55**, 129–139.

67. Laurent, E. and Delmon, B. (1994) *J. Catal.*, **146**, 281–291.

68. Bui, V.N., Laurenti, D., Delichère, P., and Geantet, C. (2011) *Appl. Catal., B*, **101**, 246–255.

69. Newman, C., Zhou, X., Goundie, B., Ghampson, I.T., Pollpck, R.A., Ross, Z., Wheeler, M.C., Meulenberg, R.W., Austin, R.N., and Frederick, B.G. (2014) *Appl. Catal., A*, **477**, 64–74.

70. Bykova, M.V., Ermakov, D.Y., Kaichev, V.V., Bulavchenko, O.A., Saraev, A.A., Lebedev, M.Y., and Yakovlev, V.A. (2012) *Appl. Catal., B*, **113-114**, 296–307.

71. Sun, J., Karim, A.M., Zhang, H., Kovarik, L., Li, X.S., Hensley, A.J., McEwen, J.-S., and Wang, Y. (2013) *J. Catal.*, **306**, 47–57.

72. Zhao, C. and Lercher, J.A. (2012) *Angew. Chem. Int. Ed.*, **51**, 5935–5940.

73. Zhao, C. and Lercher, J.A. (2012) *ChemCatChem*, **4**, 64–68.

74. Strassberger, Z., Alberts, A.H., Louwerse, M.J., Tanase, S., and Rothenberg, G. (2013) *Green Chem.*, **15**, 768–774.

75. Rachmady, W. and Vannice, M.A. (2000) *J. Catal.*, **192**, 322–334.

76. Olcay, H., Xu, L., Xu, Y., and Huber, G.W. (2010) *ChemCatChem*, **2**, 1420–1424.

77. Elliott, D.C. and Hart, T.R. (2009) *Energy Fuels*, **23**, 631–637.

78. Alotaibi, M.A., Kozhevnikova, E.F., and Kozhevnikov, I.V. (2012) *Appl. Catal., A*, **447–448**, 32–40.

79. Chen, L., Zhu, Y., Zheng, H., Zhang, C., and Li, Y. (2012) *Appl. Catal., A*, **411–412**, 95–104.

80. Kreuzer, K. and Kramer, R. (1997) *J. Catal.*, **167**, 391–399.

81. Sitthisa, S. and Resasco, D. (2011) *Catal. Lett.*, **141**, 784–791.

82. Lee, C.-L. and Ollis, D.F. (1984) *J. Catal.*, **87**, 325–331.

83. Bunch, A.Y., Wang, X., and Ozkan, U.S. (2007) *J. Mol. Catal. A: Chem.*, **270**, 264–272.

84. Romero, Y., Richard, F., Renème, S., and Brunet, S. (2009) *Appl. Catal., A*, **353**, 46–53.

85. Liu, C., Shao, Z., Xiao, Z., Williams, C.T., and Liang, C. (2012) *Energy Fuels*, **26**, 4205–4211.

86. Leung, D.Y.C., Wu, X., and Leung, M.K.H. (2010) *Appl. Energy*, **87**, 1083–1095.

87. Şenol, O.İ., Viljava, T.R., and Krause, A.O.I. (2005) *Catal. Today*, **106**, 186–189.

88. Snåre, M., Kubičková, I., Mäki-Arvela, P., Eränen, K., and Murzin, D.Y. (2006) *Ind. Eng. Chem. Res.*, **45**, 5708–5715.

89. Kubičková, I., Snåre, M., Eränen, K., Mäki-Arvela, P., and Murzin, D.Y. (2005) *Catal. Today*, **106**, 197–200.

90. Simakova, I., Simakova, O., Mäki-Arvela, P., and Murzin, D.Y. (2010) *Catal. Today*, **150**, 28–31.

91. Simakova, I., Simakova, O., Mäki-Arvela, P., Simakov, A., Estrada, M., and Murzin, D.Y. (2009) *Appl. Catal., A*, **355**, 100–108.

92. Mäki-Arvela, P., Kubičková, I., Snåre, M., Eränen, K., and Murzin, D.Y. (2006) *Energy Fuels*, **21**, 30–41.

93. Gosselink, R.W., Hollak, S.A.W., Chang, S.-W., van Haveren, J., de Jong, K.P.,

Bitter, J.H., and van Es, D.S. (2013) *ChemSusChem*, **6**, 1576–1594.

94. Bezergianni, S., Dimitriadis, A., Kalogianni, A., and Pilavachi, P.A. (2010) *Bioresour. Technol.*, **101**, 6651–6656.

95. Bezergianni, S. and Dimitriadis, A. (2013) *Fuel*, **103**, 579–584.

96. Peng, B., Yao, Y., Zhao, C., and Lercher, J.A. (2012) *Angew. Chem. Int. Ed.*, **51**, 2072–2075.

97. Peng, B., Yuan, X., Zhao, C., and Lercher, J.A. (2012) *J. Am. Chem. Soc.*, **134**, 9400–9405.

98. Mohammad, M., Kandaramath Hari, T., Yaakob, Z., Chandra Sharma, Y., and Sopian, K. (2013) *Renewable Sustainable Energy Rev.*, **22**, 121–132.

99. Zhao, C., Bruck, T., and Lercher, J.A. (2013) *Green Chem.*, **15**, 1720–1739.

100. Baker, E.G. and Elliott, D.C. (1988) in *Research in Thermochemical Biomass Conversion* (eds A.V. Bridgwater and J.L. Kuester), Springer, Dordrecht, pp. 883–895.

101. Elliott, D.C., Biller, P., Ross, A.B., Schmidt, A.J., and Jones, S.B. (2015) *Bioresour. Technol.*, **178**, 147–156.

102. Choudhary, T.V. and Phillips, C.B. (2011) *Appl. Catal., A*, **397**, 1–12.

103. Eddie, G.B. and Douglas, C.E. (1988) in *Pyrolysis Oils from Biomass* (eds E.J. Soltes and T.A. Milne), ACS, Washington, DC, pp. 228–240.

104. Elliott, D.C., Baker, E.G., Piskorz, J., Scott, D.S., and Solantausta, Y. (1988) *Energy Fuels*, **2**, 234–235.

105. Baldauf, W., Balfanz, U., and Rupp, M. (1994) *Biomass Bioenergy*, **7**, 237–244.

106. Wildschut, J., Mahfud, F.H., Venderbosch, R.H., and Heeres, H.J. (2009) *Ind. Eng. Chem. Res.*, **48**, 10324–10334.

107. Wildschut, J., Iqbal, M., Mahfud, F.H., Cabrera, I.M., Venderbosch, R.H., and Heeres, H.J. (2010) *Energy Environ. Sci.*, **3**, 962–970.

108. Venderbosch, R.H., Ardiyanti, A.R., Wildschut, J., Oasmaa, A., and Heeres, H.J. (2010) *J. Chem. Technol. Biotechnol.*, **85**, 674–686.

109. Wildschut, J., Arentz, J., Rasrendra, C.B., Venderbosch, R.H., and Heeres, H.J. (2009) *Environ. Prog. Sustainable Energy*, **28**, 450–460.

110. Elliott, D.C., Hart, T.R., Neuenschwander, G.G., Rotness, L.J., and Zacher, A.H. (2009) *Environ. Prog. Sustainable Energy*, **28**, 441–449.

111. French, R.J., Stunkel, J., and Baldwin, R.M. (2011) *Energy Fuels*, **25**, 3266–3274.

112. Elliott, D.C., Hart, T.R., Neuenschwander, G.G., Rotness, L.J., Olarte, M.V., Zacher, A.H., and Solantausta, Y. (2012) *Energy Fuels*, **26**, 3891–3896.

113. Vispute, T.P. and Huber, G.W. (2009) *Green Chem.*, **11**, 1433–1445.

114. Vispute, T.P., Zhang, H., Sanna, A., Xiao, R., and Huber, G.W. (2010) *Science*, **330**, 1222–1227.

115. Samolada, M.C., Baldauf, W., and Vasalos, I.A. (1998) *Fuel*, **77**, 1667–1675.

116. de Miguel Mercader, F., Groeneveld, M.J., Kersten, S.R.A., Way, N.W.J., Schaverien, C.J., and Hogendoorn, J.A. (2010) *Appl. Catal., B*, **96**, 57–66.

117. Hamaguchi, M., Cardoso, M., and Vakkilainen, E. (2012) *Energies*, **5**, 2288–2309.

118. Elliott, D.C. and Oasmaa, A. (1991) *Energy Fuels*, **5**, 102–109.

119. Oasmaa, A., Alén, R., and Meier, D. (1993) *Bioresour. Technol.*, **45**, 189–194.

120. Adjaye, J.D. and Bakhshi, N.N. (1995) *Fuel Process. Technol.*, **45**, 185–202.

121. Vitolo, S., Bresci, B., Seggiani, M., and Gallo, M.G. (2001) *Fuel*, **80**, 17–26.

122. Al-Sabawa, M., Chen, J., and Ng, S. (2012) *Energy Fuels*, **26**, 5355–5372.

123. de Miguel Mercader, F., Groeneveld, M.J., Kersten, S.R.A., Geantet, C., Toussaint, G., Way, N.W.J., Schaverien, C.J., and Hogendoorn, K.J.A. (2011) *Energy Environ. Sci.*, **4**, 985–997.

124. NREL http://www.nrel.gov/docs/fy13osti/58056.pdf (accessed 27 September 2015).

125. Suarez, P.A.Z., Moser, B.R., Sharma, B.K., and Erhan, S.Z. (2009) *Fuel*, **88**, 1143–1147.

126. Lima, D.G., Soares, V.C.D., Ribeiro, E.B., Carvalho, D.A., Cardoso, É.C.V., Rassi, F.C., Mundim, K.C., Rubim, J.C.,

and Suarez, P.A.Z. (2004) *J. Anal. Appl. Pyrolysis*, **71**, 987–996.

127. Maher, K.D. and Bressler, D.C. (2007) *Bioresour. Technol.*, **98**, 2351–2368.

128. Idem, R.O., Katikaneni, S.P.R., and Bakhshi, N.N. (1997) *Fuel Process. Technol.*, **51**, 101–125.

129. Leibniz-Institut für Katalyse e.V. (2009) Verfahren zur Herstellung eines Produktgemisches von Kohlenwasserstoffen aus Triglyceriden. DE Patent 10 2009 045399.

130. Maher, K.D., Kirkwood, K.M., Gray, M.R., and Bressler, D.C. (2008) *Ind. Eng. Chem. Res.*, **47**, 5328–5336.

131. Leung, A., Boocock, D.G.B., and Konar, S.K. (1995) *Energy Fuels*, **9**, 913–920.

132. Lestari, S., Simakova, I., Tokarev, A., Mäki-Arvela, P., Eränen, K., and Murzin, D.Y. (2008) *Catal. Lett.*, **122**, 247–251.

133. Roh, H., Eum, I., Jeong, D., Yi, B.E., Na, J., and Ko, C.H. (2011) *Catal. Today*, **164**, 457–460.

134. Do, P.T., Chiappero, M., Lobban, L.L., and Resasco, D.E. (2009) *Catal. Lett.*, **130**, 9–18.

135. Snare, M., Kubickova, I., Mäki-Arvela, P., Eränen, K., Wärna, J., and Murzin, D.Y. (2007) *Chem. Eng. J.*, **134**, 29–34.

136. Mohite, S., Armbruster, U., Richter, M., and Martin, A. (2014) *J. Sustainable Bioenergy Syst.*, **4**, 183–193.

137. Lappi, H. and Alén, R. (2012) Pyrolysis of Tall Oil-Derived Fatty and Resin Acid Mixtures, *ISRN Renewable Energy*, **2012**, doi: 10.5402/2012/409157, Article ID 409157.

138. Mäki-Arvela, P., Rozmysłowicz, B., Lestari, S., Simakova, O., Eränen, K., Salmi, T., and Murzin, D.Y. (2011) *Energy Fuels*, **25**, 2815–2825.

139. Akhtar, J., Aishah, N., and Amin, S. (2011) *Renewable Sustainable Energy Rev.*, **15**, 1615–1624.

140. Elliott, D.C., Hallen, R.T., and Sealock, L.J. Jr., (1983) *Ind. Eng. Chem. Prod. Res. Dev.*, **22**, 431–435.

141. Reddy, S.N., Nanda, S., Dalai, A.K., and Kozinski, J.A. (2014) *Int. J. Hydrogen Energy*, **39**, 6912–6926.

142. Shabaker, J.W. and Dumesic, J.A. (2004) *Ind. Eng. Chem. Res.*, **43**, 3105–3112.

143. Davda, R.R., Shabaker, J.W., Huber, G.W., Cortright, R.D., and Dumesic, J.A. (2005) *Appl. Catal., B*, **56**, 171–186.

144. Huber, G.W. and Dumesic, J.A. (2006) *Catal. Today*, **111**, 119–132.

145. Davda, R.R. and Dumesic, J.A. (2003) *Angew. Chem. Int. Ed.*, **42**, 4068–4071.

146. Wilhelm, D.J., Simbeck, D.R., Karp, A.D., and Dickenson, R.L. (2001) *Fuel Process. Technol.*, **71**, 139–148.

147. Martin, A., Armbruster, U., Gandarias, I., and Arias, P.L. (2013) *Eur. J. Lipid Sci. Technol.*, **115**, 9–27.

148. Metzger, J.O. (2006) *Angew. Chem. Int. Ed.*, **45**, 696–698.

149. Virent, Inc. http://www.virent.com/ (accessed 27 September 2015).

150. Aramendía, M.A., Borau, V., Jiménez, C., Marinas, J.M., Ruiz, J.R., and Urbano, F.J. (2001) *J. Mol. Catal. A: Chem.*, **171**, 153–158.

151. Bulushev, D.A. and Ross, J.R.H. (2011) *Catal. Today*, **163**, 42–46.

152. Wang, X. and Rinaldi, R. (2013) *Angew. Chem. Int. Ed.*, **52**, 1–6.

153. Johnstone, R.A.W., Wilby, A.H., and Entwistle, I.D. (1985) *Chem. Rev.*, **85**, 129–170.

154. New Oil, L.L.C http://newoilresources.com/process.html (accessed 27 September 2015).

155. Catallo, W.J. and Junk, T. (2001) Transforming biomass to hydrocarbon mixtures in near-critical or supercritical water. US Patent 6,180,845.

156. Baskis, P.T. (1993) Thermal depolymerizing reforming process and apparatus. US Patent 5,269,947.

157. Nielsen, R.P., Olofsson, G., and Søgaard, E. (2012) *Biomass Bioenergy*, **43**, 2–5.

158. Kruse, A. and Dinjus, E. (2007) *J. Supercrit. Fluids*, **39**, 362–380.

159. Arai, K., Smith, R.L. Jr., and Aida, T.M. (2009) *J. Supercrit. Fluids*, **47**, 6328–6369.

160. Kruse, A. and Gawlik, A. (2003) *Ind. Eng. Chem. Res.*, **42**, 267–279.

161. Qian, Y., Zuo, C., Tan, J., and He, J. (2006) *Energy*, **32**, 196–202.

162. Borgund, A.E. and Barth, T. (1999) *Org. Geochem.*, **30**, 1517–1526.

17
C–C Coupling for Biomass-Derived Furanics Upgrading to Chemicals and Fuels

Tuong V. Bui, Steven Crossley, and Daniel E. Resasco

17.1
Introduction

Lignocellulosic biomass is composed of cellulose and hemicellulose, two carbohydrate polymers, tightly bound to lignin, an aromatic polymer. Depending on the biomass source, the relative content of the three types of polymers can vary. Sources of lignocellulosic biomass that could be used as feedstock for the production of biofuels and chemicals typically include forest products (trees, shrubs, logging residues), crops, grasses, straw, or waste materials (urban and agricultural, saw mill and paper mill discards). Woods usually contain more lignin than other sources of biomass. For example, a typical wood composition is 25% lignin, 45% cellulose, 25% hemicellulose, and 5% others [1]. But different tree species may exhibit lignin contents from as low as 15% to as high as 40% [2].

Lignin is a complex polymeric network of monolignols (coniferyl, sinapyl, and paracoumaryl aromatic alcohols).Upon thermal decomposition, lignin evolves phenolic compounds derived from depolymerization of the network. At lower decomposition temperatures, guaiacols and syringols are the dominant products, but as temperature and residence times increase, derivatives such as phenol and catechol are formed. On the other hand, cellulose is a polysaccharide consisting of linear chains of linked glucose units, which constitute the primary cell walls of plants. Thermal decomposition of cellulose produces CO, CO_2, water, acetic acid, several furanic compounds (furfural, 5-(hydroxymethyl)-furfural (HMF), dihydromethyl furanone, etc.), and several anhydrosugars (levoglucosan, levoglucosenone, and levomannosan). Finally, hemicellulose is composed primarily of xylans linked as xylopyranosyl units; it is the least stable among the three major biomass components, and it cross-links the cellulose and lignin polymers, providing the structural support to the cell walls [3]. Thermal decomposition of hemicellulose primarily results in the evolution of mostly CO, CO_2, water, acetic acid, furfural, and xylopyranose. As the temperature increases, less acetic acid and more acetone are observed in the products [4].

Chemicals and Fuels from Bio-Based Building Blocks, First Edition.
Edited by Fabrizio Cavani, Stefania Albonetti, Francesco Basile, and Alessandro Gandini.
© 2016 Wiley-VCH Verlag GmbH & Co. KGaA. Published 2016 by Wiley-VCH Verlag GmbH & Co. KGaA.

The process of thermochemical conversion, combined with catalytic upgrading, offers the potential of high efficiency from biomass to products and relatively good economics, particularly for the production of intermediate products to be further upgraded to liquid fuels in the range of gasoline and diesel components. The products of the thermal conversion processes typically include permanent gases, water, organic vapors, and solids composed of mineral matter and char. For many years, the objective of thermochemical conversion processes has been to maximize the yields of primary liquids, expecting to conduct a subsequent catalytic upgrading, which has generally been conventional hydrotreating. However, a fundamental problem with all these existing technologies is that the complex composition of bio-oil presents different problems that cannot be solved with a single solution. Hydrotreating of the entire bio-oil results in exceedingly low yields of liquid in the desirable fuel range (C_6–C_{14}) since a large fraction of bio-oil contains C_2–C_5 oxygenates (e.g., acetic acid, pyrans, furanics, etc.).

While it is desirable to promote condensation of small oxygenates (i.e., acetic acid), it is undesirable to do it with large oxygenated molecules (phenolics). The former allows maximizing C-retention in the fuel range; the latter leads to coke formation and rapid catalyst deactivation. At the same time, hydrodeoxygenation of molecules that have reached the fuel range (C_6–C_{12} for gasoline and ~C_{10}–C_{21} for diesel) is desirable, but the opposite is true for molecules in the C_2–C_4 range that represent 40–60% of the carbon in the bio-oil. As a result, a judicious combination of prefractionation and catalytic upgrading is necessary to achieve high C-efficiency. This approach results in maximized C-retention in the fuel range via condensation of C_2–C_4 molecules, transalkylation of methoxy groups, and alkylation of small alcohols and phenolics. The net improvements in yields and catalyst lifetimes due to tailored strategies for each separate bio-oil fraction will outweigh the costs of increased process complexity.

Furfural and other furanic compounds represent a significant fraction of the pyrolysis vapors. Unfortunately, they present a major obstacle for the catalytic upgrading processes. In the first place, the number of carbons in the molecule of furanic compounds is too small to become a desirable component of gasoline upon deoxygenation (e.g., furan (4C), furfural (5C), furfuryl alcohol (5C), HMF (6C)). Conventional hydrotreating would produce light hydrocarbons (C_1–C_5), which are undesirable as fuel components. Second, and even more detrimental for the upgrading process, furanics are thermally unstable and readily polymerize forming undesirable dark brown precipitates called *humins* that are insoluble in water or organic solvents. Moreover, when the fractionation of biomass components is conducted either via staged thermal treatments (i.e., increasing temperatures in the thermochemical approach) [5–8] or via sequential condensation of pyrolysis vapors [9, 10], furanics appear in several fractions. Therefore, the upgrade and stabilization of furanics are of great technological importance.

In this contribution, we review recent attempts to upgrade furanic compounds to more stable compounds with longer carbon chain, which can either be more easily upgradable for the production of fuels or lead to chemicals of higher value. Specifically, we are reviewing recent reports related to the following processes:

1) Aldol condensation of furfural–HMF or ketones derived from furanics to dimers and trimers, which can have carbon chain lengths in the range of gasoline or diesel.
2) Hydroxyalkylation and standard alkylation of furanics with carbonyl-containing molecules, also leading to longer molecular intermediates.
3) Diels–Alder cycloaddition of olefins to furanics (furan, 2-methylfuran, 2,5-dimethylfuran (DMF)) to produce xylenes and other aromatic compounds.
4) Piancatelli ring rearrangement reactions that can lead to the production of more stable ketones, such as cyclopentanone (CPO), which can be further condensed into fuel-range molecules.
5) Selective partial oxidation of furanics yields carboxylic acid containing species with both higher stability and the potential for selective coupling reactions, for example, the conversion of furfural to maleic acid (MA), succinic acid (SA), or maleic anhydride (MAH).
6) Dimerization of furfural via oxidative coupling to produce C_{10} gasoline precursor.

17.2
Upgrading Strategy for Furanics

17.2.1
Aldol Condensation of Furfural–HMF

17.2.1.1 Mechanism of Aldol Condensation Reaction

Aldol condensation is one of the most common organic reactions in which carbonyl compounds(C=O group) are coupled, forming C–C bonds via an intermediate enol (or an enolate anion) species to form a β-hydroxycarbonyl, followed by dehydration to produce a longer carbon chain compound. The reaction can be catalyzed homogeneously or heterogeneously by both acids and bases [11, 12]. With basic catalysts, the formation of the enolate is initiated by the attraction of an acidic α-H to the base (Scheme 17.1); with acid catalysts, the formation of enol is promoted by the attack of a proton to the oxygen atom of the functional group

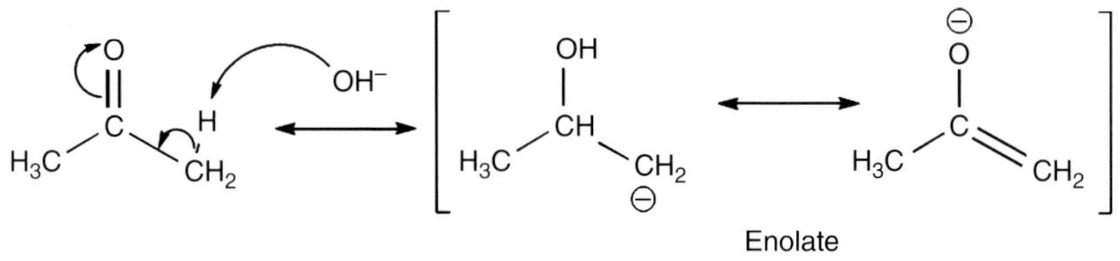

Enolate

Scheme 17.1 Base-catalyzed mechanism of enolate formation.

(C=O), followed by removal of the α-H to the conjugated base of the acid [13, 14] (Scheme 17.2).

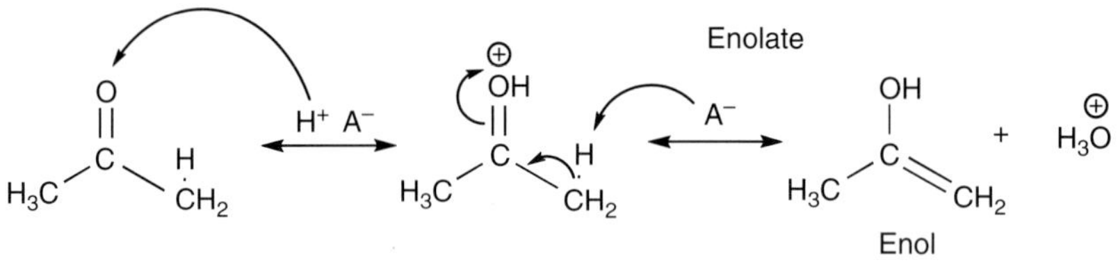

Scheme 17.2 Acid-catalyzed mechanism of enol formation in aqueous environment.

Base-Catalyzed Mechanism The conversion of furfural and acetone over a heterogeneous catalyst is of great interest in biomass conversion. A widely accepted mechanism [15] is summarized in Figure 17.1.

First, a Brønsted basic site on the catalyst surface abstracts the H atom from the α-C (adjacent to the functionality group) of the adsorbed acetone molecule to form a carbanion, which then attacks the CO group of adjacent adsorbed furfural molecules and generates the oxyanion. It has been suggested [15] that the intermediate oxyanion formed on dolomite is so unstable that it will immediately be protonated and dehydrated to form the C_8 monomer (FAc). Since the product obtained still keeps another α-H, the condensation process can continue on the other side of the monomer to generate a C_{13} dimer product (FFAc). Similar mechanisms have been recently proposed for the reaction over MgO and mixed oxide catalysts ($MgO–ZrO_2$, $MgO–Al_2O_3$) [16, 17]. For the base-catalyzed aldol condensation, the ability to abstract the α-H is the crucial property of catalysts [16].

Acid-Catalyzed Mechanism This mechanism is based on the formation of enol species. The activation of the carbonyl compound is initiated by the attack of the proton from the acid. The activated species can generate the carbocation or the enol species by giving the α-H to the conjugate base of the acid. The carbocation then becomes an electrophile, attacking the α-C of the enol to form coupling products (Scheme 17.3).

For the case of heterogeneous catalyst, there is a slight difference in which the basic site on the catalyst surface will deprotonate the α-C instead of the conjugate base of the acid as in the case of homogeneous one [18]. Therefore, the mechanism is related to the combination effect between both acid and basic sites. Climent et al. [19] have proposed a bifunctional acid–base mechanism, in which both sites are involved. They claim that the role of the acid site is to activate the carbonyl compound by protonating and polarizing the C–O bond of the carbonyl group, which

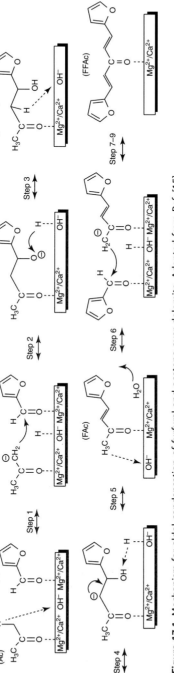

Figure 17.1 Mechanism for aldol condensation of furfural and acetone over dolomite. Adapted from Ref. [15].

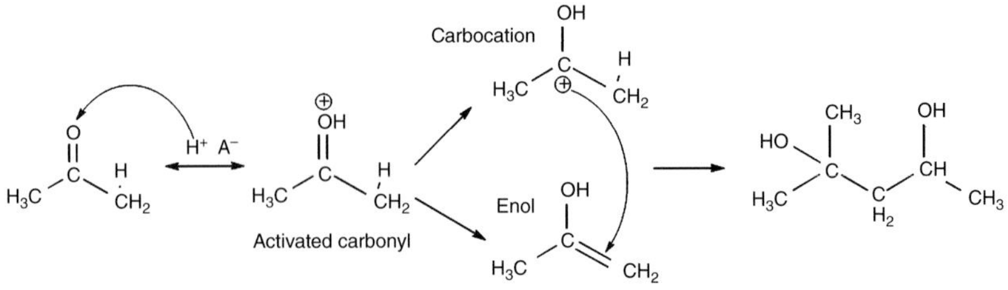

Scheme 17.3 Acid-catalyzed mechanism of aldol condensation reaction.

subsequently makes the bond more susceptible to the attack of the carbanion generated by the basic site.

17.2.1.2 Catalysts for Aldol Condensation

Homogeneous Catalysts Industrially, homogeneous bases such as NaOH, Ca(OH)$_2$, KOH, are widely used as catalysts for aldol condensation [20] due to their high activity for aldol condensation [21–23]. However, the utilization of such catalysts raises environmental and technical concerns such as high operation and separation costs [24]. Solid catalysts have many advantages over the homogeneous counterparts such as high catalytic activity and selectivity, corrosion-free, simple separation, fewer environmental issues, lower operation and maintenance costs, and lower energy requirements [20, 23]. Therefore, in the following discussion, the main emphasis will be on heterogeneous systems.

Heterogeneous catalysts
Basic Solid Catalysts Recent efforts by Dumesic's group in developing active heterogeneous catalysts for aldol condensation of biomass-derived furanics and acetone have been directed to mixed oxides of MgO with ZrO$_2$, CaO, TiO$_2$, and Al$_2$O$_3$ [25–27]. A typical example of these studies is the aldol condensation of HMF–acetone in the aqueous phase [28] over Mg–Al oxide, MgO, CaO, La–ZrO$_2$, Y–ZrO$_2$, MgO–ZrO$_2$, and MgO–TiO$_2$. The results show that mixed Mg–Al oxide, CaO, MgO–ZrO$_2$, MgO–TiO$_2$, exhibit high catalytic performance in terms of HMF conversion. Among them, MgO–ZrO$_2$ has generated remarkable interest due to its superior activity and stability. Recent investigations [16, 29, 30] have considered the effects of structure, morphology, acidity, preparation method, and composition of different MgO–ZrO$_2$ samples on its activity and selectivity. As mentioned earlier, dolomite CaMg(CO$_3$)$_2$ is another material that has resulted in improved selectivity to coupling products from furfural and acetone [15]. This catalyst especially favors the formation of dimer products at 413 K, with values of furfural conversion and dimer yield reaching 92% and 72.8%, respectively. The

active site has been attributed to Brønsted hydroxyl sites, believed to be most effective for aldol condensation [31, 32].

The performance of basic zeolites has also been investigated by Resasco et al. [33] for selective aldol condensation of acetaldehyde over NaX, NaY, and Huber et al. [34] for the aldol condensation of furfurals (and HMF) and acetone–propanal over NaY and nitrogen-substituted NaY (Nit–NaY). Huber et al. report that the incorporation of N into the framework of NaY seems to enhance the activity of the zeolite by increasing its basic strength, but it also increases the selectivity toward the monomer product. Moreover, in another study, this catalyst system has been further developed to enhance the activity and selectivity of the aldol condensation products by incorporating MgO over NaY supports [22]. The hybrid system shows superior performance compared to each individual component. The best catalyst 20% MgO–NaY achieves the conversion of furfural of 99.6% at 85 °C for 8 h in a water–ethanol solvent. The selectivity toward FAc and FFAc reaches 42.2% and 57.1%, respectively.

Acidic Solid Catalysts The study on heterogeneous acid catalyst is not as common as the base catalysts. In the study of catalyst zeolites HZSM-5, HBEA, HMOR, and USY for aldol condensation between furfural and acetone [12], it has been proposed that the wide-pore zeolite with three-dimensional crystal frameworks enhances the conversion of the reaction while medium-pore MFI zeolite and wide-pore monodimensional MOR do not. Also, by correlating the ratio C_{BAS}–C_{LAS} (Brønsted over Lewis) with the catalytic activity, it can be seen that the kinetically relevant active site for the reaction is the Brønsted acid site.

One interesting point observed on acid catalysts, typically not on base catalysts, is the formation of $(FAc)_2$ from the dimerization of FAc molecules (Scheme 17.4). The acid site strongly attracts the unsaturated bond of the monomer FAc to generate the dimerization product, while the basic site promotes the activation of α-H of the monomer to produce the dimer FFAc. Highest furfural conversions and corresponding yields, reported by Kubicka et al., of 53%, 37.4%, 9.5%, and 2.4% for FAc, $(FAc)_2$, and FFAc are obtained in reactions over HBEA at 100 °C for 8 h. Further discussion about the catalysts for aldol condensation of biomass-derived compounds can be found in the review by Pham et al. [35]. Some common aldol condensation catalysts are summarized in Table 17.1.

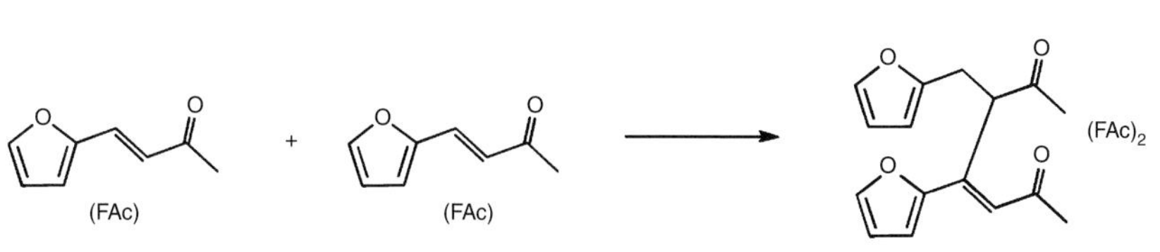

Scheme 17.4 Dimerization of FAc (Furfural-Acetone coupling product).

Table 17.1 Catalysts for aldol condensation reaction of furfural–HMF and acetone.

System	Catalysts	Reactant ratio	Reaction condition	Conversion of furfural–HMF (%)	Selectivity			Carbon balance (%)	References
					FAc (%)	FFAc (%)	FAc-OH (%)		
Basic oxides	MgO–ZrO$_2$ (Mg–Zr = 11:1)	F–Ace = 1:1	325 K, 10 bar N$_2$, 24 h Solvent: water	81.4	14.7[a]	61.5[a]	—	98.4	[16, 36]
	Mg–Al LDH (3:1)			63.8	31.1[a]	42.9[a]	—	97.8	
	CaO–ZrO$_2$ (1:3)			84.4	16.9[a]	15.4[a]	—	75	
	Mg–Al LDH	F–Ace = 1:10	373 K, 10 h, Mg–Al = 2.5	78.6	72.3	—	—	99	[24]
		F–Ace = 1:10	323 K, 3 h, Mg–Al = 3.03 (L)[b]	97	60.9	39.1	0.97	—	
		F–Ace = 1:10	323 K, 3 h, Mg–Al = 3.28 (I)	77.5	34.4	33.3	32.4	—	
	MgO–ZrO$_2$ (Mg–Zr = 1)	F–Ace = 1:10 (volume)	333 K, 3 h, Solvent: water	68.6	43.15	5.72	46.68	80	[30]
	Dolomitic rock CaMg(CO$_3$)$_2$	F–Ace = 1:1	306–413 K, 20 bar He Solvent: water–methanol (1:1.5)	75[c]	15.8	60	1.28	98.3	[15]
				92[d]	19.6	72.4	—	95.2	
Basic oxides + zeolite	MgO–ZrO$_2$ (Mg–Zr = 11:1)	HMF–Ace = 1:1	393 K; 750 psig He, 16 h	51.4	56.5	43.5	—	91.9[e]	[34]
	NaY (Si–Al = 2.55)		Solvent: water–methanol (1:1) by volume	14.1	85.9	14.1	—	89.7[e]	
	Nit-NaY			40.1	72.7	27.3	—	80.9[e]	
	MgO–ZrO$_2$ (Mg–Zr = 11:1)	F–Ace = 1:1	393 K; 750 psig He, 44 h	77.6	38.9	61.1	—	100.4[e]	[34]
	Nit-NaY (Si–Al = 2.55)		Water–methanol (2:1) by volume	73.3	57.3	42.7	—	100.4[e]	
	MgO–NaY (20%MgO)	F–Ace = 1:2	358 K, 8 h, water–ethanol (1:1)	99.4	42.2	57.1	—	99.2[e]	[22]
Acid zeolite	HBEA (SiO$_2$–Al$_2$O$_3$ = 25)	F–Ace = 1:10	373 K, 2 h	38.5	79.5	3.7	16.8[f]	—	[12]
	USY (SiO$_2$–Al$_2$O$_3$ = 80)			23	68.3	2.5	29.1[f]	—	

a) Atomic yield for each of these condensation products: $\Psi_{C8}\% = \frac{8 \times mol_{C8}}{5 \times mol_{C3=0} + 3 \times mol_{C3=0}} \times 100$ and $\Psi_{C13}\% = \frac{13 \times mol_{C13}}{5 \times mol_{C3=0} + 3 \times mol_{C3=0}} \times 100$.

b) Laboratory catalyst (L) and industrial catalyst (I); LDH: layered double hydroxides; FAc: monomer products; FFAc: dimer products; FAc-OH, alcohol product.

c) Conversion and yield at 323 K, 60 min.

d) Conversion and yield at 413 K, 60 min.

e) Selectivity to aldol products.

f) The selectivity of dimerization of FAc.

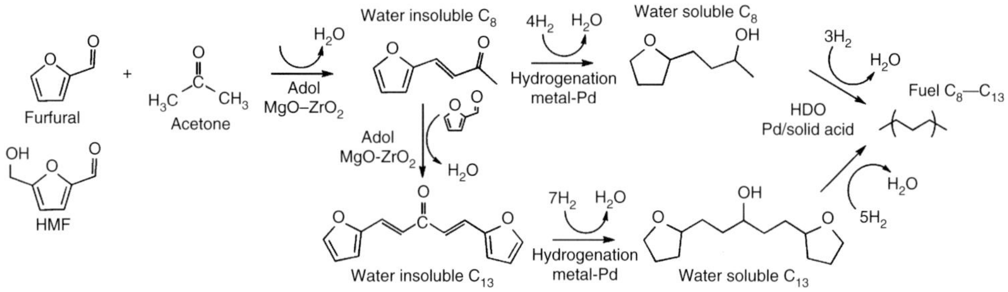

Figure 17.2 Strategy for furfural (or HMF) upgrading based on aldol condensation and hydrogenation. Adapted from Ref. [25].

17.2.1.3 Upgrading Strategy

The upgrading process of furfural (or HMF) with acetone via aldol condensation and deoxygenation is presented in Figure 17.2 including three steps:

- *Step 1:* Aldol condensation of furfural (or HMF) with acetone to generate the longer carbon chain. The catalysts could be oxides, mixed oxides, zeolites, and so on, as discussed earlier.
- *Step 2:* Hydrogenation is carried out to stabilize the coupling products to produce the alcohol with saturated ring. Common catalysts used for the hydrogenation step are noble metals, such as $Pd-A_2O_3$ [26–28].
- *Step 3:* (optional) The hydrogenated product can be hydrodeoxygenated (HDO) including hydrogenation + dehydration to completely remove oxygen content and produce linear alkanes. Some examples for HDO catalysts are as follows: $Pt-Al_2O_3$, $Pt-SiO_2-Al_2O_3$ [37], and $Pd-NbOPO_4$ [38].

Some recent examples propose conducting aldol condensation and hydrogenation in two separate reactors, followed by the HDO process to produce linear alkanes [26]. Alternatively, an integrated process can also be considered, in which aldol condensation and hydrogenation can be conducted in a single reactor over bifunctional catalysts. The bifunctional catalysts contain two components: a basic oxide for aldol condensation and a metal function for hydrogenation. Some examples of bifunctional catalysts for these reactions can be $Pd/MgO-ZrO_2$ [25, 28], $Pd/MO-Al_2O_3$ [39], Pd/TiO_2-ZrO_2, and Pd/WO_3-ZrO_2 [40]. Recently, two examples of this integrated approach have been discussed in the literature.

Single-Reactor Process for Sequential Aldol Condensation and Hydrogenation
This approach was first developed by Barrett et al. [25], using a bifunctional $Pd/MgO-ZrO_2$ catalyst. In this approach, first the aldol condensation (step 1) is conducted under mild conditions ($50-120\,^\circ$C, 10 bar). Due to the low solubility in water, the monomer and dimer products formed during the reaction precipitate

out of the aqueous phase, which complicates the recycle runs and separation of the catalysts [28]. To tackle this problem, after the first step, the authors have carried out the hydrogenation (step 2) at 120 °C at 55 bar H_2 to convert water insoluble C_8–C_{15} to water-soluble species. It is reported that the highest conversion of furfural in this method can reach 80% after 24 h with 34% and 43% selectivity of C_8 and C_{13}, respectively. Additionally, a proposed modification is to replace water in step 2 by an organic solvent (hexadecane), which can further improve the removal of the products in the solvent, resulting in higher yields. In fact, the results show that the furfural conversion and the selectivity toward C_{13} are enhanced to 100% and 85%, respectively, when the organic solvent is used. Alternatively, a biphasic system of tetrahydrofuran (THF) and water is used as solvent [21]. In this case, NaCl is added to the aqueous phase to further decrease the solubility of the coupling products, so not only the yield and carbon retention could be improved, but it also removes the need for the solvent change step. One potential limitation for this approach is the use of large amounts of NaOH, which might hinder its practical application.

Simultaneous Condensation–Hydrogenation in Water–Oil Emulsions We [41, 42] have proposed using amphiphilic nanohybrid catalysts to effectively stabilize water–oil emulsions and simultaneously conduct condensation–hydrogenation of acetone and furfural. The catalysts investigated comprise a family of bifunctional nanohybrids containing carbon nanotubes fused to different basic oxides (MgO, MgO–Al_2O_3, TiO_2, ZnO, $Ce_xZr_{1-x}O_2$, V_2O_5) with Pd clusters selectively supported on the hydrophobic nanotubes. During the process, the nanohybrids remain at the interface between the organic and aqueous phases, with the hydrophilic side in contact with the aqueous phase and the hydrophobic side with the organic phase. The aldol condensation occurring on the aqueous side of the interface produces coupling products with low solubility in water, so they are transferred to the organic phase. These products are then selectively hydrogenated over the Pd clusters incorporated on the hydrophobic nanotubes, which reside on the organic phase side. The hydrophobic nature of the supports will ensure that the Pd cannot enter the aqueous phase so that selective hydrogenation can only occur in organic phase. The illustration of the process is shown in Figure 17.3.

In general, some advantages of the water–oil emulsion system are as follows:

- Easier separation of products, reducing the complexity of the process and operation cost, allowing continuity of reaction.
- High interfacial areas leading to high reaction rate under diffusion limited condition.
- The activity is significantly higher than the reaction in single aqueous phase.
- Excellent tolerance to water leaching compared to bare oxides.

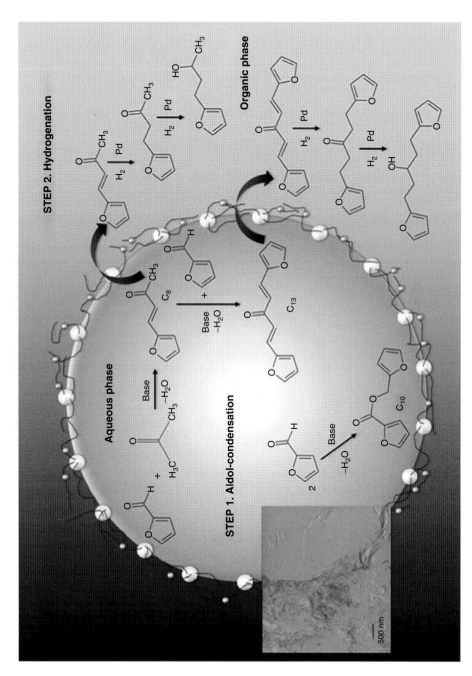

Figure 17.3 Condensation–hydrogenation in water–oil emulsions. Adapted from Ref. [41].

- High efficiency of hydrogenation step due to the hydrophobic property of the single-walled carbon nanotube support (preventing Pd from contacting with unreacted reactant).
- High flexibility due to the capability of conducting sequential reactions in a single reactor.

In the processes described earlier, HDO is one of the most energy-consuming steps, since it requires severe conditions to produce linear alkanes from the aldol condensation oxygenated products. For example, high temperatures (523–538 K) and H_2 pressures (52–60 bar) are required to hydrodeoxygenate the water-soluble C_8 and C_{13} products [26], or very high H_2 pressure (50–80 bar) is required for opening the saturated ring [43–46]. Therefore, finding alternative reactions that can operate at more moderate conditions than conventional HDO is a desirable target, which may improve the techno-economic aspects of the production of biofuels from biomass-derived furanic compounds.

An interesting alternative path, which would require milder conditions, has been recently investigated. The heterocyclic ring of the aldol condensation adducts can be readily opened at mild conditions under acid environment to produce a triketone [47], which can subsequently undergo conventional hydrogenation–dehydration to produce linear alkanes in a more energy-effective way.

For instance, by utilizing a Brønsted acid under mild conditions (80–100 °C, atmospheric pressure), the heterocyclic ring of 4-(5-methyl-2-furyl)-2-butanone (aldol condensation product between 5-methylfurfural and acetone) can be easily opened with yields near 100%. The general reaction scheme is summarized in Figure 17.4.

The ring opening products (4) can only be obtained when the substituent R is CH_3 or CH_2OH, whereas decomposition products are produced if R is H (as in **1c**-Figure 17.4). These differences clearly show that the substituents in 5-position play a very important role in the evolution of the reaction [48, 49]. Another important

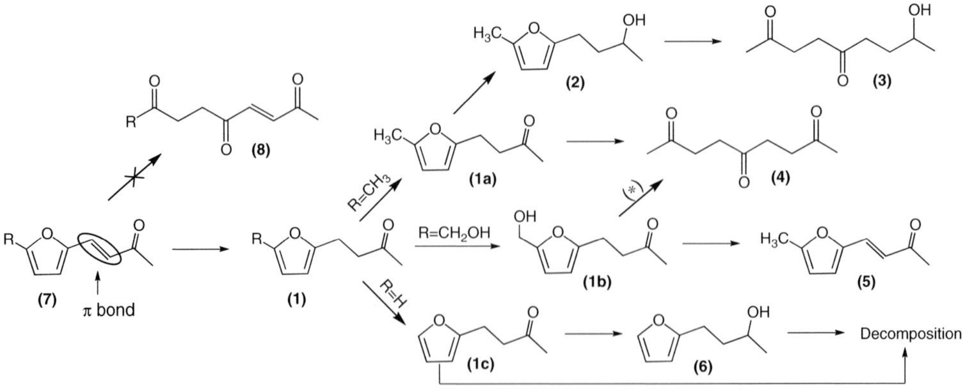

Figure 17.4 Reaction pathway for ring opening of aldol condensation products. Reaction condition: 80 °C, 24 h, water–methanol (1:1); (*)100 °C, 3 h, acetic acid–water (1:1).

Table 17.2 Yield of ring opening products over different acid catalysts (10 mol% catalyst in 1:1, water–methanol as solvent, 60–80 °C, 24 h). Adapted from Ref. [49].

Acid	Yield (%)	Acid	Yield (%)	Acid	Yield (%)
$HCl + FeCl_3$	95	HNO_3	77	HCOOH	4
HCl	94	$Fe_2(SO_4)_3 \cdot 5H_2O$	64	CH_3COOH	1.4
$Fe(NO_3)_3 \cdot 9H_2O$	94	$FeCl_2$	48	$Lu(OTf)_3$	1
$FeCl_3 \cdot H_2O$	93	$AlCl_3$	44	$NiCl_2$	0
$Fe^{III}Cl_3$	92	$Fe(SO_4) \cdot 7H_2O$	14	$CoCl_2$	0
$FeCl_3$ (Ar)	92	$CuCl_2$	10	$MgSO_4$	0
H_2SO_4	84	$CeCl_3$	9		
CF_3COOH	77	$FeCl_2$ (Ar)	4		

characteristic of this reaction is that the exocyclic unsaturated bond has to be removed before the reaction can proceed. This has been experimentally verified when no opening products are obtained but rather a darkening of the mixture, which is most probably related to the formation of humins when the compound (**7**) is heated without hydrogenation. In fact, density functional theory (DFT) calculations [49] have shown that the ring opening is thermodynamically unfavorable for the unsaturated compound.

A wide range of homogeneous acids have been tested for the ring opening of the compound (**1a**) in water–methanol (1:1) as a solvent [49] (see Table 17.2). The results show that strong Brønsted acids are the most effective catalysts for this type of reaction, whereas carboxylic acids, such as acetic or formic acid, show poor activity. However, the acetic acid in the mixture with water (1:1) is reported as an effective catalyst with 96% yield of (**4**) [47]. To our knowledge, only homogeneous catalysts have been studied for this reaction so it would be interesting to compare the behavior of solid acid catalysts.

As discussed earlier, to utilize ring opening as an upgrading step following aldol condensation, the exocyclic double bond should be removed. To conduct this saturation step, noble metal catalysts such as Pd and Pt could be utilized. However, it is important to point out that the hydrogenation process should be selective enough so that only the exocyclic unsaturated bond is removed without affecting the furanic ring. Saturation of the ring deactivates the electron delocalization, which would impede the ring opening reaction. Therefore, appropriate hydrogen pressure, temperature, and amount of catalyst should be taken into consideration.

The strategy for alkane synthesis from aldol condensation products has been proposed in a study by Sutton et al. [47] (Figure 17.5). In general, the proposed reaction sequence includes the following steps:

- Initial hydrogenation at 65 °C for 3 h under 1 atm H_2 over 5% Pd–C with acetic acid–water as solvent. The acetic acid is used because it can provide acidity for the following acid-catalyzed ring opening step.

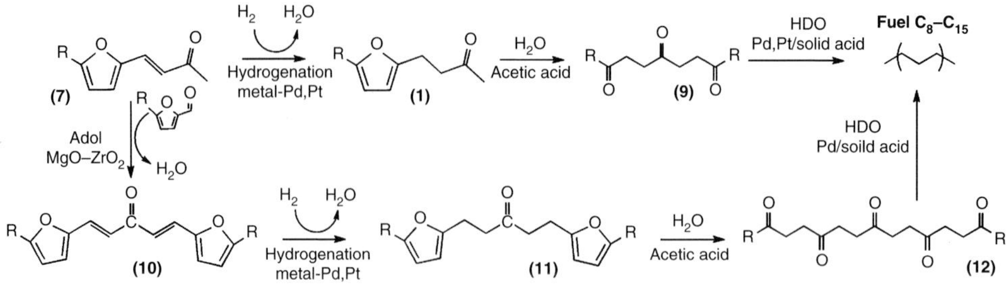

Figure 17.5 The strategy for aldol condensation products upgrading based on ring opening reaction.

- The ring opening is conducted at 100 °C for 3 h, with an excellent yield of 96% of compound (**9**) in Figure 17.5 (for the case of HMF-acetone coupling).
- The resulting ketone groups are HDO via hydrogenation–dehydration over a Pd catalyst with the addition of La(OTf)$_3$ at moderate condition 200 °C under 3.45 MPa H$_2$ for 12 h. It is reported that 90% yield of single alkane product could be achieved.

In addition, an interesting one-pot integrated process has been suggested in which the initial hydrogenation step and ring opening step are carried out simultaneously in a single step at 100 °C under 0.34 MPa H$_2$ for 3 h, without the need of changing reaction conditions or separating intermediate products. After the HDO process (at 200 °C, 2.07 MPa H$_2$), 74% yield of n-C$_9$ is achieved (for the case of HMF-acetone coupling). Compared to the severe conditions required for the conventional HDO process [43–46], the advantage of the ring opening process is obvious. An important limitation to notice for the general approach of ring opening is that it cannot be applied to furfural (i.e., the R group is H). As mentioned earlier, the absence of substituents inhibits the ring opening step, as confirmed by experimental results and DFT calculations. If ring opening could be achieved for the case of furfural, this strategy could have a much wider scope of application in biofuel upgrading.

17.2.2
Hydroxyalkylation–Alkylation: Sylvan Process

The rapid polymerization of furfural to humins could lead to significant deactivation of catalysts. One feasible strategy to upgrade furanics is to prestabilize them into key intermediates such as furan compounds (furan, 2-methylfuran (2MF), DMF) before coupling with functional molecules to produce alternative fuel precursors. In the ensuing section, the hydroxyalkylation–alkylation (HAA) followed by Diels–Alder reaction will be discussed.

17.2.2.1 Mechanism of Hydroxyalkylation–Alkylation (HAA)

A possible mechanism for the HAA reaction [50–52] is summarized in Scheme 17.5.

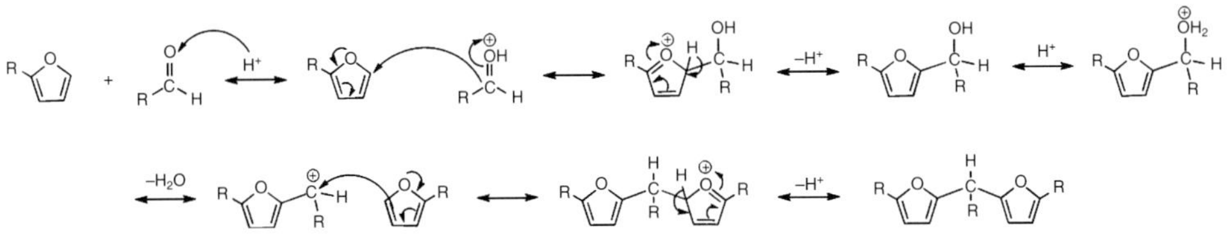

Scheme 17.5 Mechanism of furan hydroxyalkylation–alkylation (HAA).

In this mechanism, the HAA reaction starts with the protonation of the carbonyl group by a Brønsted acid. The protonated intermediate undergoes a nucleophilic attack from the molecule containing the furan ring. The oxonium ion intermediate generated from this attack leads to the production of the alcohol. When the Brønsted acid is strong to catalyze the dehydration of the alcohol, a condensation step with another furan ring molecule may occur, producing a difurylalkane product.

17.2.2.2 Catalysts for Hydroxyalkylation–Alkylation (HAA)

The catalysts that have been used for the HAA reaction of furanic compounds with ketones–aldehydes can be categorized into (a) liquid inorganic acids, such as HCl, H_2SO_4, HF, H_3PO_4 [50–56]; (b) soluble organic acids, such as acetic acid, benzoic acid, p-toluene sulfonic acid [57–60]; and (c) solid acids, including cation-exchange resins [57, 61] and zeolites [58].

Brown and Sawatzky [51] have studied the hydroxymethylation of furan and 2-methyl furan (2MF or sylvan) with various aldehydes in the presence of HCl (Scheme 17.6).

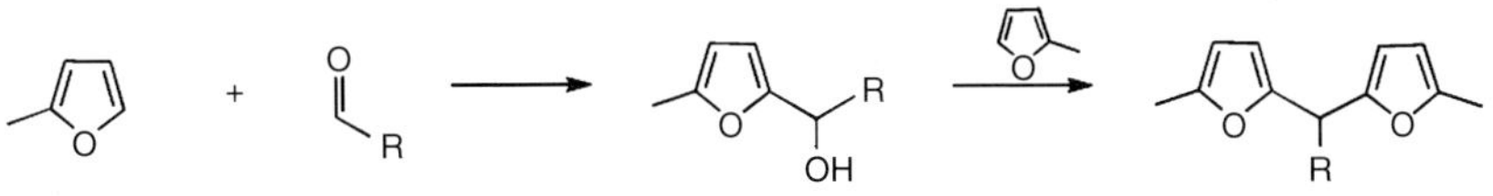

Scheme 17.6 Reaction scheme for HAA of 2MF and carbonyl compounds.

They showed that HCl is an effective homogeneous catalyst for this type of reaction. The observed yields of dimer reach up to 58% and 73% for furan and sylvan feeds, respectively. Here, again, it is worth noticing the substituent at C5 position enhances the activity of 2-methyl furan compared to furan; interestingly, its presence also inhibits the formation of polymerization products, due to the protection of the most reactive position in the ring. Additionally, the substituent R in the carbonyl compounds also plays an important role in activity and selectivity. Pennanen and Nyman [52] have shown that while for aliphatic aldehydes, the type of substituent affects the yield only slightly, the activity considerably decreases when the R group is an aromatic. For instance, with benzaldehyde, the yield of the dimer is low [52] or zero [51]. This is due to the resonance stabilization, resulting in the decrease of the amount of positive charge in the C of the carbonyl, which makes it much more difficult for the coupling with the furan ring (Scheme 17.7).

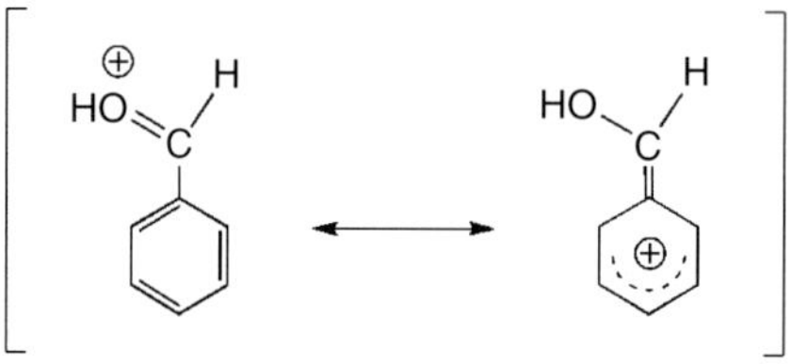

Scheme 17.7

Soluble organic acids, such as acetic acid, benzoic acid, *p*-toluene sulfonic acid, and so on, have also been investigated [57–59]. It has been proposed that not only acid strength but also lipophilicity play a key role in determining the performance of the organic catalysts. High acid strength is necessary for carbonyl activation, while sufficient lipophilicity ensures the efficient transport of cations in organic phase [50, 57]. Nevertheless, although the organic acid catalysts, especially *p*-toluene sulfonic acid, display excellent activity, the major drawback is the fast resinification of the furanic compound, which would hinder any practical application.

Compared to homogeneous catalysts, the solid catalysts exhibit obvious advantages related to separation and reusability. For example, Iovel et al. [57] have examined the performance of the H^+ form of cation-exchange resins, such as Amberlyst-15 (A15), Dowex 5OWx4 (D50), and Amberlite IRC-50 (A50) with different functionality groups, acidity, exchange capacity, and porous structure. A strong correlation between the catalyst acidity and the activity is observed for the hydroxymethylation of formaldehyde and furan (and 2-methyl furan).

The strongest acidic resins show the highest activity, while the more weakly acidic resins (A50) are only capable of catalyzing the hydroxymethylation of 2-methyl furan but do not work for furan. Understandably, the higher activity of the 2-methyl furan for HAA reaction could be ascribed to the electron donation effect of the methyl group. The strongest catalyst A15 yields 95% and 98% of difurylalkane for the case of furan and 2-methyl furan, respectively. Other resin catalysts such as macroporous sulfonic (Lewatit SPC 108, SP 120, Amberlite 15, XN 1010) are also effective catalysts, reaching over 90% difurylalkane yields [61]. It is important that these catalysts are not successful for catalyzing the furfural–formaldehyde nor the 2,5-dimethylfuran–formaldehyde system. Negligible amounts of the condensed alcohol are detected.

Strongly acidic zeolites, such as beta (BEA) and faujasite (USY), are also exploited for the HAA reaction [58]. The activity of these catalysts for the sylvan–butanal system seems to increase with increasing Si–Al ratios despite a decrease in acid density, which could be due to deactivation or an increase in acid strength. However, the governing factor seems not to be acidity, but rather the pore size of the zeolite. The intracrystalline diffusion limitations may cause a decrease in accessibility to active sites; in fact the pore diameter of zeolite beta is very close to the size of the products. Mesoporous materials with larger pore diameter (e.g., MCM-41, 3.0 nm) have been tested. However, the low acid strengths of active sites in mesoporous catalysts offset the advantage of having larger pore sizes. Delaminated zeolites with microcrystalline structure and mesoporous zeolites (fast diffusion, accessible for bulky molecules, strong acidity) could be promising.

There are a few studies applying zeolites for coupling carbonyl compounds to furfural and furfuryl alcohol system [62–65]. Strong acid sites present in the zeolite possibly cause fast polymerization of the highly reactive furanic compounds, which greatly reduces the product carbon yield. A potential mitigation of this problem is using the electron-donating 1,3-dithiolane functioning as a protecting group for the electron-withdrawing carbonyl in furanic compounds, so that the active aldehyde–ketone remains isolated during the course of the reaction. This technique makes it possible for 90% selectivity of hydroxymethylated product to be achieved [62].

A new type of hybrid catalyst, so-called sulfonic acid-functionalized ionic liquid (SAFIL), has recently attracted attention in this field [66]. It combines the advantages of homogeneous and heterogeneous catalysts with the sulfonic acid functionality providing highly acidic activity and the strongly ionic nature, which enhances the immiscibility of the catalysts in the hydrophobic product mixture. Basically, the hydrophilic nature of the ionic liquid facilitates the stabilization of a two-phase system during the course of reaction, which occurs at the interface. The water released from the reaction accumulates into the phase containing the catalysts, which could potentially simplify the downstream separation.

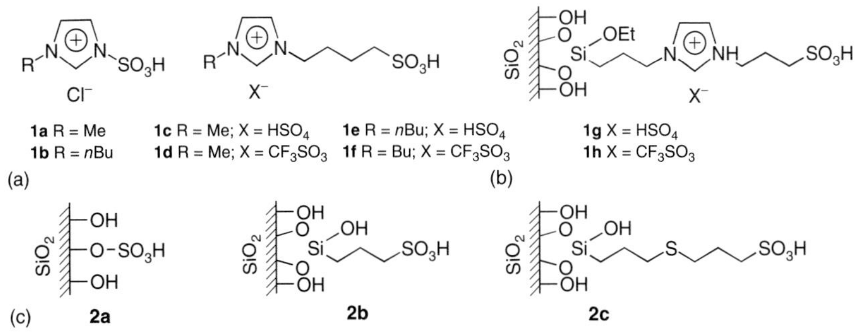

Figure 17.6 (a) Sulfonic acid-functionalized ionic liquid catalysts: Type 1 (**1a,b**), type 2 (**1c–f**). (b) Silica-supported sulfonic acid-functionalized ionic liquid catalysts (**1g–h**). (c) Silica-supported sulfonic acid catalysts (**2a–c**). Adapted from Ref. [66].

Two types of SAFIL catalysts have been investigated (Figure 17.6). One type of structural arrangement has the sulfonic acid functionality directly attached to the imidazole ring (**1a,b**). The other type has the functionality separated by an alkyl linker (**1c–f**). Since it is directly connected via an N atom, the acidity of type **1a,b** is expected to be weaker than that of type **1c–f**, as the N atom prevents the sulfonic acid group from donating a proton. This has been experimentally confirmed. That is, the catalyst **1c** reaches 89% yield of trimer products from the furfural–sylvan system and as expected is more active than the one with N linker.

A more interesting and intricate topic of discussion is the direct correlation between the performance of the catalyst and its hydrophobicity. When the hydrophobicity of the catalyst is increased by (a) replacing a polar sulfate anion by a more hydrophobic anion, such as triflate anion, or (b) replacing the R group by *n*-butyl functionalities, or (c) increasing the length of alkyl linker groups attached to the imidazole nitrogen, the catalytic activity is enhanced. However, it should be noted that a higher hydrophobicity could also increase the solubility of the catalyst in the hydrophobic phase containing the reactants so that the later separation process would be more difficult.

An alternative approach followed by the same authors is to immobilize the ionic catalyst on a silica support (**1g–h**). In this case, the catalyst phase and the reaction phase remain separated even if the catalyst is highly hydrophobic [66]. However, this system is less active compared to the free ionic liquid catalyst (61% and 92% yield of the trimer, respectively), which the authors attribute to the restrained accessibility of reactants to the active site inside the porous structure of heterogeneous silica. To tackle that problem, the sulfonic acid group is deposited directly to silica particle (**2a**) or via an alkyl linker (**2b–c**) instead of ionic liquid linker (**1g–h**) to alleviate the steric hindrance to reactants. It has been demonstrated that propane sulfonic acid-functionalized silica (**2b**) and 3-(propylthio) propane-1-sulfonic acid-functionalized silica (**2c**) are more efficient in improving the catalytic activity than the one without a linker (**2a**). The performance of **2c** is comparable to that of homogeneous catalyst *p*-toluenesulfonic acid (*p*-TsOH) and free ionic liquid catalyst with 89%, 92%, and 92% yield of the trimer achieved, respectively. Besides, the **2c** system offers high yield and a wide range of condensation products (Figure 17.7). Some examples for the performance of liquid inorganic catalysts, organic catalysts, cation-exchange resin, and zeolite catalysts are summarized in Table 17.3. It should be noted that the hydroxyalkylation can also be catalyzed by Lewis acid, typically $AuCl_3$ [56, 69, 70].

17.2.2.3 Upgrading Strategy

Although HAA of furan and 2-methyl furan has been studied for a long time, the idea of utilizing this type of reaction for biofuel upgrading has only recently been proposed by Corma and coworkers [58–60] with the so-called Sylvan process. Certainly, achieving high yields with low catalyst deactivation in the furfural hydroxyalkylation process is a difficult if not impossible task, since furfural polymerization is prevalent. On the other hand, the more stable sylvan, which could

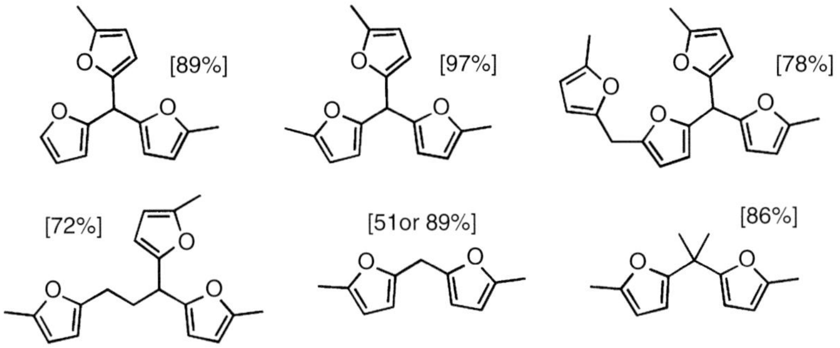

Figure 17.7 Products (with corresponding yields) formed by condensation of various carbonyl compound with 2-methyl furan using catalyst **2c**. Adapted from Ref. [66].

be industrially produced from furfural hydrogenation [71, 72], is a viable alternative. While alkylation of sylvan with alcohol is certainly possible, a more attractive way of increasing the carbon chain length in biomass-derived liquid products is the HAA since this coupling reaction could directly use acetone, which can be produced from acetic acid, an abundant component in bio-oil. In this regard, the Sylvan process would be an excellent complement to the overall bio-oil upgrading strategy described earlier. The two attractive properties that make 2-methyl furan promising candidate for the HAA reaction are the potentially high yield, selectivity and the reduced catalyst deactivation by minimization of the selectivity to undesired polymerization. Industrially, sylvan can be obtained as a side product during the production of furfuryl alcohol [58]. The sylvan upgrading strategy is summarized in Figure 17.8.

While the Sylvan process refers to only methyl furan, the original study by Corma et al. [58] has demonstrated the concept for a variety of carbonyl compounds, including acetone, butanal, methyl furfural, 4-oxopentanal (ring opening product of 2-methyl furan) and 2-pentanone (Routes 2, 3, 4, 6, 7) in Figure 17.8. The remarkable efficiency of this process has stimulated further studies attempting to broaden the scope of the Sylvan process to other lignocellulose-derived carbonyl compounds over a broad range of catalysts. For example, studies with butanal, acetone [73, 74], hydroxyacetone obtained from glycerol dehydration [75], and ethyl levulinate [74] have recently been reported. Moreover, the concept has been extended to other furanic compounds used as hydroxyalkylation agent to couple with 2-methyl furan or furanics, such as furfural–sylvan [66, 74], furfural–furan [76], 5-methylfurfural–sylvan, and HMF–sylvan [60]. The advantage of this approach is its high flexibility since a wide range of aldehydes–ketones can be used as the cofeed with sylvan.

This process is important not only for bio-oil upgrading but also for its capability to combine with other biomass-derived approaches. For example, butanal obtained from the selective oxidation of 1-butanol which is in turn produced from biomass fermentation with enzymes [77], or 5-methyl furfural production from

Table 17.3 Summary for the catalysts of HAA reaction.

Reactant	Catalyst	Aldehyde–ketone	Reaction product	Yield (%)	References
Furan	HCl	Formaldehyde	$C_9H_8O_2$	17	[51]
		Acetaldehyde	$C_{10}H_{10}O_2$	58	
			$C_{22}H_{22}O_4$	9.7	
		Propionaldehyde	$C_{11}H_{12}O_2$	24	
		Butyraldehyde	$C_{12}H_{14}O_2$	38	
		Benzaldehyde	No reaction		
		Furfural	No reaction		
		Glyoxal	No reaction		
Sylvan		Formaldehyde	$C_{11}H_{12}O_2$	34	
		Acetaldehyde	$C_{12}H_{14}O_2$	61	
		Propionaldehyde	$C_{13}H_{15}O_2$	63	
		Butyraldehyde	$C_{14}H_{18}O_2$	73	
Furan		Acetone	$C_{28}H_{22}O_4$	18.5	[55]
		Butanone	$C_{32}H_{40}O_4$	11.5	
		Methyl *n*-propyl ketone	$C_{36}H_{48}O_4$	5.5	
		Methyl isobutyl ketone	$C_{40}H_{56}O_4$	6.3	
		Methyl *n*-amyl ketone	$C_{44}H_{64}O_4$	6.1	
		Pinacolone	No reaction		
		Acetophenone	No reaction		
		3-Pentanone	$C_{13}H_{16}O_2$	17	[54]
			$C_{22}H_{28}O_3$	28	
			$C_{31}H_{40}O_4$	3.5	
Furan	H_2SO_4	Aliphatic aldehyde–ketone	Symmetric[a]	25–60	[67]
			Asymmetric	16–20	
Sylvan		5-Methylfurfural	$C_{16}H_{16}O_3$	83[b]	[58–60]
		4-Oxopentanal[c]	$C_{15}H_{18}O_3$	75[b]	
		2-Pentanone	$C_{15}H_{20}O_2$	10	[58]
			$C_{15}H_{18}O_3$	39	
			$C_{15}H_{22}O_3$	1.4	
		4-Methylpent-4-en-2-one	$C_{11}H_{16}O_2$	4.8	
			$C_{11}H_{18}O_3$	2.7	
			$C_{15}H_{18}O_3$	43.5	
			$C_{13}H_{16}O_2$	13.8	
Furan	HF	Formaldehyde	$C_9H_8O_2$	14	[68]
Sylvan	H_3PO_4		$C_{11}H_{12}O_2$	40	[50]
Sylvan	*p*-TsOH[d]	Butanal	$C_{14}H_{18}O_2$	78–88[b]	[58, 60]
		5-Methylfurfural	$C_{16}H_{16}O_3$	93[b]	[59, 60]
		Ethanal	$C_{12}H_{14}O_2$	66–87[b]	[58]
		Propanal	$C_{13}H_{16}O_2$	88[b]	
		Butanal	$C_{14}H_{18}O_2$	88[b]	
		Pentanal	$C_{15}H_{20}O_2$	87[b]	
		2-Pentanone	$C_{15}H_{20}O_2$	51	
			$C_{15}H_{18}O_3$	11.4	
			$C_{15}H_{22}O_3$	2.3	

Table 17.3 (*Continued*)

Reactant	Catalyst	Aldehyde–ketone	Reaction product	Yield (%)	References
		4-Methylpent-4-en-2-one	$C_{11}H_{16}O_2$	55.5	
			$C_{11}H_{18}O_3$	6.5	
			$C_{15}H_{18}O_3$	4.3	
			$C_{13}H_{16}O_2$	4.3	
Furan	Amberlyst-15	Formaldehyde	$C_5H_6O_2$	93 (24 h)[e]	[57]
			$C_9H_8O_2$	3	
Sylvan		Formaldehyde	$C_{11}H_{12}O_2$	86 (1 h)[e]	[57]
		5-Methylfurfural	$C_{16}H_{16}O_3$	86[b]	[60]
		Butanal	$C_{14}H_{18}O_2$	90	[58]
Sylvan	Amberlite IRC-50	Formaldehyde	$C_6H_8O_2$	90 (68 h)	[57]
			$C_{11}H_{12}O_2$	7	
Furan			No reaction	—	
Furan	Dowex 50WX4	Formaldehyde	$C_5H_6O_2$	68 (2 h)[e]	
Sylvan	Dowex 50WX4	Formaldehyde	$C_{11}H_{12}O_2$	71 (1 h)[e]	
	Dowex 50WX2[f]	Butanal	$C_{14}H_{18}O_2$	80	[58]
	Lewatit SPC 108	Formaldehyde	$C_{11}H_{12}O_2$	98	[61]
	Lewatit SPC 108	Acetaldehyde	$C_{12}H_{14}O_2$	100	
Sylvan	Beta (F−)	Butanal	$C_{14}H_{18}O_2$	16(13), 33(27), 18(47), 11(103), 0(∞)[g]	[58]
	Beta comm[h]		$C_{14}H_{18}O_2$	67[b]	
	Beta nano[i]		$C_{14}H_{18}O_2$	59[b]	
	Beta (OH−)		$C_{14}H_{18}O_2$	34(16), 9(26), 0(∞)[g]	
	MCM-41		$C_{14}H_{18}O_2$	45–60	
	ITQ-2[j]		$C_{14}H_{18}O_2$	86	
	USY		$C_{14}H_{18}O_2$	4(2.5), 9(6), 18(15), 53(20), 36(27.5)[g]	

a) Difurylalkanes with symmetric and asymmetric structure.
b) Yield after distillation.
c) Ring opening product of 2-methylfuran.
d) *para*-Toluenesulfonic.
e) Furan–CH_2O–[H+] = 1 : 3 : 0.1.
f) Gel-type sulfonic acid resins.
g) Yield and corresponding Si–Al ratio.
h) Commercial sample from Zeolyst.
i) Nanocrystalline sample.
j) Delaminated zeolite.

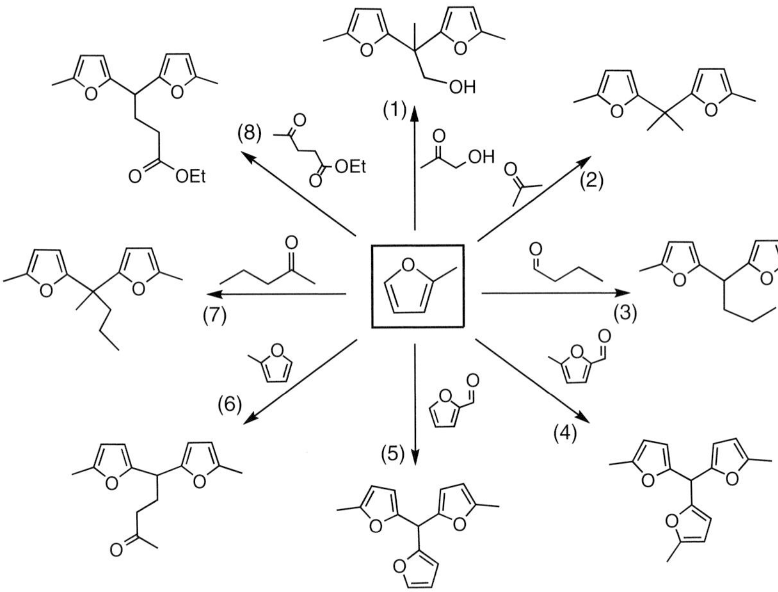

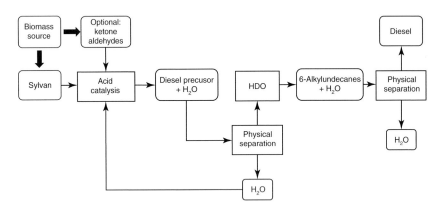

Figure 17.8 Summary of 2-methyl furan upgrading strategy into fuel application.

Figure 17.9 Sylvan process to produce diesel fuel from biomass.

cellulose [78–80] could be integrated with the Sylvan process to directly produce fuel from biomass source. In fact the Sylvan process can be applicable to any biomass-derived carbonyl compound, avoiding C–C and C–H dissociation, which provides high efficiency in terms of energy and carbon retention [81]. An example of the specific Sylvan process designed for biomass-derived fuel production is summarized in Figure 17.9 [58].

The sylvan and ketone–aldehyde obtained from biomass can be fed to an HAA reactor, which uses catalysts that could be either a soluble organic acid (such as

Table 17.4 Liquid product mixture obtained of HDO with different types of diesel precursors[a]. Adapted from Ref. [58].

Catalyst support[b] (%)						Product mixture[c] (%)		Yield[d] (%)	
C	Al$_2$O$_3$	TiO$_2$	Precursors from route (in Figure 17.8)	Feed passed-g	(W/F)[e] h^{-1}	Diesel[f]	C$_3$–C$_8$	Liquid	Diesel
100	0	0	4	91	0.7	89.3	5.8	86.3	77.1
75	0	25	6	198	1	90.5	8.2	89.1	80.7
100	0	0	7	125	0.55	89.4	5.3	79.1	70.7
80	20	0	3	113	0.55	94.6	0.5	89.6	84.7
100	0	0	4 and 6[g]	99	0.7	86.6	4.2	89.7	77.7

a) The table has been simplified. More detailed information can be found in Ref. [58].
b) Platinum (3.0 wt%) was impregnated onto the support.
c) Product distribution of the liquid organic phase recovered.
d) Calculated on a molar basis.
e) Contact time as a ratio of the amount of catalyst (g) to feed rate (g h^{-1}); reaction temperature was 350 °C and 5.0 g of catalyst was employed.
f) Diesel fraction was considered from C$_9$ to C$_{24}$ containing linear, branched, and cyclic alkanes.
g) Feed was a mixture of 4 and 6 in a 88:12 (wt%–wt%) ratio.

p-TsOH) or a solid catalyst (such as a zeolite or a cation-exchange resin). The diesel precursors produced from the HAA process can be introduced directly to an HDO process without additional fractionation. Laboratory experiments have shown that high selectivity (80–95%) and yield (75–85%) can be obtained [58] (Table 17.4). Interestingly, sylvan can be used as a single feed due to its capability of self-condensing into C$_{15}$ diesel precursors with high yields (∼65–75%) in the presence of sulfuric acid at a very mild condition [53, 60]. This could be an obvious benefit in terms of process simplification [58]. Another techno-economic advantage is that the water produced in the condensation reaction can be recycled to the HAA reactor and used as the reactant for the ring opening reaction. Although the H$_2$SO$_4$ catalyst for sylvan trimerization could not be considered a green catalyst, it has the potential to be recovered and recycled without losing its activity or selectivity, as recently shown [60].

Improvements to the original Sylvan process have recently been studied in the literature, which might make it even more attractive. For example, Wen et al. [81] have proposed a single-step fuel production process, which is based on the Sylvan process but utilizes a bifunctional heterogeneous catalyst, such as Pt-MCM-41. The dual function of this catalyst includes the acid support that catalyzes the HAA reaction and the metal site (Pt) that catalyzes the HDO reaction. This concept is tested in a plug flow reactor at 503 K. The results show 98% conversion of sylvan with 96% selectivity to C^{8+} hydrocarbons, which is significantly higher than those reported for the two-step Sylvan process [58] (Table 17.5). Similar improvements and process integration have recently been outlined by Refs. [82, 83].

Table 17.5 Catalytic performances of bifunctional catalysts for C_{8+} production from 2-methylfuran. Adapted from Ref. [81].

Catalyst	Temperature (K)	Conversion[a] (%)	Selectivity to C_{8+}[b] (diesel) (%)	C_{8+} yield (%)
0.1 wt% Pt-MCM-41	623	100	69	69
	553	100	76	76
	503	98	98	96

a) Conversion is calculated by dividing moles of converted 2MF to original moles of 2MF measured by GC-MS (Gas chromatography-mass spectrometry). The error of calculated conversion is ±4%. The butanal has the same conversion as the reactant is added with stoichiometric ratio and was measured also by GC-MS.
b) The selectivity to C_{8+} is calculated by the portion of carbon from the 2MF and butanal converted to C_{8+} hydrocarbons and is measured by GC-MS. The error for the calculated selectivity is ±7% [81].

17.2.3
Diels–Alder

17.2.3.1 Mechanism of Diels–Alder Reaction

The Diels–Alder reaction, commonly referred to as *[4+2] cycloaddition*, is a well-known C–C coupling reaction to produce a substituted cyclohexene from a dienophile and a diene [84] (Scheme 17.8). Furanics, containing two π–π bonds inside the ring, could serve as a diene that couples with a dienophile such as an olefin to form larger species. The Diels–Alder reaction has been proposed to follow either a concerted or stepwise mechanism as shown in Scheme 17.8 [87]. For the former, the C–C bonds are formed simultaneously with only one activation barrier, whereas for the latter, the bonds are formed consecutively

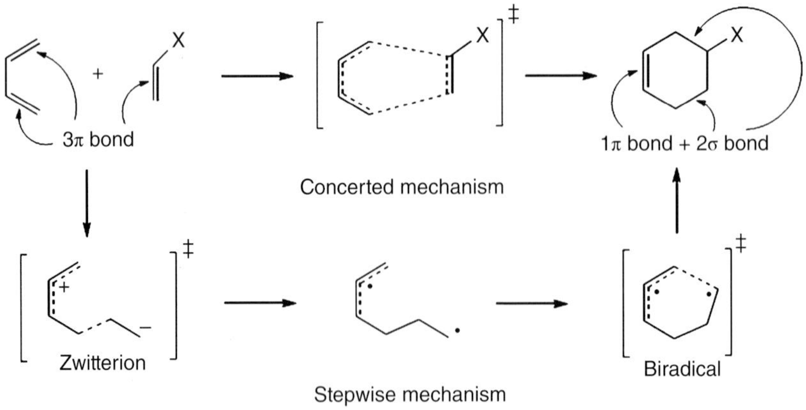

Scheme 17.8 Concerted and stepwise mechanism. Adapted from [85, 86].

with two activation barriers [85]. The concerted mechanism is more universally accepted.

According to the frontier molecular orbital theory, C–C bond formation via the Diels–Alder reaction is related to the overlapping of the HOMO (highest occupied molecular orbital) containing 4π electrons of a diene molecule with the LUMO (lowest unoccupied molecular orbital) containing 2π electrons of a dienophile [88]. In a normal demand scheme, if the dienophile possesses an electron-withdrawing group (EWG) capable of lowering its LUMO energy level, it will facilitate electron movement from the HOMO of the diene to the LUMO of the dienophile to form a bond by reducing the HOMO–LUMO gap [84, 89]. In that same manner, if the diene has an electron-donating group (EDG), the energy level of the HOMO of the diene will increase, leading to the lower energy gap between the HOMO and the LUMO. If an EWG is instead attached to the diene, with the EDG on the dienophile, this could facilitate electron movement from the HOMO of the dienophile to the LUMO of the diene. This is referred to as *inverse demand*. Both strategies reduce the energy gap between the LUMO and HOMO such that the reaction is more favorable and faster. More details regarding the mechanism and characteristics of the Diels–Alder reaction can be found in Refs. [86, 88, 90, 91].

In the normal demand scenario, furan is typically considered a weak diene for Diels–Alder reactions, which could couple only with active dienophiles such as alkenes, alkyls, and so on. [92]. For example, the Diels–Alder reaction between furan and ethylene requires Lewis acids [93] or high pressures to achieve satisfactory yields [92, 94]. On the other hand, DMF is more active due to the election-donating methyl groups. For instance, the reaction between DMF and methyl acrylate yields 66% of coupling product, whereas in the case of furan, 48% yield is obtained under the same conditions [94]. As another example, upon cofeeding DMF with ethylene over H-beta zeolites, 90% yield to *p*-xylene is produced, while the reaction of ethylene with methylfuran and furan produces only 43% yield of toluene and 25% yield of benzene, respectively [95]. The addition of an election-donating substituent such as a methyl group to an olefin (dienophile) would result in a dramatic decrease in activity [94]. The Diels–Alder reaction of furan and an olefin could follow inverse electron demand when the furan has EWGs and the olefin has EDGs [96]. However, practically, it is less common to carry out the Diels–Alder reaction in reverse electron demand due to the requirement of more powerful donor and acceptor groups [88]. DFT calculations of DMF and ethylene coupling over NaY have showed that if the DMF is bound to a Lewis acid site, the inverse mechanism is likely to occur. However, the rate of reaction under inverse demand is generally slower due to a larger HOMO–LUMO gap [96]. Furfural, which possesses an electron-withdrawing carbonyl group, is expected to be less active than furan in a normal demand scheme but could conceivably serve as a nucleophile in inverse electron demand. Thus far, attempts at furfural coupling with various compounds including both electron-rich (i.e., ethyl vinyl ether) and electron-poor (i.e., dimethyl acetylenedicarboxylate) dienophiles have not been successful [94].

17.2.3.2 Catalysts for Diels–Alder Reaction

By adding suitable substituent groups to the diene and dienophile, the rate of Diels–Alder coupling could be increased to the point that it could possibly occur even at room temperature in the absence of a catalyst [88]. For example, the self-coupling of cyclopentadienone is known to rapidly occur, leading to polymerization even at room temperature [97]. Therefore, it is reasonable to consider Diels–Alder a noncatalyzed reaction for certain combinations of dienes and dienophiles. However, in some less active systems, such as the non-substituted dienophile, the addition of a catalyst becomes necessary to accelerate the reaction rate under moderate conditions. The addition of a Lewis acid catalyst has been proposed to lower the energy of the LUMO of the dienophile by forming a complex between the dienophile and an acid [86, 98, 99]. Moreover, the addition of a catalyst has been proposed to trigger an increase in the stereoselectivity and regioselectivity of the Diels–Alder reaction when compared with the uncatalyzed case [88, 99].

Catalysts for the acceleration of Diels–Alder reactions have been reported for many decades including homogeneous Lewis acids [100] and heterogeneous Lewis acids such as silica gel [101, 102], magnesium silicate [102], alumina [102–104], clays [105], Fe^{3+}-doped K10 montmorillonite [106], $ZnCl_2$, ZnI_2, $TiCl_4$ [107], Et_2AlCl [108], copper [109], zeolites [110, 111], alkali and alkaline perchlorates (Mg^{2+}, Ba^{2+}, Li^+, Na^+) [112]. In this contribution, we focus only on the catalysts that have been investigated for the production of targeted valuable chemicals from biomass-derived furanics.

Toste and Shiramizu [113] have studied the Diels–Alder reaction of DMF and acrolein over several Lewis acids including $AlCl_3$, Et_2AlCl, $Sc(OTf)_3$, $TiCl_4$, $SnCl_2$, $BF_3 \cdot Et_2O$, $ZnCl_2$, and ZnI_2. Prior Diels–Alder attempts at coupling between furanic compounds (furan, 2-MF, DMF) and acrolein were carried out at extreme pressures of 15 000 atm at room temperature [94, 114] or at milder pressure with Lewis acids such as ferric chloride, stannic chloride, and zinc chloride [115]; Fe^{3+}-doped K10 bentonite clay [93]; or ZnI_2 [116] to achieve satisfactory yield. Toste and Shiramizu [113] have conducted the reaction at subambient temperatures ($-60\,^\circ$C) and long reaction times (68.5 h), over $Sc(OTf)_3$ to achieve 84% yield of coupling products.

One promising example of the Diels–Alder reaction with respect to biomass utilization is the production of *p*-xylene, a valuable chemical, from DMF and ethylene. The reaction is believed to include two steps: cycloaddition (Diels–Alder) and dehydration (Figure 17.10). The inclusion of zeolites has been proven to accelerate the overall rate of aromatics formation, with Brønsted sites in the zeolite serving to dehydrate the Diels–Alder cycloadduct. Recent studies have focused on Y [117–119], ZSM-5 [117, 120], and BEA [117] zeolites.

Dauenhauer et al. [117] have noted that in the absence of a zeolite catalyst, the rate-limiting step was the dehydration of the coupling product. However, Brønsted acid sites lower the activation energy of the dehydration reaction (from 57.6 to 19.13 kcal mol^{-1}) causing an enhancement in the overall rate of xylene production. As a result, good selectivity ($\sim$75%) and high conversions (95%) of

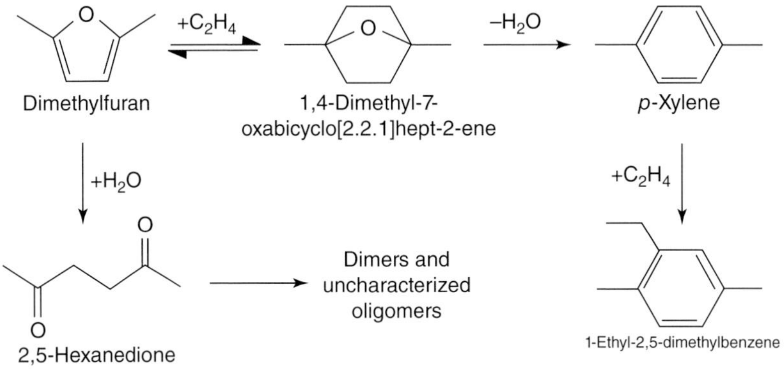

Figure 17.10 Reaction pathway of DMF and ethylene cycloaddition. Adapted from Ref. [117].

DMF are observed over HY zeolites at 300 °C, 57 bar ethylene with *n*-heptane solvent. The detailed reaction network developed by Watson and Lobo et al. [121] provides a general picture of the reaction evolution, including two main reactions, cycloaddition and dehydration, and several side reactions such as ring opening of DMF, alkylation, oligomerization. Because Lewis acid sites are also present in the zeolites and known to accelerate both dehydration and Diels–Alder coupling reactions, more work is necessary to completely separate the role of confinement on each of the active sites in these zeolites.

DFT calculations from Vlachos et al. for DMF coupling with ethylene over NaY (with no Brønsted acid sites) and HY (including both Brønsted and Lewis sites) have shown that the Diels–Alder reaction follows a concerted mechanism with only one activation barrier peak in the energy profile (Figure 17.11) [96]. They demonstrate that if the ethylene is bound to a Na^+ cation in NaY, the reaction follows normal election demand whereas the reaction proceeds via inverse election demand if DMF interacts with the Na^+. As with most cases, the normal electron demand has a lower overall barrier. DFT calculations indicate that Lewis acid sites in NaY help to reduce the barrier for the cycloaddition from 24.7 in the absence of a catalyst to 18.65 kcal mol^{-1} over NaY (Figure 17.11). The Na^+ Lewis acid sites could also catalyze the dehydration step but are far less effectively than Brønsted acid sites. In the case of NaY, dehydration is the rate-limiting step. Conversely, over HY, the rate-limiting step is the Diels–Alder reaction. The idea of utilizing a bifunctional catalyst with an appropriate ratio of Brønsted and Lewis acid sites holds promise since both cycloaddition and dehydration could be catalyzed simultaneously. It should be noted that the DFT calculations discussed earlier do not take into consideration the effect of solvation and confinement of the sites within the zeolite. Subsequent DFT studies have been published to elucidate the effect of the solvent [122] and the HY framework environment [123].

The rate enhancement of the Diels–Alder reaction in NaY has been previously reported [124]. For the Diels–Alder coupling of cyclopentadiene (CPD), the

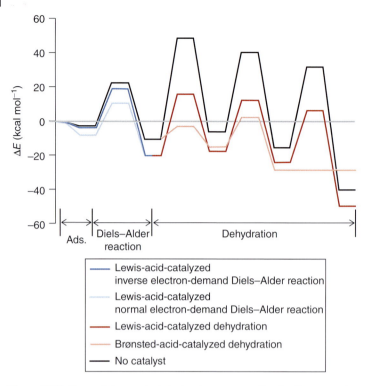

Figure 17.11 Zero-point-corrected electronic energy profiles for the conversion of DMF and ethylene to *p*-xylene relative to the reactants' energy at infinite separation. Adapted from Ref. [96]

supercage structure of hydrophilic NaY (Si/Al = 2.4) is proposed to trap the nonpolar CPD inside, generating an "internal pressure" to activate the CPD to couple with methacrylonitrile, a weakly active dienophile. High yields (91%) of coupling products are obtained.

In 1986, studying the conversion of buta-1,3-diene to 4-vinylcyclohexene over Na-Y, Na-β, Na-ZSM-20, Na-ZSM-15, and Na-ZSM-12, Dessau et al. [110] stated that the highly concentrated butadiene inside the porous structure of zeolites is the proposed reason for enhanced reaction rates. Similar enhancements in rate are observed when non-acidic highly porous carbon is used.

Huber and Cheng [120] have studied aromatics production (benzene, toluene, xylene) via the Diels–Alder reaction between furanic compounds (furfural, furfuryl alcohol, furan, 2MF, DMF) and olefins (ethylene, propylene). The reaction is carried out in a fixed-bed flow reactor at 600 °C over HZSM-5. Different types of aromatics could be obtained with a moderate yield via three main reactions: (i) decarbonylation of furan to form allene (C_3H_4) and olefins, which could undergo sequential coupling reactions; (ii) Diels–Alder coupling and dehydration of furan compounds with olefins to produce single ring aromatics (e.g., benzene, toluene, *p*-xylenes, etc.); and (iii) Diels–Alder coupling of furanic compounds with aromatics to form bicyclic products (e.g., naphthalene). Furfural and furfuryl alcohol are proposed to undergo decarbonylation to furan before participating in the coupling reaction. The authors suggested that 2MF and propylene are appropriate feedstocks for highly selective production of xylene, while for toluene, furan with propylene would be the best choice.

The solvent plays a significant role in Diels–Alder coupling rates. Dauenhauer et al. have found that H-beta is a promising catalyst for this reaction when the DMF is diluted in an *n*-heptane solvent. Yields of DMF coupling with ethylene over H-beta reach 99% conversion with 90% selectivity to *p*-xylene at 250 °C and 62 bar [95, 117]. The authors attribute the superior performance of H-BEA over other zeolites to a combination of deactivation resistance, high activity, and the ability to selectively catalyze the dehydration of the Diels–Alder coupling product without facilitating other side reactions [95]. DFT calculations have indicated that the use of *n*-heptane solvent helps to dilute the concentration of DMF in the zeolite reducing the undesired side reactions such as further condensation of the cycloaddition product or DMF ring opening [125]. Dumesic et al. have recently demonstrated that other catalysts containing primarily strong Brønsted acid sites such as $WO_x–ZrO_2$ and niobic acid can serve as effective catalysts for aromatics (*p*-xylene, toluene, benzene) production via Diels–Alder coupling of DMF, 2MF, and furan with ethylene [119]. For example, $WO_x–ZrO_2$ materials give 77% selectivity to *p*-xylene at 60% conversion of DMF.

17.2.3.3 Upgrading Strategy

The application of Diels–Alder reactions for the lignocellulosic biomass upgrading has focused mainly on the production of aromatic compounds such as *p*-xylene, benzene, and toluene from furanic compounds and olefins. Two separate patents in 2009 [126] and 2010 [127] provided a method for *p*-xylene production from DMF and ethylene gas through a single-step process. Yields of *p*-xylene as high as 30–80% are claimed at 100–300 °C and 10–50 bar of ethylene pressure over a range of catalysts including acid-washed activated carbon, zeolites, and alkali earth modified zeolites.

As previously mentioned, H-Y, H-BEA, and ZSM-5 are promising catalysts for *p*-xylene production from the coupling of DMF and ethylene. Other chemicals such as toluene and benzene could also be produced from other furanic species such as 2MF and furan. The recent promise shown with other oxide catalysts such as $WO_x–ZrO_2$ deserves further investigation.

Toste and Shiramizu [113] have proposed a different route to produce *p*-xylene from DMF and acrolein (Figure 17.12). The acrolein could be produced from glycerol dehydration. The obtained *p*-xylene is then oxidized to produce terephthalic acid (TA), a monomer used for the polyester, and polyethylene terephthalate (PET). This polymer has been widely used for the large-scale manufacture of synthetic fibers and plastic bottles with total consumption in the United States reaching 5.01 million metric tons in 2007 [128].

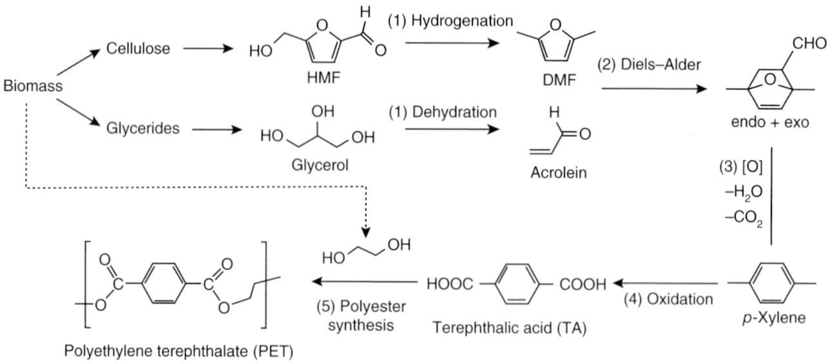

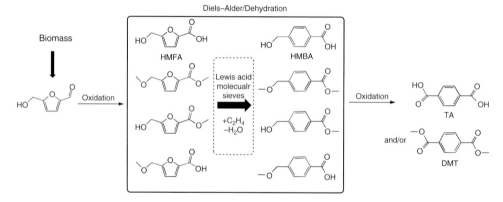

Figure 17.12 The proposed PET synthesis by using biomass-derived carbon feedstocks. Adapted from Ref. [113].

Figure 17.13 Diels–Alder pathways to TA and DMT (dimethyl terephthalate) starting from biomass-derived HMF using oxidation steps. Adapted from Ref. [129].

Due to the fact that DMF preparation from HMF hydrogenation requires an additional hydrogenation step and in some cases high hydrogen pressures, it could be economically advantageous if the hydrogenation step could be avoided. Davis and Pacheco [129] have introduced an alternative strategy in which, initially, the HMF would be partially oxidized to generate 5-(hydroxymethyl)furoic acid (HMFA). To protect the hydroxy and acid functional groups of HMFA, an alcohol (methanol) is used to generate corresponding ethers and esters. Those compounds are then reacted with ethylene to produce the aromatic products (HMBA (4-(hydroxymethyl)benzoic acid), etc.), which could be further oxidized to produce TA (Figure 17.13) [129, 130]. A yield of aromatics as high as 24% at 190 °C, 70 bar ethylene is obtained with this strategy [129]. The initial oxidation of HMF is intended to make the reactant more active for the Diels–Alder reaction and to avoid side reactions possibly caused by the unstable HMF. Pure-silica beta molecular sieves with Zr^{4+} centers (Zr-beta) and Sn-beta are the most effective catalysts used in this study with dioxane as the solvent. The authors claim that

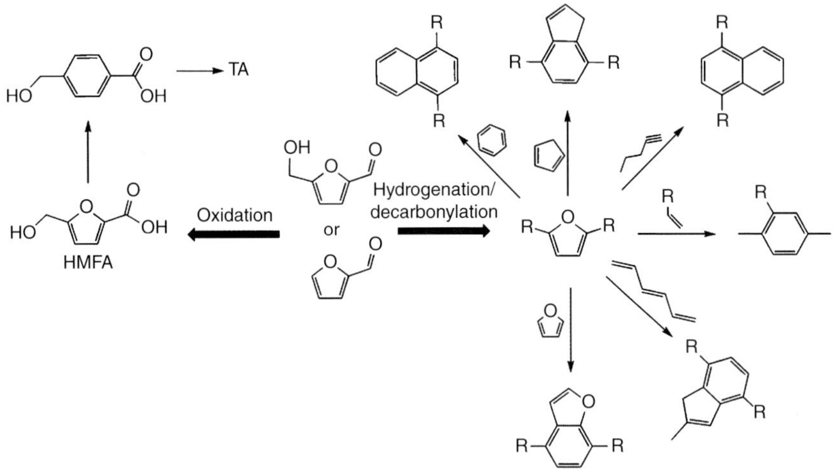

Figure 17.14 Road map for the conversion of HMF (furfural) to aromatic products via Diels–Alder reaction.

Brønsted acid sites should be prevented due to their ability to promote many undesired side reactions including coke formation, leading to low selectivity of cycloaddition products. Both Diels–Alder and dehydration steps are proposed to be catalyzed by Lewis acid sites.

Based on the reaction network of furan and propylene over ZSM-5 proposed by Huber et al. [120], a roadmap for the Diels–Alder reaction has been proposed (Figure 17.14), including two main routes: hydrogenation or decarbonylation to DMF/MF/furan and partial oxidation to HMFA. There are several possible pathways to couple DMF/MF/furan with a wide range of unsaturated compounds such as olefins and aromatic rings, which could potentially open more opportunities for the application of this approach for biomass utilization.

17.2.4
Piancatelli Reaction–Ring Rearrangement of Furfural

17.2.4.1 Mechanism of Furfural Ring Rearrangement
The ring rearrangement reaction transforms furanic species to more stable C_5 ketones through the creation of a C–C bond. This reaction has initially been reported by Piancatelli et al. upon observing the ring rearrangement of 2-furylcarbinol into a 4-hydroxycyclopent-2-enone in an acidic aqueous system [131] and has subsequently been referred to as a *Piancatelli rearrangement*. This rearrangement has been the subject of numerous additional studies with 2-furylcarbinol [132–136], furfuryl alcohol [137, 138], and a variety of substituted furanic compounds [139].

The ring rearrangement of furfural requires an initial hydrogenation step to form furfuryl alcohol. Therefore, the reaction pathway for the ring rearrangement of furfural includes two steps: hydrogenation of furfural to furfuryl alcohol, which is typically approached through the use of H_2 over metal catalysts [140, 141], and the acid-catalyzed ring rearrangement of the alcohol to CPO in the presence of water [131–135, 142, 143]. The protonation of the OH group on furfuryl alcohol to form a carbocation has been first proposed by Piancatelli as the initial step for the ring rearrangement reaction [131, 135], resulting in the formation of species A. The unstable nature of protonated H_2O promotes the decomposition of A to B, followed by the nucleophilic attack of water onto the ring to form intermediate C shown in Scheme 17.9 [131]. Intermediate C consecutively undergoes ring opening to generate species D, which can undergo 4π-conrotatory cyclization to facilitate the ring closure. This 4π-conrotatory cyclization is commonly referred to as a *Nazarov cyclization* [131, 144–147]. The successive deprotonation of this species yields 4-hydroxy-2-cyclopentenone (4-HCP). In the presence of metal catalysts and H_2, the intermediate 4-HCP could undergo hydrogenation to form 2-cyclopentene (2-PEN) and sequential CPO or cyclopentanol (CPOL) (Scheme 17.9).

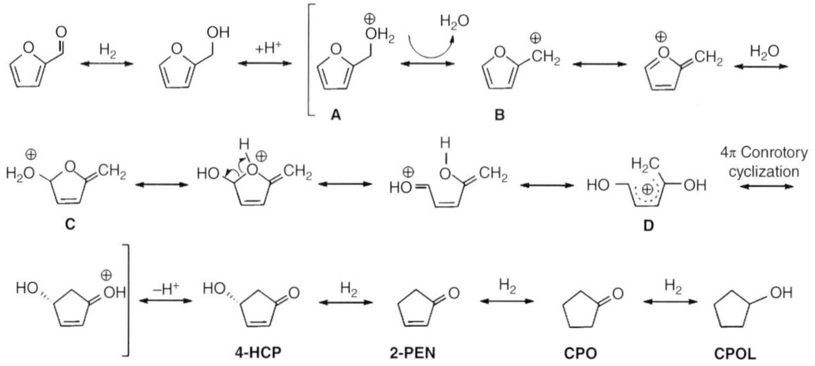

Scheme 17.9 Mechanism of the furfural conversion to cyclopentanone. Adapted from Ref. [131].

It is generally accepted that water plays an essential role in the formation of CPO since several literature sources report the necessity of aqueous media as a prerequisite for the ring rearrangement reaction [140, 141, 148]. In the presence of organic solvents, typical hydrogenation products such as tetrahydrofurfuryl alcohol (THFA), 2-methyltetrahydrofuran (MTHF), are obtained [140, 141, 148, 149]. It has been proposed that water is responsible for initiating the opening and closure of the furan ring via nucleophile attack by H_2O in the 5-position of the furan ring [138] as well as affecting the adsorption of reactants and intermediate species on the metal surface [141]. By using 97% ^{18}O abundant water as a solvent, Xu et al. [140] report 95% of the CPO product contains ^{18}O, which indicates the incorporation of oxygen from water into the ketone group of CPO.

As can be seen from Scheme 17.9, the proposed mechanism is related to the formation of 4-HCP as the intermediate. The 4-HCP and its isomers have been synthesized from the rearrangement of furfuryl alcohol and furan compounds under acidic conditions [150–153]. However, it is interesting to note that in the aqueous medium, furfuryl alcohol can spontaneously convert to 4-HCP at temperatures above 110 °C in the absence of H_2 pressure or catalyst [138, 140, 141]. It has been proposed that the ring rearrangement reaction is catalyzed by the hydronium ions generated from the auto-dissociation of water [154]. The metal catalysts and hydrogen pressure are required only for the hydrogenation of furfural to furfuryl alcohol and 4-HCP to CPO, while the hydronium ion facilitates the ring rearrangement. Since hydrogenation typically occurs in parallel with the Piancatelli rearrangement, high H_2 pressures over the metal catalysts would also enhance the formation rates of hydrogenated by-products such as THFA or MTHF, methyl furan (2-MF), resulting in a decreased selectivity to ring rearrangement products. Therefore, the reaction conditions and catalysts must be tuned to convert furfural selectively to furfuryl alcohol and 4-HCP to CPO while avoiding sequential and side reactions.

Some alternative pathways for the formation of CPO have recently been introduced. Small numbers of by-products have been reported in the product mixture including 2-cyclopentenone [140]; compounds containing one or more hydroxyl-, formyl-, or keto-groups [138]; and diol and triol compounds [149]. Ordomsky et al. have reported that, instead of 4-HCP, three different types of alcohols comprising 1,3-cyclopentanediol, 1,2-cyclopentanediol, and 1,2,3-cyclopentanetriol are rapidly formed during the hydrogenation of furfuryl alcohol in an aqueous solvent, which indicates that those alcohols could possibly be intermediates for CPO formation [149]. Based on that result, another ring rearrangement reaction pathway via the alcohol formation has been proposed as illustrated in Scheme 17.10.

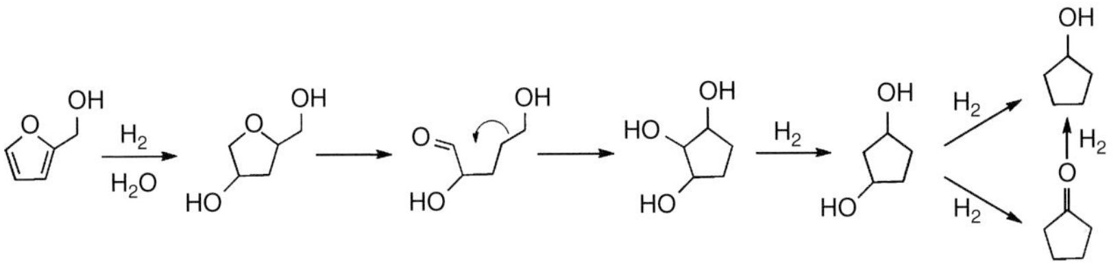

Scheme 17.10 The alternative pathway of ring rearrangement of furfural via the formation of alcohols. Adapted from Ref. [149].

17.2.4.2 Catalysts for Ring Rearrangement of Furfural

As mentioned before, the two steps in the furfural to CPO pathway require different types of catalysts including metal catalysts for hydrogenation and acid catalysts or hydronium ions formed from the dissociation of water for the ring rearrangement of furfuryl alcohol. Though chemists proposed many effective Brønsted and Lewis acid catalysts for the Piancatelli rearrangement [132–134] as early as 1978, the recent discovery of spontaneous ring rearrangement of furfuryl alcohol to

CPO in water has attracted more engineering attention due to the simplicity of the process.

The concentration of hydronium ions in water depends on the temperature and the pH. An appropriate addition of weak acids such as acetic acid has a positive effect on the conversion of furfuryl alcohol to CPO over a Ni-based catalyst, while the addition of basic sodium hydroxide shows the opposite trend [138]. Similarly, results over Pt and Pd–C catalysts show a similar correlation [140] in which weak acids such as acetic acid and NaH_2PO_4 prefer the CPO formation while Na_2CO_3 and Na_2HPO_4 favor the THFA formation as depicted in Scheme 17.11. It has been typically believed that the role of acidic medium is to enhance the ratio between the rate of ring rearrangement and that of typical hydrogenation. However, strong inorganic acids such as H_3PO_4 and H_2SO_4 over Pt and Pd–C have been reported not to favor the formation of CPO, but instead lead to rapid polymerization [141].

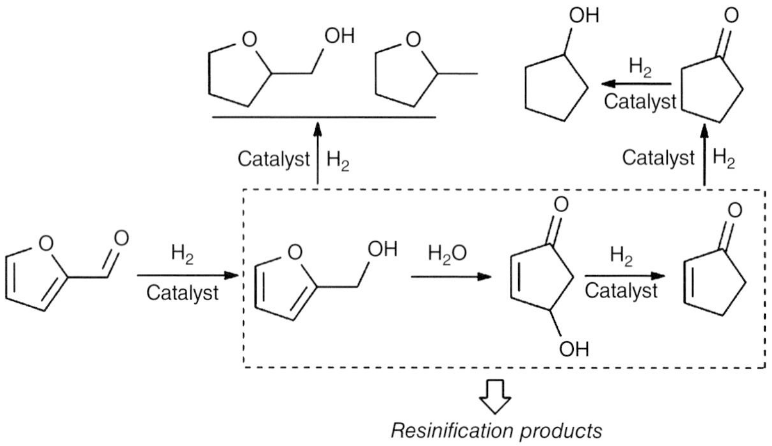

Scheme 17.11 Reaction pathway of furfural hydrogenation. Adapted from Ref. [140]

A variety of catalysts have been investigated for the hydrogenation steps in this reaction, ranging from noble metals such as Pt, Pd, Ru, and Ir to non-noble catalysts such as Ni, Cu, and so on, from mono- to bimetallic and alloy systems [155–158]. In addition to the rates of hydrogenation to ring rearrangement, the metal catalyst's selectivity obviously has a significant impact on the overall product distribution. For example, the final yield of CPO will be influenced not only by the selectivity for hydrogenation of the aldehyde C=O versus the hydrogenation of the ring but also the rate of hydrogenation to C–O hydrogenolysis or decarboxylation. Hronec et al. [138] have demonstrated this balance in competing rates by varying the partial pressure of hydrogen from 0.8 to 2.5 MPa over a commercial nickel-based catalyst at 160 °C. By tuning this ratio, they report high selectivity to CPO with yields higher than 90 mol% with a yield of by-products below 7%. The nature of the interaction of the C=O bond with the metal surface plays an important role in this selectivity as well. For example, the adsorption of furfural on Cu favors a η^1-(O) aldehyde species [159, 160] due to the interaction with the electron lone pair of oxygen. In the case of noble metals such as Pd and Pt, however, the adsorbed furfural molecule lays flat on catalyst surface with the π electron of the furan ring and the lone pair electron of the carbonyl oxygen interacting strongly with the surface [161–164]. This will inherently influence the rates of THFA formation and the resulting product distribution. Therefore, as pointed out by Hronec et al. [138, 141, 148], the catalysts and reaction conditions should be chosen appropriately to maximize the yield and the selectivity of ring rearrangement products.

The acidity of the support also influences the selectivity of ring rearrangement products (CPO + CPOL). For instance, the Pt over acidic alumina is less favorable for CPO + CPOL formation than that of neutral activated carbon, while in the case of basic MgO, the furfural alcohol is the dominant product [141]. Additionally, it has been stated that the acid–base properties of the support enhance the chemisorption of furfural on the support, which supports undesired reactions such as condensation and oligomerization of furfural and furfuryl alcohol, resulting in a low mass balance. The summary of the catalysts for the conversion of furfural to CPO is presented in Table 17.6.

17.2.4.3 Upgrading Strategy

The Piancatelli rearrangement is a promising route to stabilize furfural with high yield and selectivity. When compared with furfural and furfuryl alcohol, CPO is much more stable, even in hot water and acidic environments, which is a significant advantage in terms of storage and transportation [140]. More interestingly, the ketone group and available α-H allow CPO to couple with other ketones–aldehydes or even with itself via aldol condensation to produce heavier hydrocarbons.

The first upgrading route (Route A) as presented in Figure 17.15 is the self-aldol condensation of CPO to produce 2-cyclopentylidene-CPO (CC), which can then undergo HDO to form a C_{10} hydrocarbon (F1) with a density of $866\,g\,l^{-1}$ [169, 170], a high net heat of combustion ($42.6\,MJ\,kg^{-1}$; $37.0\,MJ\,l^{-1}$) [171], and low viscosity and freezing point, which could become appropriate transportation fuels and attractive additives for bio-jet fuels [172]. The quality of F1 has been proposed as an alternative to many jet fuels developed for use by US Air Force [173] (JP-4, JP-7) and some common civilian jet fuels such as Jet A, Jet A-1, Jet B [174].

The self-aldol condensation of CPO can be conducted over various base catalysts such as MgO, KF–Al_2O_3, CaO–CeO_2, CaO, LiAl-HT, and MgAl-HT [172]. It has been reported that magnesium–aluminum hydrotalcite (MgAl-HT) and lithium–aluminum hydrotalcite (LiAl-HT) demonstrate the best activity for the reaction in which the yield of CC reaches 86 and 80%, respectively (Figure 17.16a). The HDO step is carried out consecutively over Pd on basic, neutral, and acidic supports (Figure 17.16b). The Pd–acid supports such as Pd–SiO_2, Pd–SiO_2–Al_2O_3, Pd–Hβ, and Pd–ZrP earn excellent yield of F1 ranging from 80 to 90%, whereas low activity is obtained in the case of Pd-activated

Table 17.6 Catalytic conversion of furfural to cyclopentanone on solid catalysts[a].

Catalyst (g)	Reaction conditions	Conversion (%)	Yield (%)			References
			CPO	CPO + CPOL	TH[b]	
1.4% Pt–C (0.18)	160 °C, 80 bar H$_2$, 0.5 h	99.6	43.9	60.2	7.9	[141]
1% Pt–Al$_2$O$_3$ (0.25)	160 °C, 80 bar H$_2$, 0.5 h	97.7	44.7	48.3	5.8	[141]
1.4% Pt + 1.4% Ru–C (0.09)	160 °C, 80 bar H$_2$, 0.5 h	100	3.7	46.4	19.6	[141]
1% Pt–MgO (0.25)	160 °C, 80 bar H$_2$, 0.5 h	97.9	9.1	10	3.2	[141]
5% Pt–C (0.05)	160 °C, 30 bar H$_2$, 0.5 h	100	76.5	81.3	3.8	[141]
5% Pt–C (0.1)	175 °C, 80 bar H$_2$, 0.5 h	100	40.2	76.4	14.7	[141]
5% Pd–C (0.1)	160 °C, 30 bar H$_2$, 1 h	97.8	67	68.4	10.5	[138]
5% Ru–C (0.1)	175 °C, 80 bar H$_2$, 0.5 h	98.7	56.7	66.3	12.9	[141]
NiSAT® 320 RS (0.1)	175 °C, 80 bar H$_2$, 0.5 h	98.3	61.0	78.3	15.7	[141]
G-134 A (0.1)	175 °C, 80 bar H$_2$, 0.5 h	100	57.3	64.2	14.2	[141]
G-134 A (0.04)[c]	160 °C, 8 bar H$_2$, 1 h	100	88.5	90.1	2.13	[138]
5% Ru–C (0.1)	160 °C, 30 bar H$_2$, 0.5 h	60.1	12.7	13.3	1.69	[148]
5% Ru–C (0.1)	175 °C, 80 bar H$_2$, 1 h	100	57.3	66.8	12.6	[148]
CoMnCr (0.15)	175 °C, 80 bar H$_2$, 0.5 h	100	7.6	24	17.3	[148]
Raney Ni Actimet C (0.2)	160 °C, 30 bar H$_2$, 1 h	100	17.5	57.5	34.7	[148]
5% Ru–C (0.4)	165 °C, 25 bar H$_2$, 5 h	100	10.6	27	44.7	[149]
5% Ru–C (0.4)	165 °C, 25 bar H$_2$, 5 h [d]	100	1.1	48.3	23.7	[149]
NiCu-50-SBA-15 (0.2)	160 °C, 40 bar H$_2$, 4 h	~100	~62	65	17	[140]
10 wt% Ni-CNTs (1.5)	140 °C, 50 bar H$_2$, 10 h	94	77	79	—	[165]
30 wt% Ni-CNTs (1.5)	140 °C, 50 bar H$_2$, 10 h	95	5	88.6	—	[165]
Cu–Ni–Al hydrotalcite (1.5)	140 °C, 40 bar H$_2$, 8 h	100	95.8	98.8	—	[166]
	Cu–Ni–Al –1:14:5					
Cu-SBA-15 (0.2)	150 °C, 40 bar H$_2$, 6 h	89.6	10.5	24.2	0.2	[167]
Cu-SiO$_2$ (0.2)	150 °C, 40 bar H$_2$, 6 h	71.9	7.31	0	0.3	[167]
CuZnAl (0.2)	150 °C, 40 bar H$_2$, 6 h	97.9	60.3	62.8	0	[167]
Au–Nb$_2$O$_5$[e] (0.01)	140 °C, 80 bar H$_2$, 12 h	>99	86[f]	—	—	[168]
Au–Al$_2$O$_3$[e] (0.01)	140 °C, 80 bar H$_2$, 12 h	>99	72[f]	—	—	[168]
Pt, Pd, Ru–Nb$_2$O$_5$[e] (0.01)	140 °C, 80 bar H$_2$, 12 h	>99	28–66[f]	—	—	[168]

a) Water solvent.
b) Yield of typical hydrogenation products including THFA, 2-MeF, 2-MeTHF.
c) Feed furfuryl alcohol, 0.05 g acetic acid as additive.
d) MTHF–water solvent.
e) Feed 5-hydroxymethyl furfural.
f) Yield of 3-hydroxymethyl cyclopentanone.

carbon (36%) and Pd–MgO (0%). This could be explained in terms of the crucial role of the acid site of the support in the dehydration step. Other inexpensive catalysts such as Ni–SiO$_2$ and Cu–SiO$_2$ could also be effective alternatives with the obtained yield of F1 up to 93%, even higher than that of noble metal catalysts. A similar upgrading concept has also been reported by Yang et al. [140] where

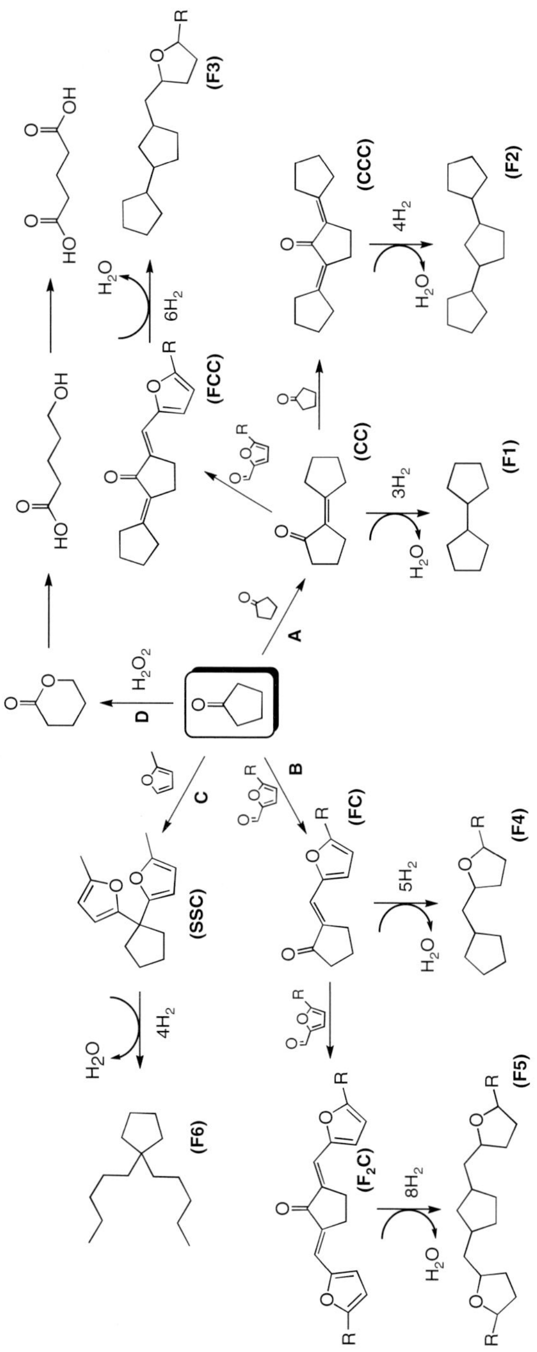

Figure 17.15 Cyclopentanone upgrading strategy via aldol condensation and hydroxyalkylation.

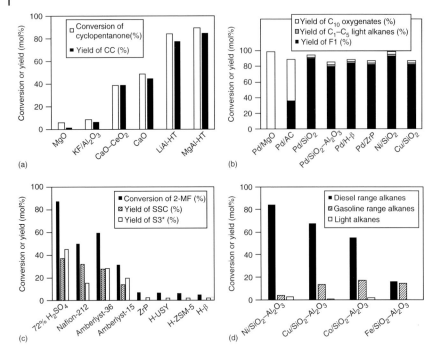

Figure 17.16 (a) Aldol condensation of cyclopentanone over different solid base catalysts at 423 K for 8 h in a batch reactor. (b) Carbon yields of F1, C_1–C_5: light alkanes and C_{10} oxygenates (2-cyclopentyl–cyclopentanone and 2-cyclopentyl–cyclopentanol) over different catalysts. Reaction conditions: 503 K, 6 MPa; liquid feedstock (CC in Figure 17.15) flow rate 0.04 ml min^{-1}; hydrogen flow rate: 120 ml min^{-1}. (c) Hydroxyalkylation of 2-MF and CPO over different solid acid catalysts. Reaction conditions: 338 K, 2 h; 2-MF/CPO molar ratio = 2. (d) Carbon yield of different alkanes obtained by the HDO of hydroxyalkylation products of 2-MF and CPO over the M/SiO$_2$–Al$_2$O$_3$ (M = Fe, Co, Ni, Cu) catalysts. Reaction conditions: 533 K; liquid flow rate = 0.04 ml-min, WHSV = 1.3 h^{-1}; H$_2$ flow rate = 120 ml min^{-1}. The diesel range alkanes, gasoline range alkanes, and light alkanes account for C_9–C_{15}, C_5–C_8, and C_1–C_4 alkanes, respectively [172, 175]. *S3 is the hydroxyalkylation product between 2MF (2-methyl furan) and 4-oxopentanal (the ring opening product of 2MF). Reproduced with permission of American Chemical Society, 2014, ref. [177]

NaOH is used for the self-aldol condensation of CPO with a yield of 65%. A subsequent HDO step over Ru-ZSM-5 yields 59% of bicyclopentyl (F1) from CPO 180 °C, 4 MPa H$_2$. Theoretically, the C_{15} jet fuel precursor (CCC – Figure 17.15) could be produced by further aldol condensation of the CC with another CPO molecule since there is still an α-H for the coupling reaction. However, until now, no study has claimed this as a significant product. The scope of CPO upgrading is expanded by another promising pathway via the aldol condensation between CPO and furan derivatives such as furfural and HMF (Route B; Figure 17.15) [176]. High yields (>96%) of solid dimer products (F$_2$C) have been reported through this route over homogeneous catalysts such as NaOH or Ca(OH)$_2$ at low temperatures

Table 17.7 Aldol condensation of furfural and HMF with cyclopentanone (CPO) [176].

No	Molar ratio		Reactant	Temperature (°C)	Reaction time (h)	Conversion (%)		Yield (mol%)	
	C_{CPO}:F	C_{CPO}:cat				CPO	F	F_2C	HF_2C
1	1:2	1:0.059	F	40	3	100	99.5	96.0	—
2	1:3	1:0.059	F	40	3	100	66.7	96.6	—
3	1:2	1:0.60	HMF	40	3	100	100	—	98.1
4	1:2	1:0.60	HMF	40	3	100	100	—	98.5
5	1:2	1:0.15	HMF	100	40	52.7	61.5	—	40.3
6	1:2	1:0.30	HMF	100	40	100	92.0	—	86.9

Table 17.8 Average pore diameters and Hammett acidity function ($-H_0$) values of the solid acid catalysts [175].

Catalyst	TON[a]	Average pore diameter (nm)	$-H_o$
H_2SO_4	7	—	—
Nafion-212	77.6	4[b]	11–13 [177]
Amberlyst-15	8.3	29[b]	2.2 [177]
Amberlyst-36	14.2	24[b]	2.2–2.65 [177]
ZrP	0.47	7.2[c]	−1.0 to about 2.8 [178]
H-USY	0	~0.7 [179]	3.0–4.4 [177, 180]
H-β	0.82	~0.6 [179]	4.4–5.7 [180]
HZSM-5	0	~0.5 [179]	5.6–5.7 [177, 180]

a) TON was evaluated by dividing the mol of SSC by the mol of acid sites over the catalyst. The acid site was measured by NH_3 chemisorption.
b) According to the information provided by supplier.
c) Measured by N_2 physical adsorption.

(<100 °C) (Table 17.7). The C_{15}–C_{17} hydrocarbons obtained after HDO lie in the diesel fuel range.

Recently, Zhang et al. [175] have converted CPO to longer chain hydrocarbons via an HAA route. This strategy, outlined as Route C in Figure 17.15, uses the carbonyl group in CPO as the coupling agent, forming a C–C bond with 2MF to produce larger hydrocarbons in the presence of Brønsted acid catalysts (Figure 17.16c). Upon comparing several homogeneous and heterogeneous acid catalysts, solid acid resins show higher activities for this reaction than zeolites. The authors attributed the lower TON (turnover numbers) values over zeolites to the small pore diameter compared to that of the dimer product (Table 17.8). The activity order of those resins follow Nafion-212 > Amberlyst-36 > Amberlyst-15 > ZrP, which matches the Hammett acidity function values of each. Among those tested, Nafion-212 has been the most promising choice for this type of reaction in terms of the turnover number and stability. The high performance of Nafion-212 is ascribed to the electron-withdrawing effect of the perfluoroalkyl backbone on the sulfonic acid site, resulting in a significant increase in acid strength when compared to the Amberlyst catalysts [181, 182]. In addition to the coupling between 2MF and CPO, the HAA between 2MF and 4-oxopentanal (the ring opening product of 2MF) could also occur under these conditions.

The liquid-diesel-range alkanes obtained after subsequent HDO (Figure 17.16d) is claimed to have a relatively high density ($0.82\,g\,ml^{-1}$), which could potentially be used as high-density fuel additives. Compared to previous studies regarding the hydroxyalkylation between furfural and sylvan [74], the CPO–sylvan route has been reported to lead to higher carbon yields of fuel-range products. Moreover, highly stable CPO with respect to furfural can be obtained, which would be a tremendous advantage in terms of storing and carbon retention enhancement. The oxidation of CPO to valerolactone (Route D) [183–188] followed by the ring opening to generate carboxylic acids is an alternative route to produce chemicals from CPO [189]. The transformation of furfural to CPO leads to a platform molecule that can undergo a variety of useful C–C upgrading strategies while also exhibiting enhanced stability, leading to decreased rates of undesirable humin formation and higher overall yields to useful products.

17.2.5
Oxidation of Furanics

17.2.5.1 Mechanism of Furanics Oxidation
It is well known that furanic compounds exhibit thermal instability leading to the formation of humins. The selective partial oxidation of furanics results in carboxylic acid containing species with both higher stability and the potential for selective coupling reactions. The oxidation of furanics has been proposed to occur via both non-radical (ion molecular) and radical-based mechanisms. In the gas phase over a vanadium(V) catalyst, for example, the conversion of furan to MA has been proposed to follow a non-radial (ion molecular) mechanism. This involves the formation of an endoperoxide, which is further oxidized to MA (Scheme 17.12) [190]. Furan homologs such as 2-MF or furfural over V_2O_5/O_2 have been proposed to first convert to furancarboxylic acid (furoic acid, FuA)

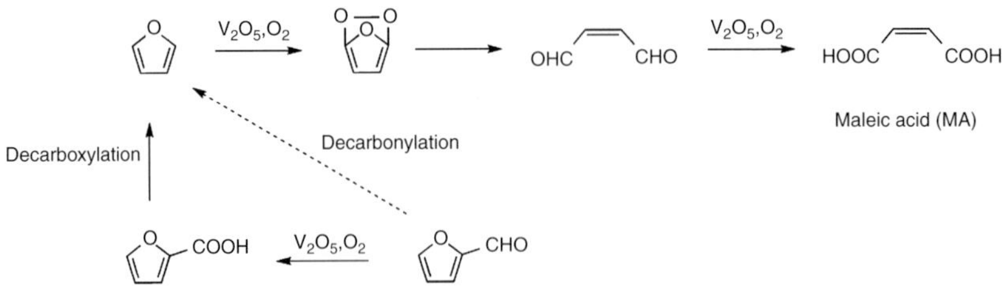

Scheme 17.12 Pathway of furanics oxidation in gas phase over V_2O_5/O_2 system. Adapted from Ref. [190].

that can undergo decarboxylation to yield furan [190, 191]. Recently, Alonso-Fagúndez et al. [192] have suggested that the oxidation of furfural to MAH with $VO_x–Al_2O_3/O_2$ forms furan as an intermediate via direct decarbonylation.

In the liquid phase, the mechanism of furfural oxidation with H_2O_2 could also follow non-radial (ion molecular) and radical mechanisms. One typical example for a non-radical mechanism is the conversion of furfural to SA and MA in the presence of H_2O_2. Under an acidic environment ($pH < 2$), the Baeyer–Villiger mechanism is prevalent (Figure 17.17), while under basic medium ($pH > 7$), the reaction proceeds via FuA formation (Figure 17.17) [193, 194]. With $pH > 8$, the decomposition of H_2O_2 is active [195]. Yet another possibility involves the formation of an epoxide by the attack of the double bond of the ring by oxidants in a H_2O_2/titanium silicalite-1 system [196].

Over $VOSO_4/H_2O_2$ or $FeSO_4/H_2O_2$, the oxidation of furan and furfural has been proposed to undergo a radical-based mechanism via interaction with an OH·[197]. Yin et al. [198] have outlined a radical-based mechanism for furfural oxidation over $H_3PMo_{12}O_{40}·xH_2O$ (phosphomolybdic acid)/O_2 in aqueous medium. The initial step is analogous to the auto-polymerization of furfural in which the furfural radical is generated via the hydrogen abstraction by oxygen at the 5-position of furan ring. The radical could lead to a chain reaction to produce resins or transfer electrons to phosphomolybdic acid to initiate pathway (A) shown in Scheme 17.13. The formation of MAH over $H_5PV_2Mo_{10}O_{40}$ combined with the Lewis acid $Cu(CF_3SO_3)_2$ in the presence of O_2 in the liquid phase has also been suggested to follow the similar radical-based mechanism (pathway B; Scheme 17.13) [199].

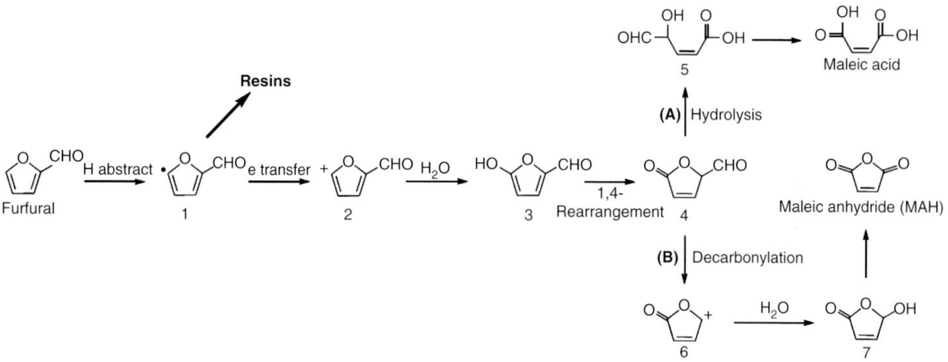

Scheme 17.13 Radical-based mechanism for maleic acid and maleic anhydride formation. Adapted from Ref. [198, 199]

17.2.5.2 Catalysts for Furfural Oxidation

The oxidation of furfural can produce a wide range of products comprising succinic acid (SA), fumaric acid (FA), MA, maleic anhydride (MAH), FuA, and

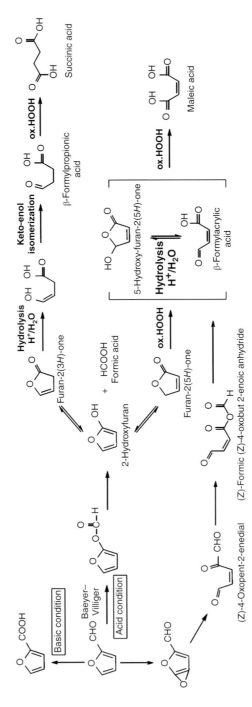

Figure 17.17 Non-radical mechanism of furfural oxidation. Adapted from Ref. [193].

formic acid (Scheme 17.14). HMF could be oxidized to 2,5-furandicarboxylic acid [200–202], which is a potential alternative chemical platform for terephthalic or isophthalic acids in the production of polyamides, polyesters, and polyurethanes [65].

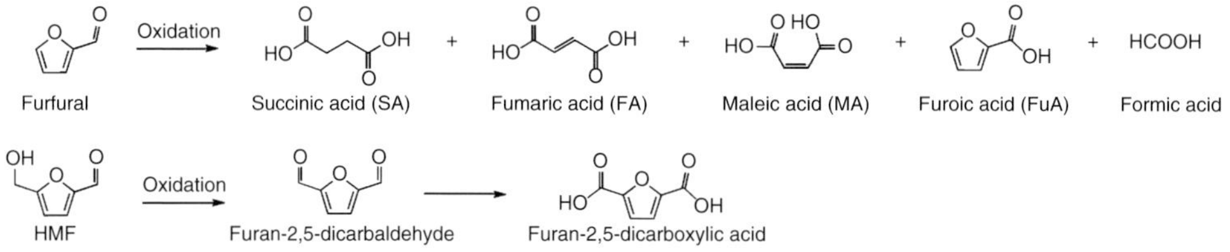

Scheme 17.14 Different products obtained from furfural and HMF oxidation.

Oxidation of Furfural to Maleic Acid (MA) MA has been reported to be formed in the gas phase with O_2 as the oxidant over V_2O_5 catalysts [191, 203], V_2O_5 and MoO_3 modified by phosphoric acid and iron molybdate [204], as well as in the liquid phase with H_2O_2/titanium silicalite [196], H_2O_2/V_2O_5, $H_2O_2/FeSO_4$, H_2O_2/SeO_2, $H_2O_2/VOSO_4$ [197], O_2/phosphomolybdic acid [198], or O_2/solid acids [205]. These reactions have been known for quite some time, with the oxidation of furfural to MAH and MA with O_2/vanadium oxides patented in 1934 [206] and the oxidation over molybdenum oxides patented in 1947 [207].

Yin et al. [205] have used a wide range of redox metal salt catalysts and O_2 as the oxidant for MA production from furfural (Table 17.9). Though a high conversion of furfural is achieved, the yield of MA is low due to the high degree of polymerization. Copper-based catalytic systems such as $Cu(OAc)_2$, $Cu(NO_3)_2$, and $CuSO_4$ show high potential for further investigation with 18.6%, 23.9%, and 19.5% yield of MA produced, respectively. Phosphomolybdic acid is found to further improve the MA production. Based on the proposed mechanism (Scheme 17.13), one feasible way to prevent the polymer formation is to accelerate the electron transfer so that the radical species will quickly transform to the carbocation intermediate (**2**) rather than condense to polymers. Copper(II) cations and phosphomolybdic acid possess high redox potential and facilitate electron transfer from intermediate (**1**) to phosphomolybdic acid [198]. This combination leads to 95.2% conversion of furfural but still produces only 49.2% yield to MA with 46% of the furfural lost to resin products [205]. No SA yield is reported in this case.

The oxidation of furfural to MA in the liquid phase with hydrogen peroxide could significantly improve yield of monomeric species. For example, by using H_2SO_4, Amberlyst-70, and soluble polystyrene sulfonic acid (PSSA) for the oxidation of furfural with hydrogen peroxide, a high carbon balance is achieved reaching up to 99% [208]. However, in addition to MA, a wide range of products are also obtained including SA, FA, 2(5*H*)-furanone, and formic acid. It has been reported that a 7:1 ratio of H_2O_2/furfural is preferable for high MA yields. The high H_2O_2/furfural ratio required is attributed to the catalytic decomposition

Table 17.9 Catalyst (with additives) scanning for the furfural oxidation to maleic acid.

Catalyst	Conversion (%)	Yield of maleic acid (%)	Catalyst Cu(NO$_3$)$_2$ with additive	Conversion (%)	Yield of maleic acid (%)
—	56.3	7.2	Ethylenediamine[a]	86.9	25.1
Mn(OAc)$_2$	26.7	0.8	Triethylamine[a]	77.2	25.8
Pd(OAc)$_2$[b]	70.0	0.3	1,10-Phenanthroline[a]	80.1	29.5
AgOAc	73.2	4.1	Bipyridine[a]	78.6	28.3
FeSO$_4$	90.0	12.1	Co$_2$(SO$_4$)$_3$	49.1	14.4
RuCl$_3$	54.8	6.1	NiCl$_2$	38.5	8.4
NiCl$_2$	35.4	3.1	RuCl$_3$	43.7	6.9
V$_2$O$_5$	72.1	6.0	Fe(NO$_3$)$_3$	36	4.6
Co(NO$_3$)$_2$	68.8	4.0	H$_3$PW$_{12}$O$_{40}$·xH$_2$O	35.8	12.1
Cu(OAc)$_2$	71.0	18.6	H$_3$PMo$_{12}$O$_{40}$·xH$_2$O	90.3	35.3
CuCl$_2$	62.1	7.4	H$_4$SiW$_{12}$O$_{40}$·xH$_2$O	36.6	9.8
Cu(NO$_3$)$_2$	85.9	23.9			
CuSO$_4$	67.0	19.5			
CuCO$_3$	49.5	1.7			

a) Cu(NO$_3$)$_2$ 0.4 mmol, additives 0.4 mmol.
b) 15.9% yield of furoic acid.
Reaction condition: water 4 ml, furfural 0.6 ml, catalyst 0.2 mmol, additive 0.2 mmol, oxygen 20 atm, reaction temperature 98 °C, and 14 h.
Adapted from Ref. [205].

and uncatalyzed thermal decomposition of H$_2$O$_2$. High MA yield (87%) and more economical utilization of H$_2$O$_2$ (H$_2$O$_2$/furfural ratio 4.4:1) have been reported over TS-1 [196].

FA, the isomer of MA, usually exists in maleic production process as a minor by-product. This isomer can be synthesized more selectively from furfural with sodium chlorate solution and V$_2$O$_5$ in water with 47% yield [209].

Oxidation of Furfural to Succinic Acid SA, one of the top 12 potential chemicals from lignocellulosic biomass reported by the US Department of Energy [210], has been investigated over many types of catalysts such as sodium molybdate [211], Mo(VI) and Cr(VI) compounds [212], VOSO$_4$ [213], palladium(II) [214], H$_2$SO$_4$ [215], and Hg(NO$_3$)$_2$ [216]. SA production over solid acid catalysts such as Amberlyst, zeolites, Nb$_2$O$_5$, ZrO$_2$, and so on has been studied by Kohki et al. (Table 17.10) [217, 218]. Catalysts containing sulfonic groups such as p-TsOH and Amberlyst-15 are demonstrated to be effective with over 70% yield to SA [217]. The authors attribute high performance of these catalysts to the moderate acidity and functionality of the tolyl group forming a π–π interaction with the furan ring [218] (Figure 17.18). While MA is obtained from furan-2(5H)-one, SA is proposed to be formed via furan-2(3H)-one type, which is enhanced via the interaction with the tolyl group in p-TsOH. Strong acids such as Nafion, HCl, and H$_2$SO$_4$, despite of

Table 17.10 Furfural oxidation in water using various acid catalysts in the presence of hydrogen peroxide[a].

Catalyst	Furfural conversion[b] (%)	SA selectivity (%)	Yield[b] (%)				Carbon mass balance[c] (%)
			SA	FA	MA	FuA	
Amberlyst-15	>99	74.2	74.2	0.1	11.0	1.9	70.1
Nafion NR50	>99	41.2	40.8	0.8	11.3	2.1	44.4
Nafion SAC13	>99	28.5	28.2	0.5	9.5	1.6	32.2
Nb_2O_5	>99	24.6	24.4	4.5	4.6	0	26.8
ZSM-5[d]	>99	16.7	16.5	1.5	2.4	0	16.3
ZrO_2	>99	16.6	16.5	2.1	5.4	0	19.2
SO_4/ZrO_2	96.5	10.4	10.1	4.3	5.9	0	16.2
HCl[e]	>99	49.4	49.0	2.9	10.5	0	49.9
Acetic acid[e]	>99	26.0	25.7	27.6	2.2	0	44.4
H_2SO_4[e]	>99	44.9	44.9	0.3	5.9	0	40.9
p-TsOH[e]	>99	72.4	72.3	0.4	11.4	1.2	68.5
Blank	68.8	1.3	0.6	1.4	4.8	18.2	23.6
Amberlyst-15[f]	34.4	0	0	0	0	0	0
Blank[f]	22.5	0	0	0	0	0	0

a) Reaction conditions: furfural (1 mmol), H_2O_2 (4 mmol, determined by iodometry titration), H_2O (3 ml), catalyst (50 mg), 353 K, 24 h, 500 rpm.
b) Furfural conversion was calculated by $[(furfural_{input} - furfural_{remain})/furfural_{input}] \times 100$ where the product yield was $(product_{detected}/furfural_{input}) \times 100$ with the HPLC (High-performance liquid chromatography) analysis.
c) Determined on the basis of observed SA, FA, MA, and FuA products.
d) $SiO_2/Al_2O_3 = 90$, JRC-Z-5-90H(1).
e) 1 mmol.
f) Without H_2O_2.
Adapted from Ref. [217].

having stronger acid strength, are less selective for SA formation due to excessive polymerization.

Indirect routes for SA production have also been examined, such as the two-step catalytic process combining the oxidation of furfural to FA over sodium chlorate/vanadium pentoxide and the hydrogenation of FA to SA over Pd/C [219, 220] as well as the hydrogenation of MA to SA using Pd/C [221].

Oxidation of Furfural to Maleic Anhydride   MAH, an important material for the fabrication of unsaturated polyester resins as well as many other chemicals, is industrially produced from petroleum-based sources such as *n*-butane over supported vanadium-based catalysts [222–224]. The vapor phase conversion of furfural to MAH has been the subject of several studies [204, 225–227]. Among those catalysts, vanadium oxides (VO_x) have engaged interest in this area. For this catalytic system, furan has been proposed as an intermediate [192] followed by its sequential oxidation to form 2-furanone, with comparable rates of MAH production reported for furfural and furan.

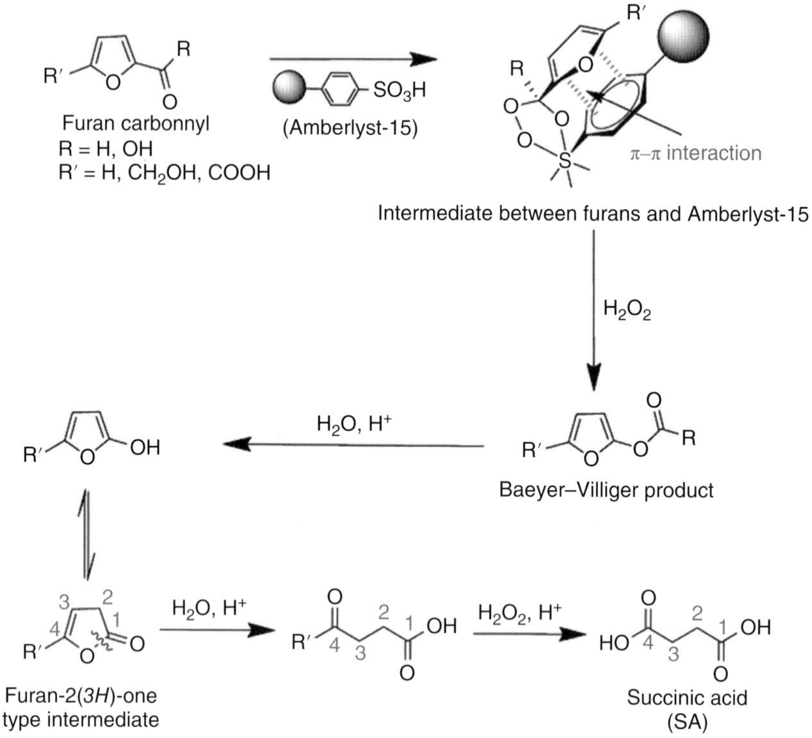

Figure 17.18 Intermediate between the catalysts and furan ring. Reproduced with permission from Ref. [218].

Liquid phase production of MAH must utilize an organic solvent to avoid the obvious hydrolysis to dicarboxylic acids that would occur in an aqueous environment [199]. It has been reported that $H_5PV_2Mo_{10}O_{40}$ combined with the Lewis acid $Cu(CF_3SO_3)_2$ used in a mixture of acetonitrile and acetic acid as a solvent results in the highest selectivity to MAH of the catalysts tested with 54% yield of MAH [228]. The enhanced performance when Cu(II) is used as an additive is ascribed to the high redox potential preferably generating a trap for organic radicals to generate organic cation intermediates [229, 230] and the ability to reoxidize the reduced $H_5PV_2Mo_{10}O_{40}$ catalyst to its original active form [199]. Other Lewis acids have also been used to produce MAH from HMF over $VO(acac)_2/O_2$ in the liquid phase [231].

17.2.5.3 Upgrading Strategy

Maleic Anhydride Utilization via Diels–Alder Coupling Lobo et al. [232] have proposed a two-step process for the production of phthalic anhydride (PA), a raw material for plasticizers, unsaturated polyesters, and alkyl resins from furan and MAH. This consists of an uncatalyzed cycloaddition of furan and MAH followed by the catalytic dehydration to form the coupling product PA (R=H) (Scheme 17.15). The MAH could be selectively produced from furfural oxidation as mentioned previously, while the furan is a well-known industrial product of furfural decarbonylation [233, 234].

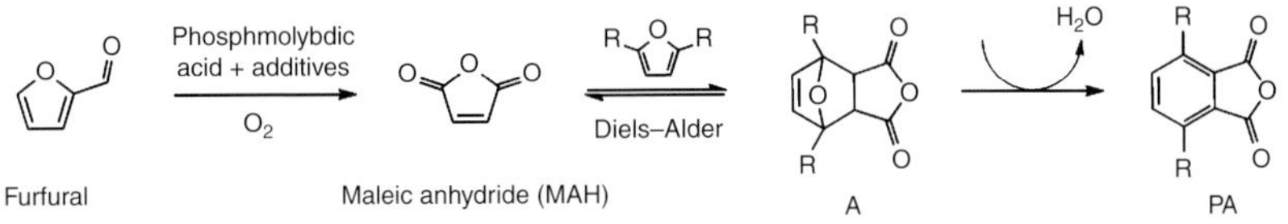

Scheme 17.15

The presence of two electron-withdrawing ketone groups on MAH (dienophile) allows the Diels–Alder reaction to be feasible even at room temperature with 96% yield of A. Some reports reveal that the addition of a small amount of water to the reaction mixture could accelerate the rate of Diels–Alder coupling [235, 236]. However, Lobo et al. [232] have observed that water also favors the hydrolysis of MAH to MA at room temperature, leading to a decrease in selectivity to cycloaddition products. Methanesulfonic acid (MSA) [232] has been utilized for the subsequent dehydration step, but this also facilitates retro-Diels–Alder and polymerization reactions. Combinations of MSA and acetic anhydride, which react to generate acetyl methanesulfonate, are found to be more effective at dehydrating A to PA [237]. This combination is capable of increasing the selectivity of PA to 80% compared to 11% in the case of pure MSA.

This production of PA can be further converted to a variety of valuable products. As an example, Figure 17.19 highlights a strategy for the production of TA (Terephthalic acid) as discussed by Tachibana et al. [238] through a combination of oxidation and Diels–Alder coupling. The oxidation of furanic species could pave the way for a variety of bio-based specialty chemicals.

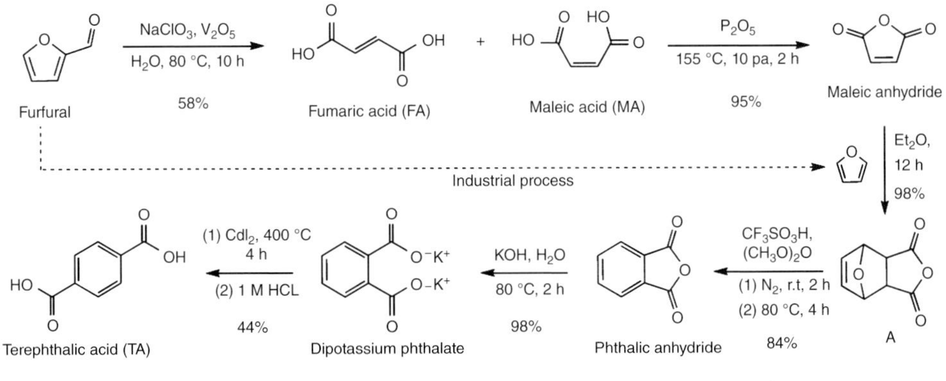

Figure 17.19 Synthesis route to bio-based TA from biomass-derived furfural. Adapted from Ref. [238].

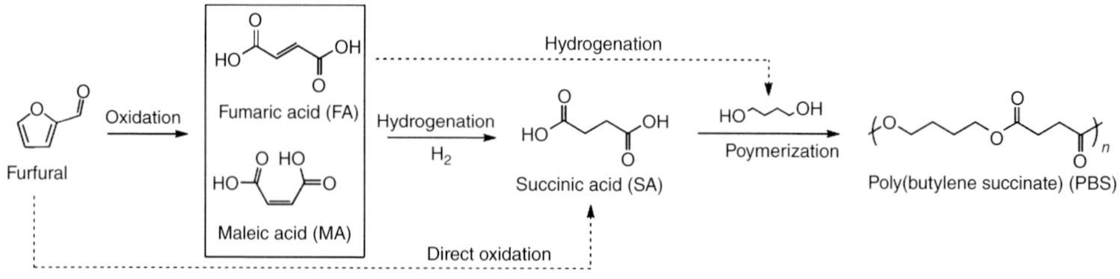

Figure 17.20 Strategy for dicarboxylic acid upgrading via polymerization.

Dicarboxylic Acid Utilization via Polymerization Another route for the use of furanic species for specialty chemicals is through the synthesis of polyesters via dicarboxylic acid. SA produced from FA/MA hydrogenation or directly from furfural oxidation could be used as the comonomer with 1,4-butanediol for the production of poly(butylene succinate) (PBS), a widely used biodegradable plastic. 1,4-Butanediol could be obtained from SA hydrogenation [239] or directly from FA/MA hydrogenation [209] (Figure 17.20). The polymerization could be conducted with Ti(i-PrO)$_4$ at 243 °C under 1 mmHg for 2 h with over 90% yield [220].

17.2.6
Dimerization of Furfural via Oxidative Coupling

The dimerization of furfural to [2,2'-bifuryl]-5,5'-dicarbaldehyde, or dimer, via palladium catalyzed-oxidative coupling is a promising route to stabilize furfural [240]. The dimer could be used as a building block for polymer synthesis [240] or undergo subsequent HDO to produce C$_{10}$ alkanes. The reaction has been reported in the presence of Pd(OAc)$_2$ [241, 242]. The proposed mechanism is presented in Scheme 17.16. The electrophilic palladium first attacks the furan ring to form the intermediate (**1**), which could couple with another furfural

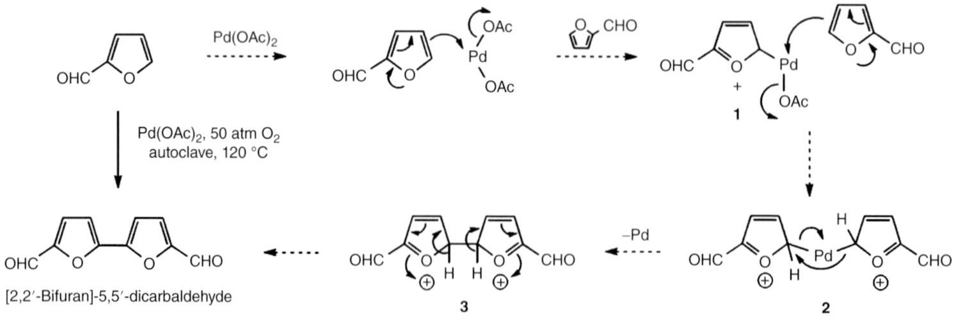

Scheme 17.16 Proposed mechanism for furfural oxidative coupling. Adapted from Ref. [240].

molecule to generate intermediate (**2**), followed by the electron delocalization to the dimer product [240].

It is worth noting that during the course of the reaction, Pd(II) will be reduced to Pd(0). The reaction becomes catalytic when co-catalysts such as Cu(II) or Fe(III) salts are utilized in the presence of O_2 to maintain the redox cycle depicted in Scheme 17.17. Kozhevnikov [241] has claimed that the dimerization of furan over co-catalysts $Pd(OAc)_2/Cu(OAc)_2$ in dimethylformamide (DMFA) and acetic acid (HOAc) solvent achieves 560% and 270% yield of 2,2'-bifuran (based on $Pd(OAc)_2$ initially used), respectively, which is a significant improvement compared to 41% yield of 2,2'-bifuran obtained when $Pd(OAc)_2$ is used alone. However, it is worth mentioning that the yields of those coupling products based on the initial amount of furan are still very low, which accounts for only 1.6% and 0.77% for DMFA and HOAc as solvents, respectively.

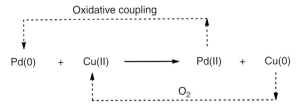

Scheme 17.17 Re-oxidation of palladium under the presence of Cu(II). Adapted from Ref. [240].

Recently, the dimerization of furfural under similar reaction conditions has provided only 1% yield of the corresponding dimer (based on the initial amount of furanic reactants) [240]. The poor activity is possibly due to a low availability of oxygen in the reaction mixture, leading to a suppression of the re-oxidation of Pd(0) to Pd(II). It has been reported that by bubbling O_2 through the reaction mixture with HOAc as the solvent, 10% yield of difurfural could be achieved over $Pd(OAc)_2/Cu(OAc)_2$ at 240 °C and 1 atm O_2 for 24 h.

Interestingly, under more severe conditions of 50 atm O_2 pressure and 120 °C, the yield of difurfural increases to 60% with the single catalytic system $Pd(OAc)_2$. Much higher amounts of coupling product compared to that of $Pd(OAc)_2$ initially introduced imply that the $Pd(OAc)_2$ is regenerated via oxidation of Pd(0) to Pd(II) without the need of a cocatalyst $Cu(OAc)_2$ under the more severe oxidizing conditions.

17.3
Summary and Conclusion

In this chapter, we have highlighted the current ideas and strategies for upgrading for furfural and other furanic compounds, which are the most unstable components in pyrolysis vapors via the utilization of C–C coupling as a backbone reaction. Thermodynamically, C–C bond formation of functional molecules is one of the efficient routes to solve two major issues in biofuel production from lignocellulosic materials: efficiently increase the energy density of feed molecules and produce heavy hydrocarbon liquid products, which closely resemble petroleum gasoline–diesel [243]. As shown in the previous discussion, the C–C coupling reactions such as aldol condensation, hydroxyalkylation, and Diels–Alder cycloaddition are not only capable of elongating the carbon chain of stated materials but also partially removing oxygen contents in a reacting mixture via dehydration catalyzed by acid–base catalysts resulting in enhancing stability, energy density of liquid products, and potentially diminishing H_2 consumption for further upgrading processes.

The direct C–C coupling reaction of furfural–HMF could be proceeded directly with functional molecules or indirectly via key intermediate compounds. A typical example of the former strategy is the aldol condensation of furfural–HMF and acetone conducted over base catalysts, typically MgO based, with high activity and selectivity. The use of an emulsion system stabilized by carbon nanotubes has shown to be a very promising strategy to technically and economically produce biomass-derived transportation fuel precursors by efficiently integrating multiple steps into a single reactor with bifunctional catalysts. The control of the transportation of each species between aqueous–organic phase could help to selectively yield hydrogenated dimer and trimer coupling products. Besides, dimerization of furfural via oxidative coupling is also a promising approach to produce C_{10} fuel precursors, which is attractive for further investigation.

However, the major obstacle of directly using furfural–HMF as coupling agent is the highly unstable nature of the carbonyl group facilitating polymerization to humins, which rapidly deactivates catalysts. Therefore, a feasible way to deal with this problem is to prestabilize furfural to key intermediates, which could be coupled subsequently with functional species. Hydrogenation–decarbonylation of furfural–HMF to furan compounds (furan, 2-methyl furan, and 2,5-dimethyl furan) or Piancatelli (ring rearrangement) of furfural to CPO is, in our opinion, an appropriate pathway. Those compounds are not only efficiently produced with excellent yield but also more stable compared to uncontrolled humin formation of furfural. While the ketone group of CPO allows it to get involved in self-aldol condensation or hydroxyalkylation, the active 5-C position of furan and 2-MF facilitates the hydroxyalkylation with carbonyl compounds or the double bond of furan, 2-MF, DMF rings favors the reaction with olefins via Diels–Alder. The concept works not just on a specific case but can be extended to many other compounds with similar structure and functional groups. For instance, furfural, which contains a carbonyl group, can play as hydroxyalkylation agent to couple with 2MF, instead of acetone. Many different bicyclic products obtained from Diels–Alder by using many different types of dienophiles are other obvious examples. Oxidation of furanics to dicarboxylic acid is a good way to stabilize furfural, which is an obvious advantage in terms of storage. Principally, those acids could be a potential platform for other types of C–C coupling reactions, such as ketonization or acylation. A network of possible upgrading routes is proposed in Figure 17.21.

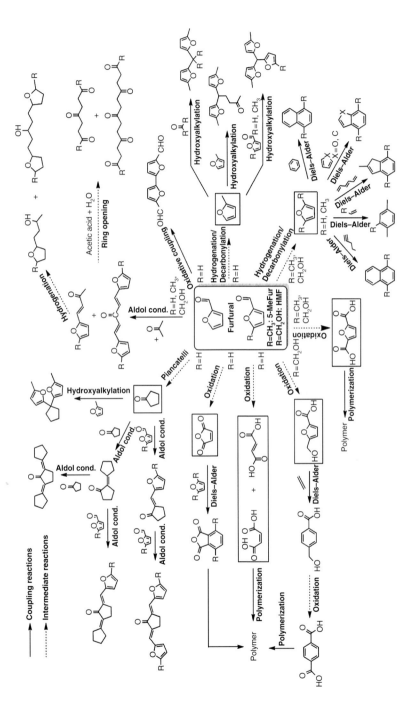

Figure 17.21 Road map for furanics upgrading strategy.

In a more general point of view, to develop upgrading strategies for a different or broader fraction of biomass pyrolysis, one should be aware of what type of compounds and useful functional groups could possibly be involved. The involvement of other types of compounds such as phenolics, alcohols, or carboxylic acid could change the whole picture of the upgrading strategy since many additional pathways could be opened. More comprehensive understanding about the functional groups, reaction mechanisms, and catalysts in complex biomass-derived stream is crucial to exploit the full potential of biomass-derived platform chemicals for C–C coupling reactions.

References

1. Sjöström, E. (1993) *Wood Chemistry: Fundamentals and Applications*, San Diego, CA, Academic Press.
2. Novaes, E., Kirst, M., Chiang, V., Winter-Sederoff, H., and Sederoff, R. (2010) Lignin and biomass: a negative correlation for wood formation and lignin content in trees. *Plant Physiol.*, **154** (2), 555–561.
3. Severian, D. (2005) *Polysaccharides. Structural Diversity and Functional Versatility*, 2nd edn, Marcel Dekker, New York.
4. Shen, D.K., Gu, S., and Bridgwater, A.V. (2010) Study on the pyrolytic behaviour of xylan-based hemicellulose using TG–FTIR and Py–GC–FTIR. *J. Anal. Appl. Pyrolysis*, **87** (2), 199–206.
5. Ciolkosz, D. and Wallace, R. (2011) A review of torrefaction for bioenergy feedstock production. *Biofuels, Bioprod. Biorefin.*, **5** (3), 317–329.
6. van der Stelt, M.J.C., Gerhauser, H., Kiel, J.H.A., and Ptasinski, K.J. (2011) Biomass upgrading by torrefaction for the production of biofuels: a review. *Biomass Bioenergy*, **35** (9), 3748–3762.
7. Prins, M.J., Ptasinski, K.J., and Janssen, F.J.J.G. (2006) More efficient biomass gasification via torrefaction. *Energy*, **31** (15), 3458–3470.
8. de Wild, P.J., Uil, H.d., Reith, J.H., Kiel, J.H.A., and Heeres, H.J. (2009) Biomass valorisation by staged degasification: a new pyrolysis-based thermochemical conversion option to produce value-added chemicals from lignocellulosic biomass. *J. Anal. Appl. Pyrolysis*, **85** (1–2), 124–133.
9. Sui, H., Yang, H., Shao, J., Wang, X., Li, Y., and Chen, H. (2014) Fractional condensation of multicomponent vapors from pyrolysis of cotton stalk. *Energy Fuel*, **28** (8), 5095–5102.
10. Pollard, A.S., Rover, M.R., and Brown, R.C. (2012) Characterization of bio-oil recovered as stage fractions with unique chemical and physical properties. *J. Anal. Appl. Pyrolysis*, **93**, 129–138.
11. Ji, W., Chen, Y., and Kung, H.H. (1997) Vapor phase aldol condensation of acetaldehyde on metal oxide catalysts. *Appl. Catal., A*, **161** (1–2), 93–104.
12. Kikhtyanin, O., Kelbichová, V., Vitvarová, D., Kubů, M., and Kubička, D. (2014) Aldol condensation of furfural and acetone on zeolites. *Catal. Today*, **227**, 154–162.
13. Heathcock, C.H. (1991) The aldol reaction: acid and general base catalysis, in *Comprehensive Organic Synthesis*, Pergamon, Oxford, pp. 133–179.
14. Hurd, C.D. (1964) Basic principles of organic chemistry, in *Science*, Benjamin, New York, p. 911.
15. O'Neill, R.E., Vanoye, L., De Bellefon, C., and Aiouache, F. (2014) Aldol-condensation of furfural by activated dolomite catalyst. *Appl. Catal., B*, **144**, 46–56.
16. Faba, L., Díaz, E., and Ordóñez, S. (2012) Aqueous-phase furfural-acetone aldol condensation over basic mixed

oxides. *Appl. Catal., B*, **113–114**, 201–211.

17. Di Cosimo, J.I., Diez, V.K., Ferretti, C., and Apesteguia, C.R. (2014) Chapter 1. Basic catalysis on MgO: generation, characterization and catalytic properties of active sites, in *Catalysis*, vol. 26, The Royal Society of Chemistry, pp. 1–28.

18. Snell, R., Combs, E., and Shanks, B. (2010) Aldol condensations using bio-oil model compounds: the role of acid–base Bi-functionality. *Top. Catal.*, **53** (15–18), 1248–1253.

19. Climent, M.J., Corma, A., Fornés, V., Guil-Lopez, R., and Iborra, S. (2002) Aldol condensations on solid catalysts: a cooperative effect between weak acid and base sites. *Adv. Synth. Catal.*, **344** (10), 1090–1096.

20. King, F. and Kelly, G.J. (2002) Combined solid base/hydrogenation catalysts for industrial condensation reactions. *Catal. Today*, **73** (1–2), 75–81.

21. West, R.M., Liu, Z.Y., Peter, M., Gärtner, C.A., and Dumesic, J.A. (2008) Carbon–carbon bond formation for biomass-derived furfurals and ketones by aldol condensation in a biphasic system. *J. Mol. Catal. A: Chem.*, **296** (1–2), 18–27.

22. Huang, X.-m., Zhang, Q., Wang, T.-j., Liu, Q.-y., Ma, L.-l., and Zhang, Q. (2012) Production of jet fuel intermediates from furfural and acetone by aldol condensation over MgO/NaY. *J. Fuel Chem. Technol.*, **40** (8), 973–978.

23. Kelly, G.J., King, F., and Kett, M. (2002) Waste elimination in condensation reactions of industrial importance. *Green Chem.*, **4** (4), 392–399.

24. Hora, L., Kikhtyanin, O., Čapek, L., Bortnovskiy, O., and Kubička, D. (2015) Comparative study of physico-chemical properties of laboratory and industrially prepared layered double hydroxides and their behavior in aldol condensation of furfural and acetone. *Catal. Today*, **241** (Pt. B), 221–230.

25. Barrett, C.J., Chheda, J.N., Huber, G.W., and Dumesic, J.A. (2006) Single-reactor process for sequential aldol-condensation and hydrogenation of biomass-derived compounds

in water. *Appl. Catal., B*, **66** (1–2), 111–118.

26. Huber, G.W., Chheda, J.N., Barrett, C.J., and Dumesic, J.A. (2005) Production of liquid alkanes by aqueous-phase processing of biomass-derived carbohydrates. *Science*, **308** (5727), 1446–1450.

27. Huber, G.W. and Dumesic, J.A. (2006) An overview of aqueous-phase catalytic processes for production of hydrogen and alkanes in a biorefinery. *Catal. Today*, **111** (1–2), 119–132.

28. Chheda, J.N. and Dumesic, J.A. (2007) An overview of dehydration, aldol-condensation and hydrogenation processes for production of liquid alkanes from biomass-derived carbohydrates. *Catal. Today*, **123** (1–4), 59–70.

29. Sádaba, I., Ojeda, M., Mariscal, R., Richards, R., and Granados, M.L. (2011) Mg–Zr mixed oxides for aqueous aldol condensation of furfural with acetone: effect of preparation method and activation temperature. *Catal. Today*, **167** (1), 77–83.

30. Sádaba, I., Ojeda, M., Mariscal, R., Fierro, J.L.G., and Granados, M.L. (2011) Catalytic and structural properties of co-precipitated Mg–Zr mixed oxides for furfural valorization via aqueous aldol condensation with acetone. *Appl. Catal., B*, **101** (3–4), 638–648.

31. Cota, I., Ramírez, E., Medina, F., Sueiras, J.E., Layrac, G., and Tichit, D. (2010) New synthesis route of hydrocalumite-type materials and their application as basic catalysts for aldol condensation. *Appl. Clay Sci.*, **50** (4), 498–502.

32. Climent, M.J., Corma, A., Iborra, S., and Velty, A. (2002) Synthesis of pseudoionones by acid and base solid catalysts. *Catal. Lett.*, **79** (1–4), 157–163.

33. Zhang, L., Pham, T.N., Faria, J., and Resasco, D.E. (2015) Improving the selectivity to C4 products in the aldol condensation of acetaldehyde in ethanol over faujasite zeolites. *Appl. Catal., A: General*, **504**, pp. 119–129.

34. Shen, W., Tompsett, G.A., Hammond, K.D., Xing, R., Dogan, F., Grey, C.P.,

Conner, W.C. Jr., Auerbach, S.M., and Huber, G.W. (2011) Liquid phase aldol condensation reactions with $MgO–ZrO_2$ and shape-selective nitrogen-substituted NaY. *Appl. Catal., A*, **392** (1–2), 57–68.

35. Pham, T.N., Shi, D., and Resasco, D.E. (2014) Evaluating strategies for catalytic upgrading of pyrolysis oil in liquid phase. *Appl. Catal., B*, **145**, 10–23.

36. Faba, L., Díaz, E., and Ordóñez, S. (2011) Performance of bifunctional Pd/MxNyO (M=Mg, Ca; N=Zr, Al) catalysts for aldolization–hydrogenation of furfural–acetone mixtures. *Catal. Today*, **164** (1), 451–456.

37. Huber, G.W., Cortright, R.D., and Dumesic, J.A. (2004) Renewable alkanes by aqueous-phase reforming of biomass-derived oxygenates. *Angew. Chem. Int. Ed.*, **43** (12), 1549–1551.

38. Xia, Q.-N., Cuan, Q., Liu, X.-H., Gong, X.-Q., Lu, G.-Z., and Wang, Y.-Q. (2014) Pd/NbOPO4 multifunctional catalyst for the direct production of liquid alkanes from aldol adducts of furans. *Angew. Chem. Int. Ed.*, **53** (37), 9755–9760.

39. Faba, L., Díaz, E., and Ordóñez, S. (2013) Improvement of the stability of basic mixed oxides used as catalysts for aldol condensation of bio-derived compounds by palladium addition. *Biomass Bioenergy*, **56**, 592–599.

40. Dedsuksophon, W., Faungnawakij, K., Champreda, V., and Laosiripojana, N. (2011) Hydrolysis/dehydration/aldol-condensation/hydrogenation of lignocellulosic biomass and biomass-derived carbohydrates in the presence of Pd/WO3–ZrO$_2$ in a single reactor. *Bioresour. Technol.*, **102** (2), 2040–2046.

41. Zapata, P., Faria, J., Pilar Ruiz, M., and Resasco, D. (2012) Condensation/hydrogenation of biomass-derived oxygenates in water/oil emulsions stabilized by nanohybrid catalysts. *Top. Catal.*, **55** (1–2), 38–52.

42. Crossley, S., Faria, J., Shen, M., and Resasco, D.E. (2010) Solid nanoparticles that catalyze biofuel upgrade reactions at the water/oil interface. *Science*, **327** (5961), 68–72.

43. Koso, S., Furikado, I., Shimao, A., Miyazawa, T., Kunimori, K., and Tomishige, K. (2009) Chemoselective hydrogenolysis of tetrahydrofurfuryl alcohol to 1,5-pentanediol. *Chem. Commun.*, (15), 2035–2037.

44. Watson, J.M. (1973) Butane-1,4-diol from Hydrolytic Reduction of Furan. *Ind. Eng. Chem. Prod. Res. Dev.*, **12** (4), 310–311.

45. Nakagawa, Y. and Tomishige, K. (2012) Production of 1,5-pentanediol from biomass via furfural and tetrahydro-furfuryl alcohol. *Catal. Today*, **195** (1), 136–143.

46. Chen, K., Mori, K., Watanabe, H., Nakagawa, Y., and Tomishige, K. (2012) C–O bond hydrogenolysis of cyclic ethers with OH groups over rhenium-modified supported iridium catalysts. *J. Catal.*, **294**, 171–183.

47. Sutton, A.D., Waldie, F.D., Wu, R., Schlaf, M., Pete Silks, L.A., and Gordon, J.C. (2013) The hydrodeoxygenation of bioderived furans into alkanes. *Nat. Chem.*, **5** (5), 428–432.

48. Montagnon, T., Noutsias, D., Alexopoulou, I., Tofi, M., and Vassilikogiannakis, G. (2011) Green oxidations of furans-initiated by molecular oxygen-that give key natural product motifs. *Org. Biomol. Chem.*, **9** (7), 2031–2039.

49. Waidmann, C.R., Pierpont, A.W., Batista, E.R., Gordon, J.C., Martin, R.L., "Pete" Silks, L.A., West, R.M., and Wu, R. (2013) Functional group dependence of the acid catalyzed ring opening of biomass derived furan rings: an experimental and theoretical study. *Catal. Sci. Technol.*, **3**, 106–115.

50. Iovel, I.G. and Lukevics, E. (1998) Hydroxymethylation and alkylation of compounds of the furan, thiophene, and pyrrole series in the presence of H+ cations (review). *Chem. Heterocycl. Compd.*, **34** (1), 1–12.

51. Brown, W.H. and Sawatzky, H. (1956) The condensation of furan and sylvan with some carbonyl compounds. *Can. J. Chem.*, **34** (9), 1147–1153.

52. Pennanen, S. and Nyman, G. (1972) Studies on the furan series, part I. The acidic condensation of aldehydes with

methyl 2-furoate. *Acta Chem. Scand.*, **26**, 1018–1022.

53. Eftax, D.S.P. and Dunlop, A.P. (1965) Hydrolysis of simple furans. Products of secondary condensation. *J. Org. Chem.*, **30** (4), 1317–1319.

54. Beals, R.E. and Brown, W.H. (1956) The condensation of 3-pentanone with furan. *J. Org. Chem.*, **21** (4), 447–448.

55. Ackman, R.G., Brown, W.H., and Wright, G.F. (1955) The condensation of methyl ketones with furan. *J. Org. Chem.*, **20** (9), 1147–1158.

56. Butin, A.V., Stroganova, T.A., and Kul'nevich, V.G. (1999) Furyl(aryl)methanes and their analogs. (Review). *Chem. Heterocycl. Compd.*, **35** (7), 757–787.

57. Iovel, I., Goldberg, Y., and Shymanska, M. (1989) Hydroxymethylation of furan and its derivatives in the presence of cation-exchange resins. *J. Mol. Catal.*, **57** (1), 91–103.

58. Corma, A., de la Torre, O., and Renz, M. (2012) Production of high quality diesel from cellulose and hemicellulose by the Sylvan process: catalysts and process variables. *Energy Environ. Sci.*, **5** (4), 6328–6344.

59. Corma, A., de la Torre, O., and Renz, M. (2011) High-quality diesel from hexose- and pentose-derived biomass platform molecules. *ChemSusChem*, **4** (11), 1574–1577.

60. Corma, A., de la Torre, O., Renz, M., and Villandier, N. (2011) Production of high-quality diesel from biomass waste products. *Angew. Chem. Int. Ed.*, **50** (10), 2375–2378.

61. Riad, A., Mouloungui, Z., Delmas, M., and Gaset, A. (1989) New synthesis of substituted difuryl or dithienyl methanes. *Synth. Commun.*, **19** (18), 3169–3173.

62. Lecomte, J., Finiels, A., and Moreau, C. (1999) A new selective route to 5-hydroxymethylfurfural from furfural and furfural derivatives over microporous solid acidic catalysts. *Ind. Crops Prod.*, **9** (3), 235–241.

63. Lecomte, J., Finiels, A., Geneste, P., and Moreau, C. (1998) Selective hydroxymethylation of furfuryl alcohol with aqueous formaldehyde in the presence

of dealuminated mordenites. *Appl. Catal., A*, **168** (2), 235–241.

64. Lecomte, J., Finiels, A., Geneste, P., and Moreau, C. (1998) Kinetics of furfuryl alcohol hydroxymethylation with aqueous formaldehyde over a highly dealuminated H-mordenite. *J. Mol. Catal. A: Chem.*, **133** (3), 283–288.

65. Moreau, C., Belgacem, M., and Gandini, A. (2004) Recent catalytic advances in the chemistry of substituted furans from carbohydrates and in the ensuing polymers. *Top. Catal.*, **27** (1–4), 11–30.

66. Balakrishnan, M., Sacia, E.R., and Bell, A.T. (2014) Syntheses of biodiesel precursors: sulfonic acid catalysts for condensation of biomass-derived platform molecules. *ChemSusChem*, **7** (4), 1078–1085.

67. Glukhovtsev, V.G., Shuikin, N.I., Zakharova, S.V., and Abgaforova, G.E. (1966) *Dokl. Akad. Nauk*, **170**, 1327.

68. Cairns, T.L., McKusick, B.C., and Weinmayr, V. (1951) The synthesis and oxidation of 2,2'-methylene-bis-(5-acetylthiophene). *J. Am. Chem. Soc.*, **73** (3), 1270–1273.

69. Hashmi, A.S.K., Schwarz, L., Rubenbauer, P., and Blanco, M.C. (2006) The condensation of carbonyl compounds with electron-rich arenes: mercury, thallium, gold or a proton. *Adv. Synth. Catal.*, **348** (6), 705–708.

70. Patalakh, I.I., Gankin, G.D., and Karakhanov, R.A. (1992) *Izv. Vyssh. Uchebn. Zaved., Khim. Khim. Tekhnol.*, **35**, 90.

71. Zheng, H.-Y., Zhu, Y.-L., Teng, B.-T., Bai, Z.-Q., Zhang, C.-H., Xiang, H.-W., and Li, Y.-W. (2006) Towards understanding the reaction pathway in vapour phase hydrogenation of furfural to 2-methylfuran. *J. Mol. Catal. A: Chem.*, **246** (1–2), 18–23.

72. Karl, J.Z. (2000) Methylfuran, in *Sugar Series*, Elsevier, pp. 229–230.

73. Li, G., Li, N., Yang, J., Wang, A., Wang, X., Cong, Y., and Zhang, T. (2013) Synthesis of renewable diesel with the 2-methylfuran, butanal and acetone derived from lignocellulose. *Bioresour. Technol.*, **134**, 66–72.

74. Li, G., Li, N., Wang, Z., Li, C., Wang, A., Wang, X., Cong, Y., and Zhang, T. (2012) Synthesis of high-quality diesel with furfural and 2-methylfuran from hemicellulose. *ChemSusChem*, **5** (10), 1958–1966.

75. Li, G., Li, N., Li, S., Wang, A., Cong, Y., Wang, X., and Zhang, T. (2013) Synthesis of renewable diesel with hydroxyacetone and 2-methyl-furan. *Chem. Commun.*, **49** (51), 5727–5729.

76. Subrahmanyam, A.V., Thayumanavan, S., and Huber, G.W. (2010) C-C bond formation reactions for biomass-derived molecules. *ChemSusChem*, **3** (10), 1158–1161.

77. Kumar, M. and Gayen, K. (2011) Developments in biobutanol production: new insights. *Appl. Energy*, **88** (6), 1999–2012.

78. Mascal, M. and Nikitin, E.B. (2008) Direct, high-yield conversion of cellulose into biofuel. *Angew. Chem. Int. Ed.*, **47** (41), 7924–7926.

79. Hamada, K., Yoshihara, H., and Suzukamo, G. (2001) Novel synthetic route to 2,5-disubstituted furan derivatives through surface active agent-catalysed dehydration of D(-)-fructose. *J. Oleo Sci.*, **50** (6), 533–536.

80. Mascal, M. and Nikitin, E.B. (2009) Towards the efficient, total glycan utilization of biomass. *ChemSusChem*, **2** (5), 423–426.

81. Wen, C., Barrow, E., Hattrick-Simpers, J., and Lauterbach, J. (2014) One-step production of long-chain hydrocarbons from waste-biomass-derived chemicals using bi-functional heterogeneous catalysts. *Phys. Chem. Chem. Phys.*, **16** (7), 3047–3054.

82. Li, G., Li, N., Yang, J., Li, L., Wang, A., Wang, X., Cong, Y., and Zhang, T. (2014) Synthesis of renewable diesel range alkanes by hydrodeoxygenation of furans over Ni/Hβ under mild conditions. *Green Chem.*, **16** (2), 594–599.

83. Balakrishnan, M., Sacia, E.R., and Bell, A.T. (2014) Selective hydrogenation of furan-containing condensation products as a source of biomass-derived diesel additives. *ChemSusChem*, **7** (10), 2796–2800.

84. Klein, D. (2012) An introduction to pericyclic reactions, in *Organic Chemistry*, vol. 783, John Wiley & Sons, Inc.

85. Houk, K.N., Gonzalez, J., and Li, Y. (1995) Pericyclic reaction transition states: passions and punctilios, 1935–1995. *Acc. Chem. Res.*, **28** (2), 81–90.

86. Sauer, J. (1967) Diels-Alder reactions II: the reaction mechanism. *Angew. Chem. Int. Ed.*, **6** (1), 16–33.

87. Wiest, O., Montiel, D.C., and Houk, K.N. (1997) Quantum mechanical methods and the interpretation and prediction of pericyclic reaction mechanisms. *J. Phys. Chem. A*, **101** (45), 8378–8388.

88. Fleming, I. (2010) *Molecular Orbitals and Organic Chemical Reactions*, John Wiley & Sons, Ltd.

89. Gujral, S.S. and Popli, A. (2013) Introduction to Diels Alder reaction, its mechanism and recent advantages: a review. *Indo Am. J. Pharm. Res.*, **3** (4), 3192–3215.

90. Brocksom, T.J., Nakamura, J., Ferreira, M.L., and Brocksom, U. (2001) The Diels-Alder reaction: an update. *J. Braz. Chem. Soc.*, **12**, 597–622.

91. Sauer, J. and Sustmann, R. (1980) Mechanistic aspects of Diels-Alder reactions: a critical survey. *Angew. Chem. Int. Ed.*, **19** (10), 779–807.

92. Oliver Kappe, C., Shaun Murphree, S., and Padwa, A. (1997) Synthetic applications of furan Diels-Alder chemistry. *Tetrahedron*, **53** (42), 14179–14233.

93. Laszlo, P. and Lucchetti, J. (1984) Easy formation of diels-alder cycloadducts between furans and α,β-unsaturated aldehydes and ketones at normal pressure. *Tetrahedron Lett.*, **25** (39), 4387–4388.

94. Dauben, W.G. and Krabbenhoft, H.O. (1976) Organic reactions at high pressure. Cycloadditions with furans. *J. Am. Chem. Soc.*, **98** (7), 1992–1993.

95. Chang, C.-C., Green, S.K., Williams, C.L., Dauenhauer, P.J., and Fan, W. (2014) Ultra-selective cycloaddition of dimethylfuran for renewable p-xylene with H-BEA. *Green Chem.*, **16** (2), 585–588.

96. Nikbin, N., Do, P.T., Caratzoulas, S., Lobo, R.F., Dauenhauer, P.J., and Vlachos, D.G. (2013) A DFT study of the acid-catalyzed conversion of 2,5-dimethylfuran and ethylene to p-xylene. *J. Catal.*, **297**, 35–43.

97. Gomes, M.G. (2007) *Diels-Alder Reactions of a Cyclopentadienone Derivative*, University of Missouri – Columbia.

98. Lutz, E.F. and Bailey, G.M. (1964) Regulation of structural isomerism in simple Diels-Alder adducts. *J. Am. Chem. Soc.*, **86** (18), 3899–3901.

99. Branchadell, V., Oliva, A., and Bertran, J. (1983) Catalytic effect on the selectivity of the Diels—Alder reaction. *Chem. Phys. Lett.*, **97** (4–5), 378–380.

100. Yamabe, S., Dai, T., and Minato, T. (1995) Fine tuning [4 + 2] and [2 + 4] Diels-Alder reactions catalyzed by Lewis acids. *J. Am. Chem. Soc.*, **117** (44), 10994–10997.

101. Conrads, M., Mattay, J., and Runsink, J. (1989) Hetero diels-alder reactions of allenes on silica gel surface and under liquid-phase conditions. *Chem. Ber.*, **122** (11), 2207–2208.

102. Veselovsky, V.V., Gybin, A.S., Lozanova, A.V., Moiseenkov, A.M., Smit, W.A., and Caple, R. (1988) Dramatic acceleration of the Diels-Alder reaction by adsorption on chromatography adsorbents. *Tetrahedron Lett.*, **29** (2), 175–178.

103. Parlar, H. and Baumann, R. (1981) Diels-Alder reaction of cyclopentadiene with acrylic acid derivatives in heterogeneous phases. *Angew. Chem. Int. Ed.*, **20** (12), 1014.

104. Hondrogiannis, G., Pagni, R.M., Kabalka, G.W., Anosike, P., and Kurt, R. (1990) The diels-alder reaction of cyclopentadiene and methyl acrylate on γ,-alumina. *Tetrahedron Lett.*, **31** (38), 5433–5436.

105. Cabral, J. and Laszlo, P. (1989) Product distribution in diels-alder addition of n-benzylidene aniline and enol ethers. *Tetrahedron Lett.*, **30** (51), 7237–7238.

106. Cativiela, C., Fraile, J.M., Garcia, J.I., Mayoral, J.A., Figueras, F., De Menorval, L.C., and Alonso, P.J. (1992) Factors influencing the k10 montmorillonite-catalyzed diels-alder

107. Fraile, J.M., García, J.I., Massam, J., Mayoral, J.A., and Pires, E. (1997) $ZnCl_2$, ZnI_2 and $TiCl_4$ supported on silica gel as catalysts for the Diels-Alder reactions of furan. *J. Mol. Catal. A: Chem.*, **123** (1), 43–47.

108. McClure, C.K. and Hansen, K.B. (1996) Diels-Alder reactivity of a ketovinylphosphonate with cyclopentadiene and furan. *Tetrahedron Lett.*, **37** (13), 2149–2152.

109. Reymond, S. and Cossy, J. (2008) Copper-catalyzed Diels—Alder reactions. *Chem. Rev.*, **108** (12), 5359–5406.

110. Dessau, R.M. (1986) Catalysis of Diels-Alder reactions by zeolites. *J. Chem. Soc., Chem. Commun.*, (15), 1167–1168.

111. Bornholdt, K. and Lechert, H. (1995) in *Catalysis By Microporous Materials*, Studies in Surface Science and Catalysis (eds H.G. Karge, I. Kiricsi, H.K. Beyer, and J.B. Nagy), Elsevier, pp. 619–626.

112. Casaschi, A., Desimoni, G., Faita, G., Invernizzi, A.G., Lanati, S., and Righetti, P. (1993) Catalysis with inorganic cations. 2. The effect of some perchlorates on Diels-Alder reaction rates. *J. Am. Chem. Soc.*, **115** (18), 8002–8007.

113. Shiramizu, M. and Toste, F.D. (2011) On the Diels—Alder approach to solely biomass-derived Polyethylene Terephthalate (PET): conversion of 2,5-dimethylfuran and acrolein into p-Xylene. *Chem. Eur. J.*, **17** (44), 12452–12457.

114. Kotsuki, H., Nishizawa, H., Ochi, M., and Matsuoka, K. (1982) High pressure organic chemistry. V. Diels-Alder reactions of furan with acrylic and maleic esters. *Bull. Chem. Soc. Jpn.*, **55** (2), 496–499.

115. Moore, J.A. and Partain, E.M. (1983) Catalyzed addition of furan with acrylic monomers. *J. Org. Chem.*, **48** (7), 1105–1106.

116. Brion, F. (1982) On the lewis acid catalyzed diels-alder reaction of furan. regio- and stereospecific synthesis of

substituted cyclohexenols and cyclo-hexadienols. *Tetrahedron Lett.*, **23** (50), 5299–5302.

117. Williams, C.L., Chang, C.-C., Do, P., Nikbin, N., Caratzoulas, S., Vlachos, D.G., Lobo, R.F., Fan, W., and Dauenhauer, P.J. (2012) Cycloaddition of biomass-derived furans for catalytic production of renewable p-Xylene. *ACS Catal.*, **2** (6), 935–939.

118. Do, P.T.M., McAtee, J.R., Watson, D.A., and Lobo, R.F. (2013) Elucida-tion of Diels–Alder reaction network of 2,5-dimethylfuran and ethylene on HY zeolite catalyst. *ACS Catal.*, **3** (1), 41–46.

119. Wang, D., Osmundsen, C.M., Taarning, E., and Dumesic, J.A. (2013) Selec-tive production of aromatics from alkylfurans over solid acid catalysts. *ChemCatChem*, **5** (7), 2044–2050.

120. Cheng, Y.-T. and Huber, G.W. (2012) Production of targeted aromatics by using Diels-Alder classes of reactions with furans and olefins over ZSM-5. *Green Chem.*, **14** (11), 3114–3125.

121. Do, P.T.M., McAtee, J.R., Watson, D.A., and Lobo, R.F. (2012) Elucida-tion of Diels–Alder reaction network of 2,5-dimethylfuran and ethylene on HY zeolite catalyst. *ACS Catal.*, **3** (1), 41–46.

122. Xiong, R., Sandler, S.I., Vlachos, D.G., and Dauenhauer, P.J. (2014) Solvent-tuned hydrophobicity for faujasite-catalyzed cycloaddition of biomass-derived dimethylfuran for renewable p-xylene. *Green Chem.*, **16** (9), 4086–4091.

123. Nikbin, N., Feng, S., Caratzoulas, S., and Vlachos, D.G. (2014) p-Xylene for-mation by dehydrative aromatization of a Diels–Alder product in Lewis and Brønsted acidic zeolites. *J. Phys. Chem. C*, **118** (42), 24415–24424.

124. Imachi, S. and Onaka, M. (2004) Entrapment of cyclopentadiene in zeolite NaY and its application for solvent-free Diels–Alder reactions in the nanosized confined environment. *Tetrahedron Lett.*, **45** (25), 4943–4946.

125. Li, Y.-P., Head-Gordon, M., and Bell, A.T. (2014) Computational study of p-Xylene synthesis from ethylene and 2,5-dimethylfuran catalyzed by H-BEA. *J. Phys. Chem. C*, **118** (38), 22090–22095.

126. Takanishi, K., Sone, S. (2009) Method for producing para-xylene. WO Patent 2009110402A1, Toray Industries, Tokyo, Japan.

127. Brandvold, T. (2010) Carbohydrate route to para-xylene and terephthalic acid. WO Patent 2010151346A1, UOP LLC, IL.

128. Kuczenski, B. and Geyer, R. (2010) Material flow analysis of polyethylene terephthalate in the US, 1996–2007. *Resour. Conserv. Recycl.*, **54** (12), 1161–1169.

129. Pacheco, J.J. and Davis, M.E. (2014) Synthesis of terephthalic acid via Diels-Alder reactions with ethy-lene and oxidized variants of 5-hydroxymethylfurfural. *Proc. Natl. Acad. Sci. U.S.A.*, **111** (23), 8363–8367.

130. Casanova, O., Iborra, S., and Corma, A. (2009) Biomass into chemicals: aerobic oxidation of 5-hydroxymethyl-2-furfural into 2,5-furandicarboxylic acid with gold nanoparticle catalysts. *ChemSusChem*, **2** (12), 1138–1144.

131. Piutti, C. and Quartieri, F. (2013) The piancatelli rearrangement: new appli-cations for an intriguing reaction. *Molecules*, **18** (10), 12290–12312.

132. Piancatelli, G., Scettri, A., David, G., and D'Auria, M. (1978) A new synthesis of 3-oxocyclopentenes1. *Tetrahedron*, **34** (18), 2775–2778.

133. Scettri, A., Piancatelli, G., D'Auria, M., and David, G. (1979) General route and mechanism of the rearrangement of the 4-substituted 5-hydroxy-3-oxocyclopentenes into the 2-substituted analogs. *Tetrahedron*, **35** (1), 135–138.

134. Fisher, D., Palmer, L.I., Cook, J.E., Davis, J.E., and Read de Alaniz, J. (2014) Efficient synthesis of 4-hydroxycyclopentenones: dyspro-sium(III) triflate catalyzed Piancatelli rearrangement. *Tetrahedron*, **70** (27–28), 4105–4110.

135. Piancatelli, G., Scettri, A., and Barbadoro, S. (1976) A useful prepa-ration of 4-substituted 5-hydroxy-3-oxocyclopentene. *Tetrahedron Lett.*, **17** (39), 3555–3558.

136. Piancatelli, G. and Scettri, A. (1977) A useful preparation of (±) t-butyl 3-hydroxy-5-oxo-1-cyclopenteneheptanoate ans its 3-deoxy-derivative, important prostaglandin intermediates. *Tetrahedron Lett.*, **18** (13), 1131–1134.

137. Minai, M. (1990) Process for preparing oxocyclopentene derivatives. US Patent 4970345 A.

138. Hronec, M., Fulajtárova, K., and Soták, T. (2014) Highly selective rearrangement of furfuryl alcohol to cyclopentanone. *Appl. Catal., B*, **154–155**, 294–300.

139. Piancatelli, G., D'Auria, M., and D'Onofrio, F. (1994) Synthesis of 1,4-dicarbonyl compounds and cyclopentenones from furans. *Synthesis*, **1994** (09), 867–889.

140. Yang, Y., Du, Z., Huang, Y., Lu, F., Wang, F., Gao, J., and Xu, J. (2013) Conversion of furfural into cyclopentanone over Ni-Cu bimetallic catalysts. *Green Chem.*, **15** (7), 1932–1940.

141. Hronec, M., Fulajtarová, K., and Liptaj, T. (2012) Effect of catalyst and solvent on the furan ring rearrangement to cyclopentanone. *Appl. Catal., A*, **437–438**, 104–111.

142. Veits, G.K., Wenz, D.R., and Readde Alaniz, J. (2010) Versatile method for the synthesis of 4-aminocyclopentenones: dysprosium(III) triflate catalyzed aza-piancatelli rearrangement. *Angew. Chem. Int. Ed.*, **49** (49), 9484–9487.

143. Nieto Faza, O., Silva López, C., Álvarez, R., and de Lera, Á.R. (2004) Theoretical study of the electrocyclic ring closure of hydroxypentadienyl cations. *Chem. Eur. J.*, **10** (17), 4324–4333.

144. Frontier, A.J. and Collison, C. (2005) The Nazarov cyclization in organic synthesis. Recent advances. *Tetrahedron*, **61** (32), 7577–7606.

145. Pellissier, H. (2005) Recent developments in the Nazarov process. *Tetrahedron*, **61** (27), 6479–6517.

146. Shimada, N., Stewart, C., and Tius, M.A. (2011) Asymmetric Nazarov cyclizations. *Tetrahedron*, **67** (33), 5851–5870.

147. Tius, M.A. (2005) Some new Nazarov chemistry. *Eur. J. Org. Chem.*, **2005** (11), 2193–2206.

148. Hronec, M. and Fulajtarová, K. (2012) Selective transformation of furfural to cyclopentanone. *Catal. Commun.*, **24**, 100–104.

149. Ordomsky, V.V., Schouten, J.C., van der Schaaf, J., and Nijhuis, T.A. (2013) Biphasic single-reactor process for dehydration of xylose and hydrogenation of produced furfural. *Appl. Catal., A*, **451**, 6–13.

150. Curran, T.T., Hay, D.A., Koegel, C.P., and Evans, J.C. (1997) The preparation of optically active 2-cyclopenten-1,4-diol derivatives from furfuryl alcohol. *Tetrahedron*, **53** (6), 1983–2004.

151. Watson, T.J.N., Curran, T.T., Hay, D.A., Shah, R.S., Wenstrup, D.L., and Webster, M.E. (1998) Development of the carbocyclic nucleoside MDL 201449A: a tumor necrosis factor-α inhibitor. *Org. Process Res. Dev.*, **2** (6), 357–365.

152. Curran, T.T., Evans, J.C., and Hay, D.A. (1998) Preparation of Cis-4-O-Protected-2-Cyclopentenol Derivatives. Mar. 17, 1998, Hoechst Marion Roussel Inc.

153. Ghorpade, S.R., Bastawade, K.B., Gokhale, D.V., Shinde, P.D., Mahajan, V.A., Kalkote, U.R., and Ravindranathan, T. (1999) Enzymatic kinetic resolution studies of racemic 4-hydroxycyclopent-2-en-1-one using Lipozyme IM®. *Tetrahedron: Asymmetry*, **10** (21), 4115–4122.

154. Bandura, A.V. and Lvov, S.N. (2006) The ionization constant of water over wide ranges of temperature and density. *J. Phys. Chem. Ref. Data*, **35** (1), 15–30.

155. Nakagawa, Y., Tamura, M., and Tomishige, K. (2013) Catalytic reduction of biomass-derived furanic compounds with hydrogen. *ACS Catal.*, **3** (12), 2655–2668.

156. Yan, K., Wu, G., Lafleur, T., and Jarvis, C. (2014) Production, properties and catalytic hydrogenation of furfural to fuel additives and value-added chemicals. *Renewable Sustainable Energy Rev.*, **38**, 663–676.

157. Sitthisa, S. and Resasco, D. (2011) Hydrodeoxygenation of furfural over supported metal catalysts: a comparative study of Cu, Pd and Ni. *Catal. Lett.*, **141** (6), 784–791.

158. Nakagawa, Y., Takada, K., Tamura, M., and Tomishige, K. (2014) Total hydrogenation of furfural and 5-hydroxymethylfurfural over supported Pd–Ir alloy catalyst. *ACS Catal.*, **4** (8), 2718–2726.

159. Sitthisa, S., Sooknoi, T., Ma, Y., Balbuena, P.B., and Resasco, D.E. (2011) Kinetics and mechanism of hydrogenation of furfural on Cu/SiO2 catalysts. *J. Catal.*, **277** (1), 1–13.

160. Sexton, B.A., Hughes, A.E., and Avery, N.R. (1985) A spectroscopic study of the adsorption and reactions of methanol, formaldehyde and methyl formate on clean and oxygenated Cu(110) surfaces. *Surf. Sci.*, **155** (1), 366–386.

161. Mavrikakis, M. and Barteau, M.A. (1998) Oxygenate reaction pathways on transition metal surfaces. *J. Mol. Catal. A: Chem.*, **131** (1–3), 135–147.

162. Davis, J.L. and Barteau, M.A. (1989) Polymerization and decarbonylation reactions of aldehydes on the Pd(111) surface. *J. Am. Chem. Soc.*, **111** (5), 1782–1792.

163. Davis, J.L. and Barteau, M.A. (1990) Spectroscopic identification of alkoxide, aldehyde, and acyl intermediates in alcohol decomposition on Pd(111). *Surf. Sci.*, **235** (2–3), 235–248.

164. Henderson, M.A., Zhou, Y., and White, J.M. (1989) Polymerization and decomposition of acetaldehyde on rutherium(001). *J. Am. Chem. Soc.*, **111** (4), 1185–1193.

165. Zhou, M., Zhu, H., Niu, L., Xiao, G., and Xiao, R. (2014) Catalytic hydroprocessing of furfural to cyclopentanol over Ni/CNTs catalysts: model reaction for upgrading of bio-oil. *Catal. Lett.*, **144** (2), 235–241.

166. Zhu, H., Zhou, M., Zeng, Z., Xiao, G., and Xiao, R. (2014) Selective hydrogenation of furfural to cyclopentanone over Cu-Ni-Al hydrotalcite-based catalysts. *Korean J. Chem. Eng.*, **31** (4), 593–597.

167. Guo, J., Xu, G., Han, Z., Zhang, Y., Fu, Y., and Guo, Q. (2014) Selective conversion of furfural to cyclopentanone with CuZnAl catalysts. *ACS Sustainable Chem. Eng.*, **2** (10), 2259–2266.

168. Ohyama, J., Kanao, R., Esaki, A., and Satsuma, A. (2014) Conversion of 5-hydroxymethylfurfural to a cyclopentanone derivative by ring rearrangement over supported Au nanoparticles. *Chem. Commun.*, **50** (42), 5633–5636.

169. Levina, R.Y., Skvarchenko, V.R., and Okhlobystin, O.Y. (1955) *Zh. Obshch. Khim.*, **25**, 1466–1469.

170. Turova-pollak, M.B., Sosnina, I.E., and Treschova, E.G. (1953) *Zh. Obshch. Khim.*, **23**, 1111–1116.

171. Gollis, M.H., Belenyessy, L.I., Gudzinowicz, B.J., Koch, S.D., Smith, J.O., and Wineman, R.J. (1962) Evaluation of pure hydrocarbons as jet fuels. *J. Chem. Eng. Data*, **7** (2), 311–316.

172. Yang, J., Li, N., Li, G., Wang, W., Wang, A., Wang, X., Cong, Y., and Zhang, T. (2014) Synthesis of renewable high-density fuels using cyclopentanone derived from lignocellulose. *Chem. Commun.*, **50** (20), 2572–2574.

173. Obaid Faroon, D.M. (1995) Hernan Navarro, Toxicological Profile for Jet Fuels JP-4 and JP-7, P.H.S. U.S. Department of health and human services, Agency for Toxic Substances & Disease Registry, Editor.

174. Hemighaus, G., Boval, T., Bacha, J., Barnes, F., Franklin, M., Gibbs, L., Hogue, N., Jones, J., Lesnini, D., Lind, J., and Morris, J. (2004) *Aviation Fuels Technical Review*, Chevron.

175. Li, G., Li, N., Wang, X., Sheng, X., Li, S., Wang, A., Cong, Y., Wang, X., and Zhang, T. (2014) Synthesis of diesel or jet fuel range cycloalkanes with 2-methylfuran and cyclopentanone from lignocellulose. *Energy Fuel*, **28** (8), 5112–5118.

176. Hronec, M., Fulajtárová, K., Liptaj, T., Štolcová, M., Prónayová, N., and Soták, T. (2014) Cyclopentanone: a raw material for production of C15 and C17 fuel precursors. *Biomass Bioenergy*, **63**, 291–299.

177. Okuhara, T. (2002) Water-tolerant solid acid catalysts. *Chem. Rev.*, **102** (10), 3641–3666.

178. Rao, K.N., Sridhar, A., Lee, A.F., Tavener, S.J., Young, N.A., and Wilson, K. (2006) Zirconium phosphate supported tungsten oxide solid acid catalysts for the esterification of palmitic acid. *Green Chem.*, **8** (9), 790–797.

179. Jae, J., Tompsett, G.A., Foster, A.J., Hammond, K.D., Auerbach, S.M., Lobo, R.F., and Huber, G.W. (2011) Investigation into the shape selectivity of zeolite catalysts for biomass conversion. *J. Catal.*, **279** (2), 257–268.

180. Cadenas, M., Bringué, R., Fité, C., Ramírez, E., and Cunill, F. (2011) Liquid-phase oligomerization of 1-hexene catalyzed by macroporous ion-exchange resins. *Top. Catal.*, **54** (13–15), 998–1008.

181. Tanabe, K., Misono, M., Hattori, H., and Ono, Y. (1989) *New Solid Acids and Bases: Their Catalytic Properties*, vol. 51, Elsevier Science Publisher, Tokyo, Japan.

182. Grot, W. (2011) *Fluorinated Ionomers*, Elsevier Inc., Waltham, MA.

183. Pillai, U.R. and Sahle-Demessie, E. (2003) Sn-exchanged hydrotalcites as catalysts for clean and selective Baeyer–Villiger oxidation of ketones using hydrogen peroxide. *J. Mol. Catal. A: Chem.*, **191** (1), 93–100.

184. Murahashi, S.-I., Oda, Y., and Naota, T. (1992) Fe2O3-catalyzed baeyer-villiger oxidation of ketones with molecular oxygen in the presence of aldehydes. *Tetrahedron Lett.*, **33** (49), 7557–7560.

185. Rios, M.Y., Salazar, E., and Olivo, H.F. (2008) Chemo-enzymatic Baeyer–Villiger oxidation of cyclopentanone and substituted cyclopentanones. *J. Mol. Catal. B: Enzym.*, **54** (3–4), 61–66.

186. Jiménez-Sanchidrián, C. and Ruiz, J.R. (2008) The Baeyer–Villiger reaction on heterogeneous catalysts. *Tetrahedron*, **64** (9), 2011–2026.

187. Lei, Z., Zhang, Q., Luo, J., and He, X. (2005) Baeyer–Villiger oxidation of ketones with hydrogen peroxide catalyzed by Sn-palygorskite. *Tetrahedron Lett.*, **46** (20), 3505–3508.

188. Ueno, S., Ebitani, K., Ookubo, A., and Kaneda, K. (1997) The active sites in the heterogeneous Baeyer-Villiger oxidation of cyclopentanone by hydrotalcite catalysts. *Appl. Surf. Sci.*, **121–122**, 366–371.

189. Fischer, J. and Hölderich, W.F. (1999) Baeyer–Villiger-oxidation of cyclopentanone with aqueous hydrogen peroxide by acid heterogeneous catalysis. *Appl. Catal., A*, **180** (1–2), 435–443.

190. Badovskaya, L.A. and Povarova, L.V. (2009) Oxidation of furans (Review). *Chem. Heterocycl. Compd.*, **45** (9), 1023–1034.

191. Milas, N.A. and Walsh, W.L. (1935) Catalytic oxidations. I. Oxidations in the furan series. *J. Am. Chem. Soc.*, **57** (8), 1389–1393.

192. Alonso-Fagúndez, N., Granados, M.L., Mariscal, R., and Ojeda, M. (2012) Selective conversion of furfural to maleic anhydride and furan with VOx/Al2O3 catalysts. *ChemSusChem*, **5** (10), 1984–1990.

193. Takagaki, A., Nishimura, S., and Ebitani, K. (2012) Catalytic transformations of biomass-derived materials into value-added chemicals. *Catal. Surv. Asia*, **16** (3), 164–182.

194. Poskonin, V.V. and Badovskaya, L.A. (1998) Catalytic oxidation of furan and hydrofuran compounds. 2.* Oxidation of furfural in the hydrogen peroxide-vanadyl sulfate-sodium acetate system. *Chem. Heterocycl. Compd.*, **34** (6) 646–650.

195. Poskonin, V.V. (2009) Catalytic oxidation reactions of furan and hydrofuran compounds 9.* characteristics and synthetic possibilities of the reaction of furan with aqueous hydrogen peroxide in the presence of compounds of niobium (ii) and (v). *Chem. Heterocycl. Compd.*, **45** (10), 1177–1183.

196. Alonso-Fagundez, N., Agirrezabal-Telleria, I., Arias, P.L., Fierro, J.L.G., Mariscal, R., and Granados, M.L. (2014) Aqueous-phase catalytic oxidation of furfural with H_2O_2: high yield of maleic acid

by using titanium silicalite-1. *RSC Adv.*, **4** (98), 54960–54972.

197. Poskonin, V.V. and Badovskaya, L.A. (1991) Reaction of furan compounds with hydrogen peroxide in the presence of vanadium catalysts. *Chem. Heterocycl. Compd.*, **27** (11), 1177–1182.

198. Guo, H. and Yin, G. (2011) Catalytic aerobic oxidation of renewable furfural with phosphomolybdic acid catalyst: an alternative route to maleic acid. *J. Phys. Chem. C*, **115** (35), 17516–17522.

199. Lan, J., Chen, Z., Lin, J., and Yin, G. (2014) Catalytic aerobic oxidation of renewable furfural to maleic anhydride and furanone derivatives with their mechanistic studies. *Green Chem.*, **16** (9), 4351–4358.

200. Gallezot, P. (2012) Conversion of biomass to selected chemical products. *Chem. Soc. Rev.*, **41** (4), 1538–1558.

201. Pasini, T., Piccinini, M., Blosi, M., Bonelli, R., Albonetti, S., Dimitratos, N., Lopez-Sanchez, J.A., Sankar, M., He, Q., Kiely, C.J., Hutchings, G.J., and Cavani, F. (2011) Selective oxidation of 5-hydroxymethyl-2-furfural using supported gold-copper nanoparticles. *Green Chem.*, **13** (8), 2091–2099.

202. Albonetti, S., Pasini, T., Lolli, A., Blosi, M., Piccinini, M., Dimitratos, N., Lopez-Sanchez, J.A., Morgan, D.J., Carley, A.F., Hutchings, G.J., and Cavani, F. (2012) Selective oxidation of 5-hydroxymethyl-2-furfural over TiO2-supported gold–copper catalysts prepared from preformed nanoparticles: effect of Au/Cu ratio. *Catal. Today*, **195** (1), 120–126.

203. Sessions, W.V. (1928) Catalytic oxidation of furfural in the vapor phase. *J. Am. Chem. Soc.*, **50** (6), 1696–1698.

204. Nielsen, E.R. (1949) Vapor phase oxidation of furfural. *Ind. Eng. Chem.*, **41** (2), 365–368.

205. Shi, S., Guo, H., and Yin, G. (2011) Synthesis of maleic acid from renewable resources: catalytic oxidation of furfural in liquid media with dioxygen. *Catal. Commun.*, **12** (8), 731–733.

206. Zumstein, F. (1934) Process for preparing maleic anhydride and maleic acid. US Patent 1,956,482, United States Patent Office.

207. Nielsen, E.R. (1947) Catalytic Oxidation or Fubfural, U.S.P. Office, Editor.

208. Alonso-Fagúndez, N., Laserna, V., Alba-Rubio, A.C., Mengibar, M., Heras, A., Mariscal, R., and Granados, M.L. (2014) Poly-(styrene sulphonic acid): an acid catalyst from polystyrene waste for reactions of interest in biomass valorization. *Catal. Today*, **234**, 285–294.

209. Kunioka, M., Masuda, T., Tachibana, Y., Funabashi, M., and Oishi, A. (2014) Highly selective synthesis of biomass-based 1,4-butanediol monomer by alcoholysis of 1,4-diacetoxybutane derived from furan. *Polym. Degrad. Stab.*, **109**, 393–397.

210. Werpy, T.A. and Petersen, G. (2004) Top Value Added Chemicals from Biomass. DOE/GO-102004-1992, U.S. Department of Energy (DOE), Golden, CO.

211. Grunskaya, E.P., Badovskaya, L.A., Poskonin, V.V., and Yakuba, Y.F. (1998) Catalytic oxidation of furan and hydrofuran compounds. 4. Oxidation of furfural by hydrogen peroxide in the presence of sodium molybdate. *Chem. Heterocycl. Compd.*, **34** (7), 775–780.

212. Badovskaya, L.A., Latashko, V.M., Poskonin, V.V., Grunskaya, E.P., Tyukhteneva, Z.I., Rudakova, S.G., Pestunova, S.A., and Sarkisyan, A.V. (2002) Catalytic oxidation of furan and hydrofuran compounds. 7. Production of 2(5H)-furanone by oxidation of furfural with hydrogen peroxide and some of its transformations in aqueous solutions. *Chem. Heterocycl. Compd.*, **38** (9), 1040–1048.

213. Poskonin, V.V. and Badovskaya, L.A. (1998) Catalytic oxidation of furan and hydrofuran compounds. *Chem. Heterocycl. Compd.*, **34** (6), 646–650.

214. Krupenskii, V.I. (1996) *Zh. Obshch. Khim.*, **66**, 1874.

215. Taniyama, M. (1954) *Toho Reiyon Kenkyu Hokoku*, **1**, 40.

216. Krupenskii, V.I. (1973) *Nauchn. Tr.-Leningr. Lesotekh. Akad.im. S. M. Kirova*, **158**, 68.

217. Choudhary, H., Nishimura, S., and Ebitani, K. (2012) Highly efficient aqueous oxidation of furfural to succinic acid using reusable heterogeneous acid

catalyst with hydrogen peroxide. *Chem. Lett.*, **41** (4), 409–411.

218. Choudhary, H., Nishimura, S., and Ebitani, K. (2013) Metal-free oxidative synthesis of succinic acid from biomass-derived furan compounds using a solid acid catalyst with hydrogen peroxide. *Appl. Catal., A*, **458**, 55–62.

219. Cross, C.F., Bevan, E.J., and Briggs, J.F. (1900) Einwirkung des Caro'schen Reagens auf Furfurol. *Ber. Dtsch. Chem. Ges.*, **33** (3), 3132–3138.

220. Tachibana, Y., Masuda, T., Funabashi, M., and Kunioka, M. (2010) Chemical synthesis of fully biomass-based poly(butylene succinate) from inedible-biomass-based furfural and evaluation of its biomass carbon ratio. *Biomacromolecules*, **11** (10), 2760–2765.

221. Yamada, S., Haramaki, H., Matsumoto, S., and Akazawa, Y. (1997) Method for producing succinic acid. JP Patent 0931 011.

222. Zhanglin, Y., Forissier, M., Sneeden, R.P., Vedrine, J.C., and Volta, J.C. (1994) On the mechanism of n-butane oxidation to maleic anhydride on VPO catalysts: I. A kinetics study on a VPO catalyst as compared to VPO reference phases. *J. Catal.*, **145** (2), 256–266.

223. Wachs, I.E., Jehng, J.-M., Deo, G., Weckhuysen, B.M., Guliants, V.V., Benziger, J.B., and Sundaresan, S. (1997) Fundamental studies of butane oxidation over model-supported vanadium oxide catalysts: molecular structure-reactivity relationships. *J. Catal.*, **170** (1), 75–88.

224. Centi, G., Trifiro, F., Ebner, J.R., and Franchetti, V.M. (1988) Mechanistic aspects of maleic anhydride synthesis from C4 hydrocarbons over phosphorus vanadium oxide. *Chem. Rev.*, **88** (1), 55–80.

225. Wang, S., Leng, Y., Lin, F., Huang, C., and Yi, C. (2009) Catalytic oxidation of furfural in vapor-gas phase for producing maleic anhydride. *Chem. Ind. Eng. Prog.*, **28** (6), 1019.

226. Murthy, M.S. and Rajamani, K. (1974) Kinetics of vapour phase oxidation of furfural on vanadium catalyst. *Chem. Eng. Sci.*, **29** (2), 601–609.

227. Slavinskaya, V.A., Kreile, D.R., Dzilyuma, é.E., and Sile, D.é. (1977) Incomplete catalytic oxidation of furan compounds (review). *Chem. Heterocycl. Compd.*, **13** (7), 710–721.

228. Hjelmgaard, T., Persson, T., Rasmussen, T.B., Givskov, M., and Nielsen, J. (2003) Synthesis of furanone-Based natural product analogues with quorum sensing antagonist activity. *Bioorg. Med. Chem.*, **11** (15), 3261–3271.

229. Yin, G., Piao, D.-g., Kitamura, T., and Fujiwara, Y. (2000) Cu(OAc)2-catalyzed partial oxidation of methane to methyl trifluoroacetate in the liquid phase. *Appl. Organomet. Chem.*, **14** (8), 438–442.

230. Kochi, J.K. and Subramanian, R.V. (1965) Kinetics of electron-transfer oxidation of alkyl radicals by copper(II) complexes. *J. Am. Chem. Soc.*, **87** (21), 4855–4866.

231. Du, Z., Ma, J., Wang, F., Liu, J., and Xu, J. (2011) Oxidation of 5-hydroxymethylfurfural to maleic anhydride with molecular oxygen. *Green Chem.*, **13** (3), 554–557.

232. Mahmoud, E., Watson, D.A., and Lobo, R.F. (2014) Renewable production of phthalic anhydride from biomass-derived furan and maleic anhydride. *Green Chem.*, **16** (1), 167–175.

233. Singh, H., Prasad, M., and Srivastava, R.D. (1980) Metal support interactions in the palladium-catalysed decomposition of furfural to furan. *J. Chem. Technol. Biotechnol.*, **30** (1), 293–296.

234. Hoydonckx, W.M.V.R.H.E., Van Rhijn, W., De Vos, D.E., and Jacobs, P.A. (2007) *Ullmann's Encyclopedia of Industrial Chemistry*, Wiley-VCH Verlag GmbH & Co.

235. Windmon, N. and Dragojlovic, V. (2008) Diels-Alder reactions in the presence of a minimal amount of water. *Green Chem. Lett. Rev.*, **1** (3), 155–163.

236. Narayan, S., Muldoon, J., Finn, M.G., Fokin, V.V., Kolb, H.C., and Sharpless, K.B. (2005) "On water": unique reactivity of organic compounds in aqueous suspension. *Angew. Chem. Int. Ed.*, **44** (21), 3275–3279.

237. Mazur, Y. and Karger, M.H. (1971) Mixed sulfonic-carboxylic anhydrides. II. Reactions with aliphatic ethers and amines. *J. Org. Chem.*, **36** (4), 532–540.

238. Tachibana, Y., Kimura, S., and Kasuya, K.-i. (2015) Synthesis and verification of biobased terephthalic acid from furfural. *Sci. Rep.*, **5** 1–5.

239. Colonna, M., Berti, C., Fiorini, M., Binassi, E., Mazzacurati, M., Vannini, M., and Karanam, S. (2011) Synthesis and radiocarbon evidence of terephthalate polyesters completely prepared from renewable resources. *Green Chem.*, **13** (9), 2543–2548.

240. Taljaard, B. and Burger, G.J. (2002) Palladium-catalysed dimerisation of furfural. *S. Afr. J. Chem.*, **55**, 56–66.

241. Kozhevnikov, I.V. (1976) Oxidative coupling of 5-membered heterocyclic compounds catalyzed by palladium(II). *React. Kinet. Catal. Lett.*, **4** (4), 451–458.

242. Itahara, T., Hashimoto, M., and Yumisashi, H. (1984) Oxidative dimerization of thiophenes and furans bearing electron-withdrawing substituents by palladium acetate. *Synthesis*, **1984** (03), 255–256.

243. Simonetti, D.A. and Dumesic, J.A. (2008) Catalytic strategies for changing the energy content and achieving C-C coupling in biomass-derived oxygenated hydrocarbons. *ChemSusChem*, **1** (8–9), 725–733.

Part V
Biorefineries and Value Chains

Chemicals and Fuels from Bio-Based Building Blocks, First Edition.
Edited by Fabrizio Cavani, Stefania Albonetti, Francesco Basile, and Alessandro Gandini.
© 2016 Wiley-VCH Verlag GmbH & Co. KGaA. Published 2016 by Wiley-VCH Verlag GmbH & Co. KGaA.

18
A Vision for Future Biorefineries

Paola Lanzafame, Siglinda Perathoner, and Gabriele Centi

18.1
Introduction

The concept of biorefinery has become quite popular but controversial at the same time recently [1–10]. Biorefinery concept is not driven from economy, at least in the traditional sense of possibility to produce at lower costs, but from an ensemble of factors: (i) geopolicy and security of supplies, (ii) preserve rural developments and recently also employment in otherwise closing traditional refineries, (iii) lower impact on environment of mobility and reduced greenhouse gases (GHGs), (iv) sustainability, and so on. These driving forces have been translated to a mix of actions such as mandate fuel mix, subsidies, renewable targets, and carbon targets to comply with which biorefineries have been fast spread over the world. Only in few cases, such as the case of bioethanol for Brazil, the economic factor is relevant but cannot explain alone the fast expansion in the production, perhaps showing a contraction recently. It is necessary to understand how these sociopolitical drivers (SPDs) [9, 10] determine the evolution in the sector, because new elements and a better understanding of the effective impact of biofuel production and use are pushing a revision of these external (e.g., not techno-economical) drivers:

- New possibilities exist today to produce energy vectors not based on fossil-fuel raw materials.
- The production of renewable energy (not based on biomass) increased far beyond expectations.
- New routes are opened for nonenergy use of biomass, higher added-value production, and rural valorization.
- Biomass transport and intensive monoculture agriculture necessary for biofuel production are bottlenecks for large biorefinery plants.
- Growing and processing aquatic biomass, which show better efficiencies and lower impact on environment with respect to intensive terrestrial production of biomass (on equivalent production of fuels), are becoming a feasible possibility to produce biofuels.

Chemicals and Fuels from Bio-Based Building Blocks, First Edition.
Edited by Fabrizio Cavani, Stefania Albonetti, Francesco Basile, and Alessandro Gandini.
© 2016 Wiley-VCH Verlag GmbH & Co. KGaA. Published 2016 by Wiley-VCH Verlag GmbH & Co. KGaA.

- Cost of fossil fuels (FFs) has been maintained to low values to make costs of second-generation biofuels from lignocellulosic unacceptable.
- The need to produce large amounts of H_2 to synthetize drop-in biofuels (hydrocarbons) from oxygen-rich biomass substrates (sugars and platform molecules deriving from biomass) has been underestimated in terms of impact.
- New models of symbiosis and value chain integration require to reconsider the model of biofuel production.

The predicted scenario of biofuel expansion and production of a decade ago has largely failed. For example, the International Energy Agency (IEA) in the recent reports remarked how the world biofuel production showed the end of rapid expansion and the growth rate is currently significantly lower than expected [11]. The target of 8% biofuel share in transport energy for year 2025 is currently revised to be within the 4–5% range, for example.

There is thus the demand to reconsider the model of biorefineries and analyze the new directions and scenario to evaluate possibilities and anticipate needs for research. As evident for an area as that of biorefineries, with large interest and investments made, there are contrasting ideas for the future and dynamics of change. It is not our aim to discuss strengths and weaknesses of these different thoughts. There are too many uncertainties and energy is a so pervasive area that multiple models will coexist simultaneously, due also to the different relevance given to drivers such as GHG emissions, energy security, agriculture production, and so on in different world regions. We thus focus discussion here on our vision to future of biorefineries, although in the general context of dynamics of evolution of this sector. The aim is to evidence new opportunities and remark area on which research can open new perspectives.

We are in a transitional economy, for example, during the transition from one to another economic cycle. This transition is characterized typically in process industry from different pushing (new raw materials, technologies, and processes) and pulling forces (new social demands, environmental forces, sustainability) [12]. One of the key elements characterizing the transition is the emerging of new companies looking at the new technologies/processes as a mechanism to gain market position through innovation. This is well evident in the area of biorefineries, with the fast explosion of new models and technologies/processes. However, often the indications appear driven more from the need to gain visibility (and raise funds) in an emerging possible market rather than from a true innovation and a clear analysis of the market perspectives. The US example of governmental push to develop hydrocarbons from biomass has led to the creation of over 100 venture capital companies, most of which now closed or having markedly turned their strategies. This lesson is limitedly recognized in Europe, where some of the ideas about new model of biorefineries and circular bioeconomy may be questioned. In Europe it was a large effort about a decade ago to promote the use of large thermal biorefineries (pyrolysis, gasification, etc.) as the best method for biomass utilization [13], but the limits of this idea are evident today.

New models of biorefineries are presented, for example, by Matrica, a joint venture between Versalis (eni group) and Novamont. The project in Porto Torres (Sardinia, Italy) plans to produce various chemicals starting from an upstream process of oxidative cleavage with ozone of vegetable oils. This plant has a planned production capacity of 35 kt per year (thus small size), with the driving element of the initiative being to decommission an old industrial site (petrochemical plant) to develop a cleaner and with better market perspective production. The social factor is thus driving the creation of the new biorefinery model strongly linked to territory and as a recovery strategy for local economy. The same concept is planned to be applied in other similar situations in Italy. This is certainly a positive element, which can also raise governmental or international funds to help the start of the project, but the question is to understand when the elements for a long-term industrial sustainability are also present. Some question mark may exist on this aspect. This example shows one of the key questions about the future of biorefineries: the mixing of different driving elements that do not allow identifying clearly the long-term sustainability and when the proposed models have a general validity or depend strongly on local situations. For this reason, a broader-view effort is necessary to assess the future of bioeconomy and to identify some general guidelines for this analysis.

There are converging ideas that the future of biorefineries will be centered on four elements:

- Smaller size (to adapt better at regional level and minimize biomass transport) and lower capital investments with short-time amortization.
- More flexible production, for example, possibility to fast switch between chemical and fuel production to adapt better to market.
- Production of higher added-value products but in an integrated full biomass use.
- Value chain integration and symbiosis.

Although the concept of biorefinery already includes the idea of biomass processing into a spectrum of products (food, feed, chemicals, fuels, and materials), it is evident that it is strongly linked with oil refinery concept: (i) large scale, (ii) relatively limited range of products with low added value, and (iii) limited flexibility in terms of products. We thus prefer to indicate this new model of biorefinery as biofactory [9, 10] to remark the aspect that chemical production is becoming the main target, with energy products (mainly higher-value additives rather than direct components for fuel blending) as a side element. This model focuses at integrating within current oil-centered refinery and chemical production, rather than to substitute it, as hidden in many current models and discussion of biorefinery and biobased economy. Oil will be still abundant in the next half century, and a successful transition to a more sustainable economy is possible only when the barrier to change (e.g., the necessary economic investments but also other elements which delay the introduction of new market products) is reduced. This requires mandatory that the new processes/technologies can be inserted readily (or with minor changes) within the current infrastructure (drop-in products and processes/technologies).

The process design should allow a flexible switch between chemical and energy products, in order to adapt better to the fast-changing market opportunities. The new model is thus not only different in terms of targets but also in size, plant design, and reference market. It is likely that old and new models will coexist, as well as a number of other biorefinery concepts (indicated as integrated biorefineries [8, 14, 15], although mixing together often gives different aspects). This differentiation between biorefineries and biofactories, however, helps to clarify the diverse objectives, visions, and plant design. New biorefinery models (to address the energy–food–water nexus) were discussed in detail also by Zhang *et al.* [1].

18.2
The Concept of Biorefinery

The concept of biorefinery is apparently simple (as defined by IEA – Bioenergy Task 42): the biorefinery is the sustainable processing of biomass into a spectrum of marketable products (food, feed, materials, chemicals) and energy (fuels, power, heat). However, the realization of this concept determines many different practical possibilities of biorefineries. According to the Joint European Biorefinery Vision for 2030 and European Biorefinery Joint Strategic Research Roadmap [16], a first classification of the different types of biorefineries can be made according to four main criteria:

- Integration of biorefining into existing industrial value chains (bottom-up approach) or development of new industrial value chains (top-down approach)
- Upstream or downstream integration
- Feedstock: biomass produced locally or imported biomass or intermediates
- Biorefinery scale.

These criteria evidence how the specific model of biorefinery to be developed is based on specific sector requirements or geographical constraints: biomass type, quality and availability, infrastructure, local industries, and so on. The choice of technology options (processes, feedstocks, location, and scale) for each biorefinery will be specific. Elements for the decisions are available industrial equipment, technological and industrial know-how or access to biomass, and so on. Biorefinery development will be driven by the demands of leading players from sectors such as the agro-industries, the forest-based industry, the energy and biofuel sectors, and the chemical industry. The integration of biorefineries into industrial value chains will be driven by either upstream players (producers and transformers of biomass) or downstream customers (producers of intermediates and final products). In principle, each biorefinery is somewhat unique even if the input and output can be formally similar. This is also what occurs in oil refineries. Nevertheless, the different biorefinery concepts could be lumped into some main classes [16], based on the type of feedstock:

- *Starch and Sugar Biorefineries*: processing starch crops, such as cereals (e.g., wheat or maize) and potatoes, or sugar crops, such as sugar beet or sugar cane.
- *Oilseed Biorefineries*: currently produce mainly food and feed ingredients, biodiesel, and oleochemicals from oilseeds such as rape, sunflower, and soybean.
- *Green Biorefineries*: processing wet biomass, such as grass, clover, and lucerne or alfalfa. They are pressed to obtain two separate product streams: fiber-rich pressed juice and nutrient-rich press cake.
- *Lignocellulosic Biorefineries*: processing a range of lignocellulosic biomass via thermochemical, chemiocatalytic, or biochemical routes.
- *Aquatic (marine) Biorefineries*: processing aquatic biomass – microalgae and seaweeds.

Some of these biorefinery concepts are schematically shown in Figure 18.1 (sugar biorefinery and lignocellulosic biorefinery – biochemical approach) and Figure 18.2 (green biorefinery and oilseed biorefinery) [9a]. The simplified schemes reported in Figures 18.1 and 18.2 differentiate between primary and secondary refining and indicate main raw materials, intermediate and final products (chemicals and energy products) as well as type of technologies.

Although these schemes are general, they evidence the complexity of treatments necessary to convert different types of biomass substrates. This complexity is much higher with respect to oil which is composed to several thousands of chemicals but which can be lumped into a limited number of classes of chemicals. In addition, each type of biomass substrate requires different procedures and technologies. As a consequence, many different models and concepts of biorefineries can be foreseen, with a large variety of technologies necessary to develop

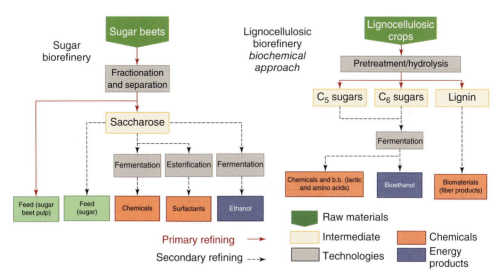

Figure 18.1 Simplified scheme of two biorefinery concepts: sugar biorefinery and lignocellulosic biorefinery (biochemical approach). (Adapted from ref. [9a].)

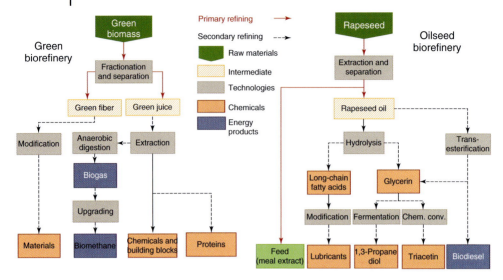

Figure 18.2 Simplified scheme of two biorefinery concepts: green biorefinery and oilseed biorefinery. (Adapted from ref. [9a].)

affecting in turn the fixed costs and profitability of the different specific routes. Only a part of these concepts will be effectively implemented on commercial scale. Sustainability is another critical element which will drive the effective implementation on a large scale of the different biorefinery concepts [17].

As remarked before, the commercial implementation of bioroutes of production appears today less promising than few years ago, for a number of reasons from economics to sustainability [9, 10]. It is thus necessary to revise the concept of biorefineries to identify priority paths and technologies.

18.3
The Changing Model of Biorefinery

One of the evolutions observed in this area is the change of the concepts of biorefinery as outlined in the section before. Although biorefinery concept is associated to multiple types of products, the current implemented models are centered on large-volume energy products, with an eventual production of chemicals as side element. We already pointed out in the introduction that new models of biorefinery will be instead focused on some keywords:

- Small scale
- Flexible production
- High added-value products
- Chemical production as core element.

The reasons is the need to integrate better at regional level and to have a production more suited to follow a fast evolving scenario, low capital investment, more efficient intensification of the processes, and overall lower impact on environment. Although some of the elements are present in the concept of biorefinery outlined in the previous section, we prefer to indicate as biofactory this new model to remark the difference [9, 10]. One of the key elements in the new models of biorefineries is also the need to integrate CO_2 reuse and introduction of renewable energy in the biorefinery value chain. These are aspects not typically considered in current biorefinery models but which are a key element to move to a sustainable, low-carbon economy [10]. Sharifzadeh *et al.* [18] showed the possibility to increase biomass to fuel conversion from 55% to 73%, when CO_2 utilization is integrated in biorefinery. In biorefineries where chemicals are the main product (biofactories), the scope is instead different, and the target is the optimal integration of CO_2 utilization within the value chain. Symbiosis with near lying factories is another emerging element characterizing the new models of bioeconomy. There are different possibilities of efficient symbiosis, but an interesting option is the use of waste from other productions (wastewater and CO_2, e.g., in advanced microalgae processes) to enhance the energy efficiency and reduce environmental impact and CO_2 emissions of a productive district. There are various other examples of possible synergies, for example, between biorefineries and fertilizers production, urea in particular. The latter finds an increasing use as selective reductant for NO_x emissions in heavy tracks and other applications, besides that as fertilizer.

There are two interesting new models of advanced biorefinery/biofactory models, which are emerging as a new opportunity [9, 10]:

1) Bioproduction of olefin and other base raw materials
2) Development of flexible production of chemicals and fuels.

The first is centered on the production of base raw materials for chemical production, while the second focuses attention on intermediate and high added-value chemical products, including monomers for polymerization, but with flexible type of production for a rapid switch to produce eventually fuel additives, depending on the market opportunities. Two of the elements characterizing both models are the full use of the biomass and process intensification (for efficient small-scale production). We have to remark that various authors, a limited selection of which with focus of catalytic processes is reported in Refs. [19–26], have discussed the production of chemicals from biomass, including some aspects related to the routes indicated previously. However, the focus of their discussion was on the technical aspects, such as the type and performances of the catalysts, rather than on the analysis of the role of the different routes in the frame of the future scenario of sustainable biofactories. We try instead to make a scenario analysis, although rather concise, of the likely dominant routes in the future bioeconomy and of the enabling factors and target objectives to realize this sustainable future.

As commented earlier, the bioeconomy is highly dependent on external factors rather than pure techno-economic considerations. It is thus necessary, in perspective analysis, to go beyond the feasibility and economic considerations only. Although they are clearly relevant, it is necessary to attempt to make a longer-term strategic evaluation. This is not a simple effort, because it is necessary to forecast from one side macroeconomic and social trends and to try to anticipate from the other side technological evolution, an aspect often critically missing in scenario analyses. One of the points shortly commented earlier is that there is a clear regionalization of the bioeconomy role in future scenario, being so strictly depending on SPDs. Therefore, in this area a deglobalization is observed. The following discussion is thus mainly focused on Europe, but different world regions (Asia, United States, etc.) may not share some of the conclusions, because other factors will drive decisions [9a].

Due to a combination of different factors (closing of various steam crackers, need of dedicated production, etc.), there is a renewed interest in producing light olefins from biobased raw materials [10, 27−34]. Although often questioned that cost is still high, it is necessary to consider the problem on a longer-term perspective. It is necessary to diversify the raw materials for chemical production to decouple from the energy production which has different dynamics, dimensions, and added values of the products. Low-carbon footprint production in Europe pushes toward a change to nonfossil-fuel sources, while preserve rural development and developed alternative routes to enhance competitiveness are further relevant elements. There are thus various motivations to consider, but economy and technical feasibility still remains relevant elements.

A balance between these aspects and how they will change in the future should be thus considered, but often this is a weakness of various estimations. There are thus some general aspects which is useful to remark. Biotechnologies have been made remarkable developments over the recent years, but productivity and separation of the mixture are two critical aspects for industrial implementation. It is possible to make engineered microorganisms to tailor the type of products obtained, but the main difference between those producing ethanol from sugars with respect to other types of products is that the strains in industrial use for bioethanol production are scarcely sensible from product inhibition, for example, they produce relatively high concentrations of ethanol, which greatly facilitate separation and enhance productivity (both key factors in costs). Most of the alternative possible products do not meet this aspect. Microorganisms operate under stressed conditions and are greatly inhibited from the products of reactions. This makes costly production and separation. Sensitivity of microorganisms to contaminants, changes in feed and operation conditions, and so on are further relevant problems. Therefore, relevant progresses have been made in terms of the possibility to control the type of products in microorganisms, but industrial implementation is still often a large issue. Separation costs in the recovery of the product, stability and flexibility in operations, quality of the product, and so on are other relevant aspects to consider.

18.3.1
Olefin Biorefinery

The analysis of the different routes in light olefin production should be thus based on the analysis of different parameters. An attempt in this direction was made by Lanzafame *et al.* [10b] and is summarized in Figure 18.3. We may identify four main lumped drivers to evaluate comparatively different possible routes to synthetize light olefins [10b]:

- *Economic drivers* (EDs), where key elements are the (i) cost of raw material versus product value, (ii) process complexity, (iii) investment necessity, and (iv) integration with other processes.
- *Technological and strategic drivers* (TSDs), composed of different aspects: (i) technological barriers to develop the process, (ii) time to market, (iii) flexibility of the process, (iv) requirements for reaction and separation steps, and (v) synergies with other process units.
- *Environmental and sustainability drivers* (ESDs), which include aspects such as (i) energy efficiency, (ii) resource efficiency, (iii) environmental impact, and (iv) GHG impact.
- *Socio-political drivers* (SPDs), where key elements are (i) social acceptance, (ii) political drivers, and (iii) public visibility.

Figure 18.3b evidences the different routes compared in this multiranking approach. Currently, light olefins are produced principally by steam cracking of naphtha or, in minor amount of natural gas, for example, using fossil fuels (FFs). This is a very energy-intensive process. The pyrolysis section of a naphtha steam cracker alone consumes approximately 65% of the total process energy and accounts for approximately 75% of the total energy loss. The specific emission factor (CO_2 Mt/Mt light olefin) depends on the starting feedstock but ranges between 1.2 and 1.8. Light olefins can be produced from different sources (crude oil, natural gas, coal, biomass and biowaste such as recycled plastics, and CO_2). Those considered in the comparative analysis are indicated in the bottom of Figure 18.3. Olefins are also a side product of the fluid catalytic cracking (FCC) process in refinery but are usually utilized inside the refinery for alkylation or oligomerization processes. The alternative processes and raw materials considered here include the dehydration of ethanol produced from biomass fermentation and the production via syngas (through the intermediate synthesis of methanol), with the syngas deriving from biomass pyrolysis/gasification. Methanol can be converted to olefins (MTOs – methanol to olefins) or even selectively to propylene (MTP – methanol to propylene). Olefins can be interconverted, for example, via metathesis reaction. New routes under development are based on the Fischer–Tropsch to olefin (FTO) reaction, by using syngas or directly CO_2/H_2 mixtures. To make sustainable the process, H_2 should be produced using renewable energy sources (rH_2). The possibility of using organic waste such as discarded plastics, used rubber, and so on via recycling pyrolysis is also considered. The potential advantage is to avoid the costs of disposal of

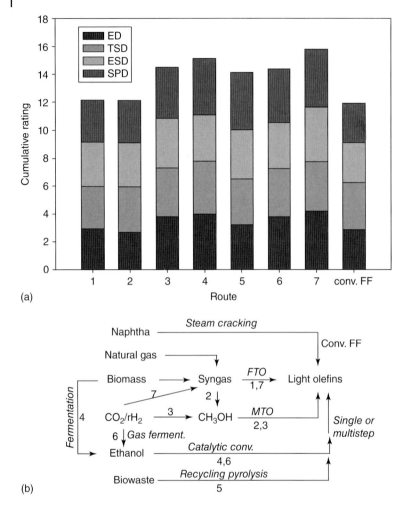

(a)

(b)

Figure 18.3 (a) Multicriteria analysis and ranking (see text for description of the parameters ED, TSD, ESD, and SPD) of different routes to produce olefins in relation to the future scenario for sustainable chemical production. (b) Indication of the different routes analyzed with respect to conventional naphtha steam cracking (conversion of fossil fuels (conv. FFs)). (Adapted from ref. [10b].)

these wastes, but at the same time the use of waste does not guarantee a constant feed composition. There are also problems in terms of process and separation costs, as well purification. One of the routes considered is based on the first step of reverse water–gas shift (RWGS) from CO/rH_2 mixtures, followed by a gas fermentation of the $CO/H_2/CO_2$ mixture to produce ethanol. LanzaTech has developed already some semicommercial units for the second step of ethanol production, although productivity is still limited and ethanol has to be recovered from solution. Ethanol could be then dehydrated to ethylene. Renewable H_2 could be derived also from processes using microorganisms, such as cyanobacteria, which are able to produce H_2 using sunlight. The last process considered is the conversion of CO_2/rH_2 mixtures by FTO process. This case assumes that a yield to C_2-C_4 olefins >75% could be obtained. Although current data are still lower, this yield seems a reasonable target that can be reached [35]. These routes are compared with steam cracking of naphtha, the current most used process in Europe to produce light olefins. This route is considered as reference for the olefin production by the conversion of fossil fuel (conv. FF).

The comparative assessment reported in Figure 18.3 evidences that the use of alternative raw materials to produce light olefins is worthwhile (e.g., a higher cumulative rating) with respect to conventional (actual) route from fossil fuels. In particular, this is even more interesting in small- to medium-scale (dedicated) productions, because economics for steam cracking are less favorable for this type of productions. There are thus motivations to develop novel sustainable olefin production routes.

Ethanol to ethylene is currently the main practical possibility [31]. The process is utilized by Braskem, Dow, and Solvay Indupa in Brazil. Industrial catalysts for this reaction are mainly based on zeolites, which should be doped to tune acidity. Selectivity is already quite high (over 98%), but an improvement of space yields is necessary. The transformation of bioethanol to C_3 and C_4 olefins/diolefins via the so-called Lebedev process (oxidative dimerization to butadiene) is more challenging, for example, the selectivities are still not enough, even if good [30, 32]. Propylene may be instead produced from ethylene via metathesis [36]. Butadiene can be also produced from biobutanol and 2,3-butandiol obtained by fermentation [32]. Propylene could be synthetized from 1- or 2-propanol or from 1,2-propanediol [37], the latter being obtained by catalytic conversion of sugars or other platform molecules (glycerol, lactic acid). The chemiocatalytic route is preferable over the fermentation paths [38], because the production by fermentation of alcohols and glycols other than ethanol has still too-low productivities, being the enzyme sensitive to inhibition by product concentrations and operating under stressed conditions.

There are various companies quite active in this area [10b], for example:

- Polyol Chemicals, Inc. (propylene and ethylene glycols from sorbitol/glucose).
- LanzaTech (2,3-butanediol by gas fermentation and in joint venture with INVISTA to produce butadiene).
- Versalis in joint venture with Genomatica to produce butadiene via 2,3-butandiol obtained by fermentation. Genencor in joint venture with Goodyear is developing a bioisoprene production process.
- Glycos Biotechnologies is developing a process to produce isoprene from crude glycerine.
- Global Bioenergies and Gevo/Lanxess are developing processes to produce isobutene from glucose or from isobutanol, respectively. The latter is produced by fermentation.

An alternative possibility is to produce syngas from biomass and convert methanol to olefins: MTO process or variations on this process such as MTP. MTP differentiates from MTO for the type of zeolite catalyst and operative conditions used [39]. The processes starting from syngas are well established and operate on a large scale, although starting on coal or other sources rather than biomass. Various issues related to gas purification, biomass variability, transport cost, and so on are present in using biomass as raw material biomass-to-olefin (BTO) [28]. The direct conversion of syngas to olefins (FTO) is alternative to the methanol route [28, 40], but selectivity and yields in olefins should be improved. Notwithstanding the past interest and development up to pilot units, the thermal routes to olefins via syngas are less promising than those based on fermentation/chemiocatalysis (low temperature).

18.3.2
Biorefinery for Flexible Production of Chemicals and Fuels

The second new emerging model of biofactories to discuss shortly, in addition to the bioproduction of olefin discussed previously, is that of the development of flexible production of chemicals and fuels. A relevant example regards the use of 5-hydroxymethylfurfural (5-HMF) as platform molecule [41–43]. There are many questions regarding whether or not this compound could be a true industrial platform molecule. In fact, in water the further conversion to levulinic acid (LA) (plus formic acid) occurs at high conversion, and thus several authors consider LA as the platform on which to build a process [44, 45]. While this is true in water, the use of organic solvents changes significantly the situation [42]. In dimethylsulfoxide (DMSO), dimethylformamide (DMF), and dimethylacetamide (DMA), yields >90% from fructose are possible. The use of biphasic systems with continuous extraction of HMF allows to obtain yields even higher, and the process is on commercial scale [42, 45]. Heterogeneous catalysts such as Amberlyst 15 catalysts allow to obtain nearly 100% yield at 120 °C/2 h from fructose. However, performances with glucose or cellulose are still unsatisfactory. Today the market incentives to start from these raw materials are limited, because fructose is available in relatively large amounts. For a future sustainability of the process, the direct use of cellulose will be important, particularly when alternative sources to terrestrial biomass could be used, for example, aquatic biomass (macroalgae).

HMF offers a wider range of possibilities as platform molecule. Some of the key possibilities to produce either chemicals or fuels are presented in Figure 18.4. It is worthwhile to mention that some of these routes may be realized in a flexible way essentially with the same type of equipments. This is the case of the oxidation to 2,5-furandicarboxylic acid (2,5-FDCA) (see later) and the etherification/coupling to produce diesel additives. This is an example of the possible process intensification in the area, because this may be designed as an industrial production process flexibly producing from the same raw materials chemicals or fuels, depending on the market demand. The flexibility is an added value in this type of production. Another example of process intensification in this area will be discussed later.

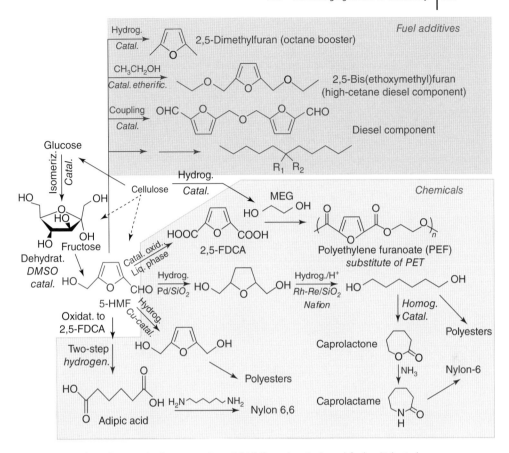

Figure 18.4 Selected routes in the conversion of 5-HMF to chemicals and fuels. (Adapted from ref. [9a].)

The other relevant element in process design is that all types of by-product formed from the initial raw material should be valorized. Humins are formed as by-product in the fructose dehydration to 5-HMF. They are derived from acid-catalyzed reactions involving sugars and derivate [46]. Although humins form in minor amount, their utilization is relevant in overall process economics. However, there are few studies in this direction, particularly from an industrial perspective [47, 48]. Another by-product is furfural derived from C_5 sugars [44]. Some possibilities are derived from the utilization of CO_2, as commented later. To note that this C_5 intermediate (furfural) could be converted to 2,5-FDCA (C_6 chemical), but a complex multistep procedure is necessary [49]: (i) catalytic oxidation of furfural into furoate, (ii) catalytic disproportionation of furoate to furan and 2,5-furandicarboxylic acid, and (iii) separation of the products. Furan can be then converted to 1,4-butanediol (1,4-BDO) by catalytic hydrogenation and hydrolysis and the latter polymerized with 2,5-FDCA to polyester poly(butylene 2,5-furandicarboxylate) (PBF). The best results of catalytic disproportionation of furoate are a yield to 2,5-FDCA of about 52% [49]. There is thus the need to improve the performances in this key step to at least over 80%, but the key issue is the complexity of the operations which lead to high production cost.

2,5-FDCA is one of the interesting products which can be obtained from 5-HMF. It is produced by catalytic oxidation of HMF with noble metal-based (Au, Pt, Ru) catalysts in liquid phase [50–52]. To maximize yields and limit side reactions, it is preferable to work with the ester equivalent of both the reactant and product. Over 90% yields are possible in industrial conditions, better than often reported in literature [42].

2,5-FDCA is polymerized with another monomer obtained from biobased raw materials (ethylene glycol, MEG) (see Figure 18.4) to produce polyethylene furanoate (PEF), a substitute to polyethylene terephthalate (PET), one of the largest volume polymers used for bottles, films, carpets, and textiles. PET is a 100% biobased substitute of PET and is one of the solutions to replay to consumer demand for "greener" products. In addition, PEF has the advantage over PET of better properties, such as a higher resistance to oxygen diffusion. It may be observed that the two monomers for PEF synthesis (2,5-FDCA and MEG) are derived both from sugars but one via oxidation processes and one via hydrogenation processes. An interesting approach for process intensification is thus to combine these two reactions in a single electrocatalytic device. This new direction, which involves the need to design and develop new type of electrodes and electrocatalytic devices, is explored by a new EU project (TERRA) in the frame of process intensification SPIRE calls of H2020. This example remarks how a combination between advances in technologies, market opportunities, and product properties can make successful new biobased production route and in turn proceed along the indicated future opportunities for a biobased economy. Some examples of using electrocatalysis in this area have started to be published [53, 54], but it is a largely unexplored emerging scientific area.

It should be also remarked that there are various large-volume chemicals, which can be potentially replaced by FDCA:

- Terephthalic acid in the production of PET, PBT, and polyamides
- Bisphenol A in polycarbonates
- Adipic acid in polyester polyols and plasticizers
- Phthalic anhydride in polyester polyols and plasticizers
- Isophthalic acid in the production of modified PET.

In addition, FDCA finds potential interesting application also in the area of polyamide. The opening of new routes of FDCA production and utilization can have then a larger positive impact in moving to biobased monomers for polymerization. These biobased polymers have a positive and relevant impact on low-carbon bioeconomy. For example, PEF production from corn-based fructose reduces about 40–50% the nonrenewable energy use and 45–55% GHG emissions compared to PET, in LCA cradle to grave estimations [55]. These reductions are higher than for other biobased plastics, such as polylactic acid (PLA) or polyethylene (PE).

HMF is a quite versatile platform molecule and finds potential application in the synthesis of other large-scale monomers, although with economics and technologies still to improve [42, 56] (Figure 18.4):

- Caprolactam [56] (commercially produced in a multistep process from benzene; it is the monomer for nylon-6). HMF hydrogenation yields selectively (>99%) 2,5-bishydroxymethyl-tetrahydrofuran, which can be then further hydrogenated to 1,6-hexanediol (86% yield), by using a Rh−Re/SiO$_2$ catalyst in the presence of Nafion. The diol can be then converted into caprolactone using Oppenauer oxidation catalyzed by a ruthenium complex yield. Overall selectivity to caprolactone is about 86% [55]. The critical step is the conversion of ε-caprolactone into caprolactam, requiring ammonia at 170 bar and 300−400 °C.
- 1,6-Hexanediol [57, 58] (commercially prepared by hydrogenation of adipic acid; it is widely used for industrial polyester and polyurethane production).
- Adipic acid [59] (commercially produced in a multistep process from benzene; it is the monomer for nylon-66).
- Adipic acid may be produced by hydrogenation of FDCA in a two-step process [60]: (i) FDCA hydrogenation to 2,5-tetrahydrofuran-dicarboxylic acid (88% yield at 140 °C for 3 h in acetic acid, catalyzed by Pd on silica) and (ii) further hydrogenation of the latter compound to adipic acid (yields up to 99% at 160 °C for 3 h in acetic acid in the presence 0.2 M of HI and 5% Pd on silica catalyst).

There are some alternative biobased routes to produce adipic acid developed by companies such as Rennovia (the HMF-based process discussed earlier), Verdezyne, BioAmber, Celexion, and Genomatica. Verdezyne uses genetically modified enzymes to ferment glucose directly to adipic acid. Rennovia converts catalytically glucose to glucaric acid, followed by hydrodeoxygenation to convert glucaric acid to adipic acid. Commercial perspectives in these processes are interesting but strongly depend on the cost of glucose feedstock. Techno-economic feasibility should be further demonstrated, as well as sustainability.

Although likely some of these processes will not enter into commercialization, if not significant advances in productivity and cost reduction can be made (e.g., the production of caprolactam from HMF), these examples evidence the concept that a single successful production may open the possibility to a larger set of opportunities, when proper choice of platform molecules is made. We have to remark that several platform molecules have been proposed in the past and even recently [25, 61−63], but likely only a part of these will be used on a large scale, because often they are missing the techno-economic conditions. Differently in part to conclusions of the cited DOE report "Top Value-Added Chemicals from Biomass" [63] and other recent reviews [25, 61−69], we believe that the main role will be played by the use of simple alcohols such as methanol/ethanol, acetone, or higher alcohols produced using biochemical routes, in addition to a limited number of platform molecules (HMF/LA, succinic acid, 3-HPA; some of them were shortly discussed previously).

18.4

Integrate CO_2 Use and Solar Energy within Biorefineries

Previous sections have introduced the concept that the integration of CO_2 utilization within bioeconomy value chain is one of the challenges for the future sustainable bioeconomy and to move to a low-carbon bioeconomy. It is also the preferential path to introduce renewable energy within the process industry [40, 70]. For conciseness, we limit the discussion here to some examples of integration of CO_2 within biorefinery schemes, but there are many other possibilities, as well as other routes of valorization of CO_2 [71, 72], for example, to produce polycarbonate or other polymers.

One of the aspects to remark is that these routes have the potential of a significant contribution to the reduction of GHG emissions, in contrast to what often remarked that CO_2 utilization does not has a relevant impact, being a minor volume production with respect to the amount of CO_2 produced in energy-related processes. This idea is erroneous, because it does not consider that CO_2 use not only avoid directly emissions but introduce renewable energy in the energy/chemical chain and avoids the use of fossil fuels. A detailed discussion is reported elsewhere [73]. It is thus not the amount used, but the global effect to consider. It is also erroneous to consider long-term storage, because every time CO_2 is used to form energy vectors through the use of renewable energy, the latter is introduced in the energy system (deduced from the energy losses in the process). Thus short cycles of CO_2 conversion and use of the derived energy vectors have a much higher cumulative effect over a certain timescale (e.g., 20 years) than CO_2 sequestration or use in long-term sequestration products (e.g., mineral carbonate).

The use of CO_2 to establish on world scale the trade of renewable energy (via conversion to methanol realized with the use of unexploited renewable energy sources) has a potential impact equal to carbon sequestration or the use of biofuels but at a lower cost [74]. This section evidences how CO_2 reuse is a facilitator for bioeconomy and in this sense boosts the transition to a low-carbon economy [10b]. Therefore, it is necessary to evaluate the impact from these perspectives and not limited to the specific amount of CO_2 which can be used.

The production of urea, which is derived from the reaction of two molecules of NH_3 with CO_2, can be used to explain further the concepts indicated in the preceding text. Urea is by far the larger consumer of CO_2 (over hundred millions of tons), but the latter is derived from the production of H_2 necessary for ammonia synthesis. Producing ammonia using H_2 that is not derived from fossil fuels may reduce up to a factor of about 9 the CO_2-equivalent emissions (on LCA bases) with respect to conventional process based on natural gas. Two molecules of ammonia are needed for each molecule of CO_2, and in addition, when ammonia is derived from nonfossil fuels, CO_2 derived also from other sources can be utilized. The global reduction is thus one order of magnitude higher with respect to the production of urea itself. The production of urea 100% from nonfossil-fuel sources thus amplifies by one order of magnitude the effect on mitigation of CO_2-equivalent emissions with respect to the amount of CO_2 utilized. On the contrary, the impact of carbon sequestration and storage (CSS) technologies is about half the amount of CO_2 utilized, because it is necessary to account the energy necessary to capture and sequestrate carbon dioxide. CO_2 utilization may thus largely enhance the effect accounted from the simple amount of CO_2 utilized.

Another common question regards the availability of CO_2 and the need to recover from flue gas. This is not necessary, at least in an initial phase. We can consider two main sources of nearly pure CO_2 in biobased production: (i) deriving from fermentation (about 1 ton CO_2 per ton bioethanol is produced in sugar fermentation) [75] and (ii) associated to biogas. Both sources emit rather pure CO_2 (>99%) and a virtual zero cost. In Europe, it may be estimated that about $10-15$ Mt nearly pure CO_2 is available from these bioresources. This amount is rapidly increasing, due to the expansion in the fermentation and biogas production processes.

There are some main options to integrate solar (renewable) energy within a biorefinery/biofactory scheme:

- Use of this energy to produce chemicals (or intermediates needed to produce the final chemicals, such as H_2), either locally (in particular, exploiting the electrical energy which is in excess to that which can be introduced to the grid) or remote, where unexploited resources can be utilized.
- Use of light-active microorganisms. There are two possibilities. The first regards synthetic enzymatic pathways for CO_2 utilization powered by electricity, as proposed, for example, by Zhang *et al.* [76, 77], and the second regards the use of microalgae [78–80].

Although still under debate, microalgae may be considered a future solution to increase sustainability of chemical/energy production. Microalgae have photosynthesis efficiency about one order of magnitude higher than plants. There are already large-scale microalgae cultivations (in open ponds) in operations by many companies, and even if still various improvements are necessary to make economically profitable the process, these are good prospective that technologies under actual development (particularly, for oil recovery from wet biomass and full biomass utilization) will drive costs to exploitable region. Therefore, even if issues to solve still remain, there are many indications about the sustainability of microalgae process. One of the challenges is to analyze the opportunities offered by its integration within the biobased production scheme [81] and how the use of CO_2 and wastewater streams to grow the microalgae can combine improved economics, lower GHG emissions, and improved water and waste management. This concept of symbiosis bioproduction defines the possibility to symbiosis operations between biorefineries/biofactories and other nearby industries. A simple option is the use of wastewater (and/or CO_2) produced in these companies to operate a microalgae unit. However, a range of different other situations are possible, from the utilization of by-products and their introduction in the biobased production value chain to exporting some of the products/by-products. Industrial symbiosis is a concept of increasing relevance, and the integration of bio- and solar refineries offers new possibilities to proceed in this direction.

There are various additional options for using CO_2 within biofactories to make added-value chemicals [82, 83]. Between these routes, the following may be cited:

- The production of succinic acid from glucose and CO_2 via biotechnological routes; various companies (BASF, DSM, Rochette, BioAmber, Purac, etc.) are working on pilot-/demo-scale production of biosuccinic acid. Succinic acid can be then hydrogenated to butanediol/tetrahydrofuran. Global succinic acid market is expected to reach about 400 M€ by 2016–2017. The applications for succinic acid expected to grow in the near future are plasticizers, polyurethanes, bioplastics, and chemical intermediates.
- The production of acetone from CO_2 and H_2 also via enzymatic routes (Evonik process); acetone, together with ammonia, methane, and methanol is a raw material for the Evonik process to produce methyl methacrylate (MMA) via acetone cyanohydrin. It is thus part of the general strategy to produce 100% fossil-fuel-free MMA. Differently from succinic acid, this reaction and the full production of "renewable" MMA is still quite far from application.
- The production of high added-value chemicals from the reaction of CO_2 (or products of its conversion) with some of the by-products of biofactory. An example is the catalytic oxidative esterification of by-product furfural, derived from dehydration of pentoses to form methyl-2-furoate (a perfuming agent), or the methylation to form methyl furfural (a fragrance agent in alcohol beverages) or the methylation on ketonic position 2-acetylfuran (a flavoring agent in bakery). Another example is instead the direct use of CO_2 for the carboxylation of biomass products. Although the use of carboxylation reaction to prepare a large variety of chemicals is known [84], but starting from activated substrates, less info are available for the direct use of by-products from biobased production. For example, lignin carboxylation leads to interesting resins/adhesives/binders, but this reaction is investigated very limitedly.

These few examples give an idea of the different options available to integrate utilization of CO_2 within biofactories, with the choice depending on the specific market targets for the biofactory. We may remark that also many possibilities arise from the use of bioelectrodes (and related processes) within biorefineries [85, 86]. This is an area largely unexplored but of great relevance for the future. In general, integration of electro- and photocatalytic routes within biorefinery/biofactory production schemes is still challenging [87–93] but can open rather interesting new opportunities.

18.5
Conclusions

The concept of biorefineries is evolving, because the nexus between chemistry and energy is changing driven from the demand for a sustainable energy and chemistry production. We have discussed here some of the possibilities for new sustainable biorefineries, with focus on the production of olefins from biomass sources, the novel concept of biofactory as opposed to that of biorefinery, the possibility of exploitation and valorization of CO_2 within biorefineries, and the opportunities to integrate renewable energy sources in the biorefinery production (solar biorefineries). Only some glimpses are given, because the aim was not a systematic state of the art, because this is an emerging scientific area, without established directions. Nevertheless, we believe that it is necessary to look at the priorities and emerging directions and issues to develop the science necessary to enable this sustainable future. To achieve this target it is necessary to have a vision and discuss the possible new scenario in biorefineries. Although this is in part a personal view and different ideas may be present, its critical analysis may help in stimulating the analysis of the pro/cons of the different options going beyond the only technical considerations. For this reason, we have highlighted some of the possible new models (olefin and intermediate/high added-value chemical production), the options to integrate CO_2 and solar energy utilization within bioeconomy, the opportunities for industrial symbiosis. Ten years ago these possibilities were considered unrealistic; today they begin to be considered a possibility even at industrial level, and we hope that 10 years from now, they will be implemented. However, science should be advanced and it is thus necessary to investigate now what could be only later implemented. It is thus necessary to have a broader and long-term view, which in this specific area is more difficult being many factors (beyond the techno-economic ones) determining the use or not of some specific routes.

There is thus a changing panorama for biobased production, with new opportunities offered by the integrated utilization of CO_2 and renewable energy to move to a sustainable and low-carbon bioeconomy. There are technologies to develop to enable this future, and legislative regulation and incentives should better recognize the possibilities offered. However, this is a clear path necessary to follow for substituting fossil fuels and to develop a low-carbon society.

Acknowledgments

The authors acknowledge the PRIN10-11 projects "Mechanisms of activation of CO_2 for the design of new materials for energy and resource efficiency" and "Innovative processes for the conversion of algal biomass for the production of jet fuel and green diesel" for the financial support and the EU IAPP project nr 324292 BIOFUR (BIOpolymers and BIOfuels from FURan based building blocks), in the frame of which part of this work was realized.

References

1. (a) Zhang, Y.-H.P. (2013) *Energy Sci. Eng.*, **1**, 27; (b) Zhang, Y.-H.P. (2011) *Process Biochem.*, **46**, 2091; (c) Chen, H.-G. and Zhang, Y.-H.P. (2015) *Renewable Sustainable Energy Rev.*, **47**, 117.

2. Kamm, B. (2014) *Pure Appl. Chem.*, **86**, 821.

3. Morais, A.R.C., da Costa Lopes, A.M., and Bogel-Lukasik, R. (2015) *Chem. Rev.*, **115**, 3.

4. McCormick, K. (2014) *Biofuels*, **5**, 191.

5. Ioanna, V. and Posten, C. (2014) *Biotechnol. J.*, **9**, 739.

6. Kazmi, A., Kamm, B., Henke, S., Theuvsen, L., and Hoefer, R. (2012) *RSC Green Chem. Ser.*, **14**, 1.

7. Waldron, K.W. (ed.) (2014) *Advances in Biorefineries. Biomass and Waste Supply Chain Exploitation*, Woodhead Publishing.

8. Stuart, P.R. and El-Halwagi, M.M. (eds) (2012) *Integrated Biorefineries: Design, Analysis, and Optimization*, CRC Press.

9. (a) Lanzafame, P., Centi, G., and Perathoner, S. (2014) *Catal. Today*, **234**, 2; (b) Centi, G., Lanzafame, P., and Perathoner, S. (2011) *Catal. Today*, **167**, 14.

10. (a) Centi, G. and Perathoner, S. (2014) *J. Chin. Chem. Soc.*, **61**, 712; (b) Lanzafame, P., Centi, G., and Perathoner, S. (2014) *Chem. Soc. Rev.*, **43**, 7562.

11. International Energy Agency (IEA) (2014) *Renewable Energy Medium-Term Market Report 2014*, IEA, Paris.

12. Cavani, F., Centi, G., Perathoner, S., and Trifiro, F. (2009) *Sustainable Industrial Chemistry: Principles, Tools and Industrial Examples*, Wiley-VCH Verlag GmbH, Weinheim.

13. Haro, P., Ollero, P., Villanueva Perales, A.L., and Vidal-Barrero, F. (2013) *Biofuels, Bioprod. Biorefin.*, **7**, 551.

14. Kamm, B., Gruber, P.R., and Kamm, M. (2010) *Biorefineries – Industrial Processes and Products: Status Quo and Future Directions*, Wiley-VCH Verlag GmbH.

15. Aresta, M., Dibenedetto, A., and Dumeignil, F. (2012) *Biorefinery: From Biomass to Chemicals and Fuels*, de Gruyter.

16. Luguel, C. (2011) Joint European Biorefinery Vision for 2030, and European Biorefinery Joint Strategic Research Roadmap/Star-Colibri, http://www.star-colibri.eu/publications (accessed 27 September 2015).

17. Zinoviev, S., Müller-Langer, F., Das, P., Bertero, N., Fornasiero, P., Kaltschmitt, M., Centi, G., and Miertus, S. (2010) *ChemSusChem*, **3**, 1106.

18. Sharifzadeh, M., Wang, L., and Shah, N. (2015) *Renewable Sustainable Energy Rev.*, **47**, 151.

19. Wang, K., Ou, L., Brown, T., and Brown, R.C. (2015) *Biofuels, Bioprod. Biorefin.*, **9**, 190.

20. Ma, R., Xu, Y., and Zhang, X. (2015) *ChemSusChem*, **8**, 24.

21. Wang, J., Xi, J., and Wang, Y. (2015) *Green Chem.*, **17**, 737.

22. Jacobs, P.A., Dusselier, M., and Sels, B.F. (2014) *Angew. Chem. Int. Ed.*, **53**, 8621.

23. Sousa-Aguiar, E.F., Appel, L.G., Zonetti, P.C., Fraga, A.d.C., Bicudo, A.A., and Fonseca, I. (2014) *Catal. Today*, **234**, 13.

24. FitzPatrick, M., Champagne, P., Cunningham, M.F., and Whitney, R.A. (2010) *Bioresour. Technol.*, **101**, 8915.

25. Climent, M.J., Corma, A., and Iborra, S. (2014) *Green Chem.*, **16**, 516.

26. Alonso, D.M., Wettstein, S.G., and Dumesic, J.A. (2013) *Green Chem.*, **15**, 584.

27. Raju, S., Moret, M.-E., and Klein Gebbink, R.J.M. (2015) *ACS Catal.*, **5**, 281.

28. Torres Galvis, H.M. and de Jong, K.P. (2013) *ACS Catal.*, **3**, 2130.

29. Van de Vyver, S., Geboers, J., Jacobs, P.A., and Sels, B.F. (2011) *ChemCatChem*, **3**, 82.

30. Angelici, C., Weckhuysen, B.M., and Bruijnincx, P.C.A. (2013) *ChemSusChem*, **6**, 1595.

31. Zhang, M. and Yu, Y. (2013) *Ind. Eng. Chem. Res.*, **52**, 9505.

32. Makshina, E.V., Dusselier, M., Janssens, W., Degrève, J., Jacobs, P.A., and Sels, B.F. (2014) *Chem. Soc. Rev.*, **43**, 7917.

33. Dutta, S. (2012) *ChemSusChem*, **5**, 2125.

34. Chieregato, A., Velasquez Ochoa, J., Bandinelli, C., Fornasari, G., Cavani, F., and Mella, M. (2015) *ChemSusChem*, **8**, 377.

35. Centi, G., Iaquaniello, G., and Perathoner, S. (2011) *ChemSusChem*, **4**, 1265.

36. Behkish, A., Wang, S., Candela, L., and Ruszkay, J. (2010) *Oil Gas*, **36**, 29.

37. Rodriguez, B.A., Stowers, C.C., Pham, V., and Cox, B.M. (2014) *Green Chem.*, **16**, 1066.

38. Marinas, A., Bruijnincx, P., Ftouni, J., Urbano, F.J., and Pinel, C. (2015) *Catal. Today*, **239**, 31.

39. Gupta, R.P., Tuli, D.K., and Malhotra, R.K. (2011) *Chem. Ind. Dig.*, **24**, 86.

40. Centi, G., Quadrelli, E.A., and Perathoner, S. (2013) *Energy Environ. Sci.*, **6**, 1711.

41. Teong, S.P., Yi, G., and Zhang, Y. (2014) *Green Chem.*, **16**, 2015.

42. van Putten, R.-J., van der Waal, J.C., de Jong, E., Rasrendra, C.B., Heeres, H.J., and de Vries, J.G. (2013) *Chem. Rev.*, **113**, 1499.

43. Wang, T., Nolte, M.W., and Shanks, B.H. (2014) *Green Chem.*, **16**, 548.

44. Lange, J.-P., van der Heide, E., van Buijtenen, J., and Price, R. (2012) *Chem-SusChem*, **5**, 150.

45. Saha, B. and Abu-Omar, M.M. (2014) *Green Chem.*, **16**, 24.

46. Rasmussen, H., Soerensen, H.R., and Meyer, A.S. (2014) *Carbohydr. Res.*, **385**, 45.

47. Hoang, T.M.C., van Eck, E.R.H., Bula, W.P., Gardeniers, J.G.E., Lefferts, L., and Seshan, K. (2015) *Green Chem.*, **17**, 959.

48. Hoang, T.M.C., Lefferts, L., and Seshan, K. (2013) *ChemSusChem*, **6**, 1651.

49. Pan, T., Deng, J., Xu, Q., Zuo, Y., Guo, Q.-X., and Fu, Y. (2013) *ChemSusChem*, **6** (1), 47–50.

50. Ardemani, L., Cibin, G., Dent, A., Isaacs, M., Lee, A.F., Parlett, C.M.A., Parry, S.A., and Wilson, K. (2015) *Chem. Sci.* **6**, 4941.

51. Ait Rass, H., Essayem, N., and Besson, M. (2015) *ChemSusChem*, **8**, 1206.

52. Albonetti, S., Pasini, T., Lolli, A., Blosi, M., Piccinini, M., Dimitratos, N., Lopez-Sanchez, J.A., Morgan, D.J., Carley, A.F., Hutchings, G.J., and Cavani, F. (2012) *Catal. Today*, **195**, 120.

53. Chadderdon, D.J., Xin, L., Qi, J., Qiu, Y., Krishna, P., More, K.L., and Li, W. (2014) *Green Chem.*, **16**, 3778.

54. Kwon, Y., Birdja, Y.Y., Raoufmoghaddam, S., and Koper, M.T.M. (2015) *Chem-SusChem*, **8**, 1745.

55. Eerhart, J.J.E., Faaij, A.P.C., and Patel, M.K. (2012) *Energy Environ. Sci.*, **5**, 6407.

56. Buntara, T., Noel, S., Phua, P.H., Melian-Cabrera, I., de Vries, J.G., and Heeres, H.J. (2011) *Angew. Chem. Int. Ed.*, **50**, 7083.

57. Tuteja, J., Choudhary, H., Nishimura, S., and Ebitani, K. (2014) *ChemSusChem*, **7**, 96.

58. Buntara, T., Noel, S., Phua, P.H., Melian-Cabrera, I., de Vries, J.G., and Heeres, H.J. (2012) *Top. Catal.*, **55**, 612.

59. Van de Vyver, S. and Roman-Leshkov, Y. (2013) *Catal. Sci. Technol.*, **3**, 1465.

60. Boussie, T.R., Dias, E.L., Fresco, Z.M., and Murphy, V.J. (2010) International Patent WO 2010144873, Assigned to Rennovia, Inc. Production of adipic acid and derivatives from carbohydrate-containing matrials.

61. Luterbacher, J.S., Martin Alonso, D., and Dumesic, J.A. (2014) *Green Chem.*, **16**, 4816.

62. Serrano-Ruiz, J.C., Luque, R., and Sepulveda-Escribano, A. (2011) *Chem. Soc. Rev.*, **40**, 5266.

63. Werpy, T. and Petersen, G. (2004) *Top Value Added Chemicals from Biomass*, US Department of Energy.

64. Pinazo, J.M., Domine, M.E., Parvulescu, V., and Petru, F. (2015) *Catal. Today*, **239**, 17.

65. Vennestrom, P.N.R., Osmundsen, C.M., Christensen, C.H., and Taarning, E. (2011) *Angew. Chem. Int. Ed.*, **50**, 10502.

66. Gerssen-Gondelach, S.J., Saygin, D., Wicke, B., Patel, M.K., and Faaij, A.P.C. (2014) *Renewable Sustainable Energy Rev.*, **40**, 964.

67. Gallezot, P. (2011) *Catal. Today*, **167**, 31.

68. Gallezot, P. (2010) *Top. Catal.*, **53**, 1209.

69. Hill, K. and Hoefer, R. (2009) *RSC Green Chem. Ser.*, **4**, 167.

70. Perathoner, S. and Centi, G. (2014) *ChemSusChem*, **7**, 1274.

71. Ampelli, C., Perathoner, S., and Centi, G. (2015) *Philos. Trans. R. Soc. London, Ser. A: Math. Phys. Eng. Sci*, **373**, 20140177.

72. Quadrelli, E.A., Centi, G., Duplan, J.-L., and Perathoner, S. (2011) *Chem-SusChem*, **4**, 1194.

73. Centi, G. and Perathoner, S. (2014) in *Green Carbon Dioxide* (eds G. Centi and S. Perathoner), John Wiley & Sons, Inc., p. 1, Chapter 1.

74. Barbato, L., Centi, G., Iaquaniello, G., Mangiapane, A., and Perathoner, S. (2014) *Energy Technol.*, **2**, 453.

75. Centi, G. and Perathoner, S. (2013) in *The Role of Catalysis for the Sustainable Production of Bio-Fuels and Biochemicals*, Chapter 16 (eds K. Triantafyllidis, A. Lappas, and M. Stöcker), Elsevier Science, p. 529.

76. Zhang, Y.-H.P. and Huang, W.-D. (2012) *Trends Biotechnol.*, **30**, 301.

77. Zhang, Y.-H.P., Xu, J.-H., and Zhong, J.-J. (2013) *Int. J. Energy Res.*, **37**, 760.

78. Razzak, S.A., Hossain, M.M., Lucky, R.A., Bassi, A.S., and de Lasa, H. (2013) *Renewable Sustainable Energy Rev.*, **27**, 622.

79. Wu, Y.-H., Hu, H.-Y., Yu, Y., Zhang, T.-Y., Zhu, S.-F., Zhuang, L.-L., Zhang, X., and Lu, Y. (2014) *Renewable Sustainable Energy Rev.*, **33**, 675.

80. Vanthoor-Koopmans, M., Wijffels, R.H., Barbosa, M.J., and Eppink, M.H.M. (2013) *Bioresour. Technol.*, **135**, 142.

81. Uggetti, E., Sialve, B., Trably, E., and Steyer, J.-P. (2014) *Biofuels, Bioprod. Biorefin.*, **8**, 516.

82. Centi, G. and Perathoner, S. (2014) *Green Carbon Dioxide: Advances in CO_2 Utilization*, John Wiley & Sons, Inc..

83. De Falco, M., Iaquaniello, G., and Centi, G. (2013) *CO_2: A Valuable Source of Carbon*, Springer.

84. Maeda, C., Miyazaki, Y., and Ema, T. (2014) *Catal. Sci. Technol.*, **4**, 1482.

85. Walcarius, A., Minteer, S.D., Wang, J., Lin, Y., and Merkoci, A. (2013) *J. Mater. Chem. B: Mater. Biol. Med*, **1**, 4878.

86. Tran, P.D., Artero, V., and Fontecave, M. (2010) *Energy Environ. Sci.*, **3**, 727.

87. Centi, G., Ampelli, C., Genovese, C., Marepally, B.C., Papanikolaou, G., and Perathoner, S. (2015) *Faraday Discuss.* **183**, 125–145

88. Lu, Q., Rosen, J., and Jiao, F. (2015) *ChemCatChem*, **7**, 38.

89. Qiao, J., Liu, Y., Hong, F., and Zhang, J. (2014) *Chem. Soc. Rev.*, **43**, 631.

90. Chen, D., Zhang, X., and Lee, A.F. (2015) *J. Mater. Chem. A: Mater. Energy Sustainability.* **3**, 14487.

91. Highfield, J. (2015) *Molecules*, **20**, 6739.

92. Sahara, G. and Ishitani, O. (2015) *Inorg. Chem.*, **54**, 5096.

93. Zhang, J., Wu, S., Li, B., and Zhang, H. (2012) *ChemCatChem*, **4**, 1230.

19
Oleochemical Biorefinery

Matthias N. Schneider, Alberto Iaconi, and Susanna Larocca

19.1
Oleochemistry Overview

19.1.1
Introduction

In nature a variety of biomolecules, that is, carbohydrates, proteins, or lipids, to name only a few examples, provide structural building blocks, energy reserves, and other functional units for the organism. The resulting value chain in nature is a prototype for the efficient use of raw materials from multiple sources to supply for intermediates and products, as most sidestreams are further used, and educts as well as products are at least partially interconvertible via platform metabolites to obtain substances as needed from available raw materials. In this context, nature's products have always been used by mankind for technical purposes. Oleochemistry, for example, has always been using renewable fats and oils to obtain standard products and specialties based on triacylglycerides and corresponding derivatives. However, with increasing demand from a variety of industries for renewable and sustainable chemical and energy feedstocks, oleochemistry nowadays not only faces the challenge to efficiently use available raw materials but also to secure a continuous, economic, and sustainable supply of fats and oils as basis to provide high-quality products, trying to cope with the increasing usage of oleochemical raw materials as substitute for petrochemicals [1, 2]. Therefore, it is of great interest for oleochemistry to enhance the supply chain for raw materials and gain independence from classical fat and oil sources that could compete with the food and feed chain, might be locally restricted, already face peak consumptions, or are challenged by new uses for the raw materials available. One potential approach toward these challenges is biorefinery concepts that mimic nature's approach to interconvert raw materials available according to the de facto demand [3, 4]. Combination of state-of-the-art technologies, ranging from mechanical processing

Chemicals and Fuels from Bio-Based Building Blocks, First Edition.
Edited by Fabrizio Cavani, Stefania Albonetti, Francesco Basile, and Alessandro Gandini.
© 2016 Wiley-VCH Verlag GmbH & Co. KGaA. Published 2016 by Wiley-VCH Verlag GmbH & Co. KGaA.

operations to biotechnological conversions, allows to get to alternative supply channels for a variety of chemicals and to improve established value chains. Such approaches ideally process biomass sidestreams, which are not efficiently used nowadays, so that the supplier benefits from added value and the downstream user profits from new raw material sources. Raw materials supplied can then be entered into various stages of the established value chain for oleochemicals, making use of existing production units, as substitute for starting materials used at present.

19.1.2
Value Chain for Oleochemistry

19.1.2.1 Raw Materials

The classical value chain of oleochemistry processes triacylglycerides, that is, esters of glycerine with fatty acids, as main raw material [5, 6]. These compounds include naturally occurring oils and fats that are classically extracted from vegetable biomasses, often farmed for oil production or from animal-based biomasses, resulting as sidestream of animal processing in the value chain of food and feed industries. In addition, recent approaches also include the extraction of oils from microorganisms such as algae that, similar to oil plants, might be selected or genetically modified to obtain raw materials with designed oil composition. Independently from source and origin, the value chain basically comprises the processing steps depicted in Figure 19.1.

Extraction of Crude Oils/Fats The oil- and fat-containing biomass is pretreated, and contaminants, such as soil residues, polymer tags of animals, or similar foreign matter from production, are separated (Figure 19.1, section A). From the pretreated biomass oils and fats are then extracted by a variety of techniques, depending on the source and targeted quality of the fat or oil. For animal fats, such as tallow, lard, poultry, or fish oils, rendering in general is carried out by cooking the grinded raw material in a wet or dry process to break up fat cells while sanitizing the resulting products (see Regulation 142/2011/EC [7]) before separating fats from the solid residues, that is, mostly proteins. The resulting intermediate is then cleaned and dried before further processing [8]. Vegetable oil grades are obtained from fruit pulp or kernel by slightly varying processes, depending on the oil type. For fruit oils, such as olive or palm oil, fruits are sterilized to inactivate enzymes and microorganism and separated from bunch stalks and other residues, before the grinded or milled material is pressed to extract the oil. For seed oils, such as soybean, sunflower, rapeseed, or palm kernel oil, the seeds are either dehulled or directly preconditioned and flaked before oil is extracted by pressing. Independent of the oil type, solvent extraction might be used to support the process and to enhance yields by further extracting remaining oil quantities in the filter cake of the pressing operation. Both for fruit and seed oils,

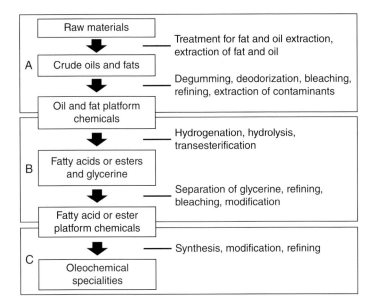

Figure 19.1 Schematic depiction of the value chain of oleochemistry: section A corresponds to the extraction of oils and fats from raw materials to obtain suitable primary platform chemicals for further processing, section B comprises the conversion of oils and fats to fatty acids and their esters as secondary platform chemicals, and section C indicates the transformation to oleochemical specialties.

the intermediates obtained are then filtered, clarified, and dried before further processing.

Refining of Crude Oils/Fats After crude fats or oils have been obtained, further processing is necessary to obtain raw materials in grades suitable for usage in oleochemical processes (Figure 19.1, section A). As natural products, the extracted crude intermediates contain a variety of substances together with the triacylglycerides of interest. The exact composition of the crude oil varies with the raw material and extraction process used, but in general, a cascade of treatment steps is necessary to improve oil quality and stability. Phospholipids, as residue of the cell content, are hydrolyzed by chemical or enzymatic treatment and removed during degumming. Naturally occurring free fatty acids or oxidation products resulting during storage and treatment, as well as pigments and metal complexes in the crude oil and fat, are extracted by physical or chemical refining, during bleaching, by adsorption over bleaching clays, or during deodorization, by stripping, in general, done by steam distillation. Additionally, waxes and other ester residues might be removed by precipitation during winterization [9]. These steps allow to obtain fats and oils in edible or nonedible qualities, with defined specifications, which can be directly used in a variety of applications or which are further processed as platform chemical in the oleochemistry value chain.

19.1.2.2 Processing

In a typical oleochemical company, the following core processes for treatment of the corresponding oil or fat are aimed at obtaining the main building blocks of oleochemistry, that is, fatty acids or their short-chain alcohol esters and glycerol. Depending on the setup of the process, the order of the most common processing steps might vary, and typical sequences are:

- Hydrogenation – splitting – separation of glycerol and distillation of fatty acid products
- Splitting – separation of glycerol and distillation of fatty acid products – hydrogenation of fatty acid products.

The resulting products might either be further processed or can be directly commercialized for a variety of applications. Some details of the essential core processes are given in the next sections.

Hydrogenation Animal fat (mostly tallow) or vegetable oil is hydrogenated in the presence of a catalyst (Ni, Pd) at approximately 160 °C and 12 bar hydrogen pressure for at least 20 min in an exothermic reaction, increasing the temperature to about 200–220 °C (Figure 19.1, section B). Similar conditions are applied also in case hydrogenation is carried out on distilled unsaturated fatty acids. In both cases the degree of saturation of the resulting products is controlled by reaction conditions, and the reaction can be performed partially, in order to get to the desired degree of saturation, in general identified by the iodine value of the product. The hydrogenated product obtained is then filtered to remove the catalyst for recycling. The raw material for hydrogenation step – which is often referred to as *hardening* – should be free of contaminants containing phosphor or sulfur groups, which could poison the catalyst during hydrogenation. Therefore, impurities like phosphatides, carotenoids, sterols, and residual proteins are removed by suitable bleaching and refining steps before hydrogenation (typically by absorption with activated bleaching earths at 100–130 °C for 40–120 min under vacuum followed by filtration), as already mentioned previously. The products of hydrogenation are fats and oils, fatty acids, or their derivatives with specific degree of saturation.

Splitting The typical splitting process in general lies in hydrolysis of fats and oils, yielding fatty acids as product for further processing or for commercialization (Figure 19.1, section B). However, separation of fatty acid groups from the glycerol backbone might also be done by alcoholysis, especially if the resulting esters are directly further processed. This procedure, however, will not be further discussed in detail here. For the production of fatty acids, typically fat/oil and process water are introduced into a splitting column by high-pressure metering pumps. Fat is fed at the bottom of the column, water at the top. The reaction takes place between the upper and lower heat exchange zone. The splitting column is heated to 250 °C by high-pressure steam resulting in a column pressure of about 50 bars. At these conditions, hydrolysis of the triacylglycerides occurs without any catalyst. A pressure regulator controls the unloading of fatty acids at the top, and an interface regulator automatically adjusts the flow of the resulting glycerol–water mixture (10–18%) from the bottom of the column. This way, fatty acids and glycerol are initially separated and can be further refined.

Crude fatty acids from splitting might still contain a number of high-boiling impurities, such as unreacted partial glycerides, soaps, glycerol, sterols, phosphatides, as well as lower-boiling materials, like water, low-molecular mass hydrocarbons, aldehydes, and ketones. In addition, their chain length distribution depends on the initial raw materials. To remove impurities and to alter the chain length distribution, if desired, distillation is carried out at more than 200 °C under vacuum to allow for desired separations. The resulting products are fatty acids of high quality with respect to color and stability.

The most important sidestream of the splitting operations is crude glycerol in water, which is initially concentrated from 11% to 20% up to minimum 80% in a multiple-effect evaporator. Concentrated crude glycerol is then treated with alkali and distilled at 155–158 °C under vacuum (typical: 4 mbar). In this stage glycerol is separated from low-boiling and high-boiling fractions to obtain glycerol of higher quality for a variety of purposes.

19.1.2.3 Oleochemical Substances

Starting from simple fatty acid esters of short-chain alcohols or fatty acids, a variety of oleochemicals can be produced in a plethora of available processes developed and introduced by oleochemical companies which are either back integrated into the depicted value chain or purchaser of the corresponding raw materials for further processing. The description of the processes needed to obtain the possible products of the value chain (Figure 19.1, section C) is beyond the scope of this chapter. However, as indicated in Figure 19.2, resulting products of oleochemistry are very versatile and are important for a huge number of industries and applications.

19.2
Applications and Markets for Selected Oleochemical Products

19.2.1
Applications for Products from Oleochemistry Value Chain

As already indicated, oleochemistry processes renewable fats and oils to a variety of standard and speciality chemicals (Figure 19.2). In principle, triacylglycerides

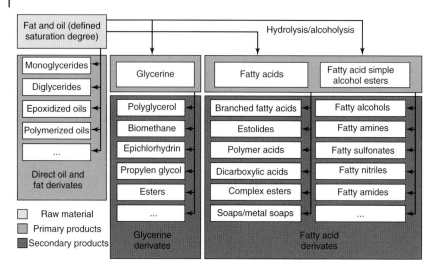

Figure 19.2 Schematic depiction of the main products of oleochemistry and their origin; the position in the value chain is indicated with different gray scales.

are the raw materials for the first chemical platform for renewable-based, sustainable materials. Direct modifications of oils/fats might include epoxidation or polymerization to obtain designed properties. In addition to the usage of triacylglycerides in food and nutrition, as well as pharmaceutical industries, technical applications of nonmodified or modified oils and fats include the usage as biodegradable base fluids for lubricants or fracturing fluids, usage in coatings, or application as hydrophobic agents in a variety of products [10–15]. Transesterifications of fats and oils with glycerol are one of the approaches to obtain mono- or diglycerides that are widely used as emulsifiers. Esters of short-chain alcohols are obtained by corresponding transesterification reactions. These are starting material to obtain fatty alcohols by reduction but can also be used for a variety of other purposes, of which the most prominent one in the last years is probably their use as biofuels. Upon hydrolysis of corresponding fats and oils, fatty acids with varying chain length distribution, specific degree of unsaturation, or with additional functional groups are obtained. Depending on the fatty acid, a variety of industry sectors find applications for these products. For example, fatty acids are directly used as nutritional supplements, as basis for candles, or in the function of lubricants in the production of textiles or polymers. Modification of the fatty acid chains yields special products, such as branched fatty acids (not naturally occurring), polymer fatty acids, and estolides [16–18], which are used in lubricants and greases or find application in a variety of coatings. Modification of the carboxylic acid group in fatty acids gives rise to sustainable, biodegradable, high-performance surfactants, for example, used in health care products, or as processing aids in a variety of industries. Soaps and metal soaps can be obtained by salt formation of fatty acids with selected (metal) cations, for which Baerlocher and Sogis are, for example, experts [19–21]. Such fatty acid salts can exemplarily be used as acid scavenger or stabilizer systems in polymer industries, exhibit anti-tacking properties of interest in rubber processing, or thicken a variety of oils and solvents, which is of interest in cosmetic applications or in designing lubricants and greases, to name only a few examples.

As described earlier, the triacylglycerides, as starting point to obtain oleochemical specialties, also represent the raw material to obtain glycerol, an important sidestream of fatty acid or fatty alcohol production. Glycerol is, for example, used as raw material in health care applications, and with growing interest in bioderivable raw materials, it gains an increasing interest to access a variety of derivates by further processing [22, 23].

19.2.2
Fats and Oil as Raw Materials for Oleochemicals

The main raw materials for oleochemistry have been, since this industry sector arose, rendered animal fats. Rendered animal fats available worldwide are more than 8 million tons per year. It used to be so for a long time, as meat consumption stayed stable during the last years. After Bovine spongiform encephalopathy (BSE) crisis, leading to an unbending ruling in the EU (see mainly Regulation 1069/2009/EC [24]), rendered animal fat supply has been split into categories, depending on specified risk materials (SRMs) separation, which is not practiced by non-EU countries, so that import of this raw material in the EU, as well as availability on the market, fell down. Production in the EU remained stable at more than 2500 kt. Traditionally, tallow has been used essentially for animal feeding, oleochemistry, or fertilizers. At the time the EU Commission started incentivizing energy from renewable sources, the tallow market was heavily challenged, as this raw material was more and more used for direct burning or for producing biodiesel. Nowadays more or less 40% of the total EU tallow supply goes into energy. Oleochemical consumption, including toilet soap manufacture, dropped from about 35% to less than 25%.

Natural oil production worldwide is about 180 million tons. The main output for this market is food. During the last two decades, we experienced an increasing demand for natural oils in energy and biofuels production. The natural substitute of tallow for oleochemicals is palm oil. This is the leading source of natural oils worldwide. The global production of palm oil is more than 55 million tons to which about 6 million tons of palm kernel oil has to be added. As the tremendous increase in palm oil production has driven to rainforest depletion, it is now more and more needed that palm oil for industrial uses has to be declared "sustainable." It is therefore expected that, besides mandatory requirements for biofuel sustainability, in a close future we will have to face mandatory requirements even for sustainability of chemicals. Another key source of vegetable oils for oleochemistry is coconut oil. Demand for medium-chain-length fatty acids is more and more increasing, correspondingly is coconut oil production, reaching more than 3.3 million tons in the last years. The other major vegetable oil sources for oleochemistry are soybean oil (more than 42 million tons per year), rapeseed oil (more than 24 million tons per year), and sunflower oil (more than 13 million tons per year) [25]. The EU oleochemistry is more and more making use of fatty acids as sidestream from vegetable oil refining. The availability of these by-products from local EU sources is supposed to be about 500 kt, while palm oil fatty acid distillate (PFAD) from palm oil accounts for about 2 million tons worldwide. It is to be noticed that biodiesel EU production is estimated to be more than 9 million tons at present, more than 24 million tons worldwide, so that this downstream use is the main driver for oil and fat prices.

19.2.3
Fatty Acid Market: History, Present, and Prognosis

Industrial oleochemistry was born in the sixteenth century, when in the EU the first industrial companies for the production of stearic acid for candles were set up. In the seventeenth century, in France, they started with the production of soaps, both toilet and laundry soaps, as Colbert, Louis XIV minister, was committed in creating new jobs. Starting from the first half of the nineteenth century, as meat consumption stepped up, a surplus of animal fats was on the market so that they started to become raw materials for the production of chemicals. The first modern companies producing fatty acids were founded in the UK and in the North of Europe.

After the Second World War, with the growth of the industrial production of polymers and detergents, oleochemistry experienced a booming time. This industry sector has been featured, till the 1990s, by the vertical integration with mass market producers in the detergent field. More or less at that time, a progressive transfer of EU assets and productions toward Far East countries was pursued, as these countries, besides profiting by a wide availability of raw materials (mainly palm oil), could take advantage from low-cost manpower, state aid, and a reduced bureaucracy.

During the last years oleochemistry had to cope with biodiesel producer competition in accessing raw materials, due to EU policy on renewable energy (Cfr Directive 28/2009/EC [26]), indirectly incentivizing the use of tallow to produce biofuels.

Traditionally, the major raw material for EU producers has been tallow, but as it used to be subsidized for the production of biodiesel or for direct burning, a consistent rise in this raw material price was experienced. While at the beginning of this century, the consumption of tallow in oleochemistry was higher than 800 kt, it is, at present, lower than 450 kt (Figure 19.3).

In the first 2000s, the production of fatty acids in the EU was about 1 million tons. In 2008–2009 it fell down to less than 800 kt, the same quantities as in 1990s, while in the last years grew again to about 900 kt.

Within the main oleochemicals produced in the EU, fatty alcohols account for more than half a million tons, fatty esters even more, while fatty acid metal soap (different from toilet and laundry soaps) and fatty nitrile production is about 100 000 tons each.

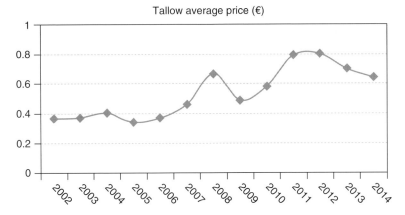

Figure 19.3 Tallow average price, Cat III, years 2002–2014.

A huge number of industry sectors are affected by oleochemistry, that is, soaps and detergents, additives for plastic materials, resins and coatings, rubber and tires, paper, drilling, food and feed, lubricants, candles, personal care, cosmetics, pharmaceuticals, medical devices, building industry, textiles, leather, fibers, firefighting, or agricultural products.

According to the information given by the EU Commission, EU biobased chemicals account at present for 5.5% of total turnover for chemicals produced in the EU, and they are expected to grow up by over 5% per year, till reaching a total proceeding of sales of about €40 billion in 2020 [27, 28]. Oleochemistry plays a pivoting role in the future growth of biobased chemistry, even if not for polymers.

19.2.4
Glycerine Market: History, Present, and Prognosis

Glycerine used to be a rare and expensive molecule, historically derived from two main industry sectors, oleochemistry and soap manufacturing. In 1995, before biodiesel appearance on the market, these two sources accounted for more than 95% of glycerine generation. The total glycerine appearance worldwide at that time was about 700 kt. In the second half of the 1990s and in the first decade of this century, the "explosion" of biodiesel production led to a huge increase in glycerine generation, up to about 3 million tons in 2014. Today biodiesel accounts for almost 80% of glycerine production. Of course this increase in glycerine generation has been followed by a drop in price, a large increase in glycerine refining capacity, an impressive enlargement in refined glycerine uses, and an extensive growth of direct uses of crude glycerine. As a direct consequence, the second most important feature, besides the large availability of this versatile molecule, is price evolution, opening a wide range of opportunities and making the once rare and expensive substance now largely available and cheap.

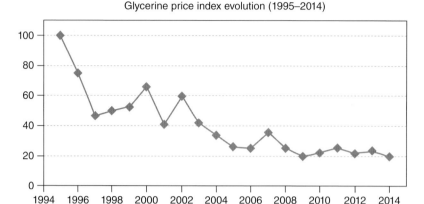

Figure 19.4 Evolution of the actualized price index for glycerine, years 1995–2014.

From the graph shown in Figure 19.4, it's clear we are experiencing a situation where price is today about 20% of the one it used to be before biodiesel explosion. Biodiesel mandatory consumption started in Europe not only as part of the program aimed at fighting climate changes but also to find alternatives to agricultural extra production (at the beginning, it was born only as a rapeseed derivative obtained by set-aside crops). It than quickly became a way to "rule" vegetable oil prices, and it is now supported by many countries worldwide, mainly where the production of oleaginous crops is a consistent contribution to the local economy (Argentina, Indonesia, Malaysia, etc.).

This kind of support is gradually taking biodiesel out of the – sometime hysteric – European politics and is making it a more global and stable product. This attitude brings stability to the glycerine market. Even if the growth rate of biodiesel capacity is slowing down, we can easily expect that biodiesel production and the consequent glycerine appearance will stay on the high side in the years to come.

Despite biodiesel is today the main source of glycerine, the growing demand for renewable products is strongly leading to research and development. Some projects have already been scaled up, and they are now implemented in the so-called integrated biorefineries. Whenever a biorefinery is fed by oils or fats as raw materials, we are fully back to what has always been referred to as *oleochemistry*, where the final objective is to process, modify, and functionalize fatty acids. Even though via innovative processes, getting fatty acids out of a triglyceride means generating glycerine as a coproduct.

The combined effect of the interest toward biofuels and biobased chemical building blocks creates the necessary confidence that glycerine will be available in large volumes at low prices in the coming future. It is then wise and economically extremely interesting to look at this molecule not only as a great ingredient for food, pharmaceutical, and cosmetic products but also as a fully renewable, harmless, biodegradable, and hyperversatile building block for a downstream, high added value green chemistry.

19.3
Future Perspectives of Oleochemistry in the View of Bioeconomy

19.3.1
Potentials for Oleochemistry

Oleochemistry has been studied for a very long time, being developed before mineral oil chemistry, but it now represents a niche industry sector due to mineral oil competitive price. It is the most versatile biobased chemistry, as a number of reactions are very well known, plants and procedures for the main processes have been in place for a long time, and oils and fats do not contain a lot of oxygen so that they are very efficient raw materials, able to replace mineral oil in a huge number of applications. As oleochemistry is mostly based on renewable raw materials, it is clearly a sustainable chemistry, with the potential to represent a prototype for the integration of bioeconomy concepts and industrial chemistry.

In this context, access to raw materials is the main key point. In endorsing substitution of petrochemical products with products from biomasses, it would be worth not repeating the same mistakes EU made with biofuels: industrial projects are to be favored only in case raw material availability has been ascertained.

The new agricultural policy opens the door to different scenarios. EU farmers are looking for solutions, in order to be able to face international markets. A new deal between industry and agriculture is essential to cope with the needs of fighting climate changes and, at the same time, finding local raw materials to feed local value chains. Oil and fat prices are now very strictly related to mineral oil price, due to their use in biofuels. In case no mandatory rules are put in place, the future development of oleochemistry depends, besides on oil and fat availability, on the delta price between mineral oil and natural oils and fats. Therefore synergies between the long-time established concepts of oleochemistry and bioeconomy concepts, for example, including biorefinery making use of low-value biomasses to get to oleochemical raw materials, can be expected, providing that the following points are considered:

- Securing continuous supply of raw materials matching economical and quality requirements and being able to substitute classically used oils and fats (e.g., tallow, palm oil, sunflower, or rapeseed oil).
- Enhancing independence from existing fat and oil supply chains, including regional and regulatory factors.
- Making EU independent on oils that can only be obtained in non-European, tropical countries but have specific properties (e.g., coconut oil, palm oil).
- Access to supply chains that are not in competition with other value chains, for example, nutrition or fuels.
- Access to sustainable supply chains ensuring "green oleochemistry" while allowing for a renewable and hence continuous supply logistic of needed amounts.

Furthermore, there are expectations about oils from algae and single-cell oils, which can be engineered to match specific demands of the market. In case such kind of raw materials would become available in a reasonable time, perspectives, both for chemical and for biofuel market, would become very interesting. Such oils could, for example, reduce efforts to fractionate the fatty acids resulting from the value chain into specific chain length distribution or allow to obtain specifically functionalized fatty acids for further processing. The most widely used raw materials for oleochemistry are tallow and palm oil, with limited quantities of oleochemical substances obtained from sunflower, soybean, rapeseed, and castor oil. An extensive use of oils and fats featured by a functional group in the carbon chain could lead to a widespread availability of chemicals from renewable sources now scarcely available. First steps in this direction have already been done. For example, high oleic oils from soybean or sunflower selected cultivars, producing suitable oils, have already been introduced [29]. Although they represent classical oleaginous plants modified to produce special oils, this approach can be taken on by biorefinery and bioeconomy concepts and might widen the range of available raw materials.

However, in addition to a large potential for bioeconomy concepts as access route to more or less classical oleochemical raw materials, research and development might as well also combine oleochemical processes to further diversify the current range of products available, beyond the typical value chain. Some potential can be expected from the following exemplary developments:

- *Development of Biological Processes*: Enzymatic hydrolysis and other enzymatic reactions have been studied since the 1980s, but till now they never found an extensive industrial implementation. As these methodologies would allow to save energy, they are becoming more and more interesting [30].
- *Reactions on the Carbon Atom Chain*: Although a number of reactions on the carbon atom chain are already known, the bulk of fatty acid oleochemistry has been focused on the carboxylic functional group reactions. There is now more and more interest in looking into the carbon atom chain, with a particular focus on the possibility of reacting double bonds. This has driven to a lot of interesting ongoing developments. Oxidation, epoxidation, and metathesis are only possible examples.
- *Short-Chain Fatty Acids/Alcohols*: Some industry sectors, like detergents, rely on short-chain fatty acids/fatty alcohols. To switch from mineral oil to renewable sources, it would be very important to have a wide availability of these raw materials at a reasonable price.
- *Branched Fatty Acids/Alcohols*: The demand for such building blocks has recently increased, as these molecules – although biodegradable – are more resistant to oxidation than the corresponding unsaturated homologs with the same physical properties.
- *Multifunctional Molecules for Polymers*: Although sugars are better candidates to be converted in monomers for polymers, there is a wide interest in producing multifunctional substances to be used as monomers for polymers, starting from oils and fats. Depending on the availability of the raw materials, this could become an interesting industry sector [11, 31].

- *Cracking of Oils and Fats*: Oils and fats could be used as well as mineral oil in catalytic cracking reactions. They are already extensively used in hydrocracking processes to obtain hydrotreated vegetable oils (HVOs) to produce high-quality biodiesel [32, 33].
- *Glycerol*: Glycerol is a very interesting molecule with three functional groups. A lot of developments aimed at transforming glycerol into other useful molecules are now very close to implementation. In case the availability of this molecule will stay stable, as well as price, it will become more and more one of the key building blocks for biobased economy.

19.3.2
Legal and Regulatory Background

Since EU signed the Kyoto Protocol, in April 1998, a strong policy endorsing the fight against climate changes has been put in place in the Union. Although it is hard to understand if this process was really driven by the will to limit greenhouse gas (GHG) emissions or by economic interests, at last, with Directive 28/2009/EC Renewable Energy Directive (RED) [26], a common EU policy for energy from renewable sources has been definitively issued. Transposition in all EU member states is now almost completed.

New industry sectors have been started up, driven by legislation, but a lot of efforts are needed in order to make them profitable without subsidies, avoiding market distortions, as raw materials for chemicals from renewable sources, biofuels, and energy production from biomasses are the same. Oleochemistry has been facing challenging times for a long period, as oleochemical companies had to cope with an intensified competition for tallow sourcing against biodiesel producers, as this traditional raw material for EU oleochemistry, enabling EU companies to compete with the Far East producers, that benefit from a wide availability of palm oil, was indirectly subsidized for use in biofuels.

Also, a clear policy fostering local raw materials and productions has to be supported, in order to avoid import of subsidized biofuels, moving enormous resources swept up from EU citizen extra taxes outside the EU and not solving the problem of reducing the dependence of EU energy demand from foreign countries, one of the key issues EU government claims they intend to tackle.

As biomasses were a preferred option for renewable energy, it is debatable if it was wise or not to start ruling on low-carbon economy with a focus on energy, as chemistry needs less input in terms of raw materials, it has enough skill to find out suitable technical solutions, and it is much easier to handle. With Communication (2010) 2020 [34], in 2010, the EU Commission has launched a program in order to foster a "smart, sustainable and inclusive growth," based on three mutual reinforcing priorities, five headline targets, and seven flagship initiatives. This program includes also biobased economy, dealt with mainly in the flagship initiatives, "resource efficient Europe," "an industrial policy for the globalization era," and "Innovation Union."

Biobased economy has been than extensively introduced mainly in communications: COM (2012) 60 [35], "Innovating for sustainable growth: A bioeconomy for Europe," and COM (2011) 112 [36], "A roadmap for moving to a competitive low carbon economy in 2050."

A key concept, which is becoming more and more popular in all EU preparatory laws and was already there in the Waste Framework Directive (Directive 98/2009/EC) [37], is the one of "hierarchy of use," this means a clever use of raw materials where the number of production steps is increased so that the value in the value chain goes up, as well as labor intensity.

Circular economy is another fundamental concept for the biobased economy, developed mainly in COM (2014) 398 [38], "Towards a circular economy: A zero waste programme for Europe." Besides waste legislation, biobased economy is strongly affected by this policy, due to waste oils and fats, other food wastes, and agricultural residues.

After these preparatory documents, there are now many expectations on binding rules able to establish a common policy within EU member states.

19.4
Conclusions

The final consumption of petrol, diesel, and biofuel for transport in the EU in 2012 was 272.967 kt, 31.8% of the total energy demand in the EU (1104.5 million tons) [39]. This means that the full current worldwide production of vegetable oils would not be able to cover even the EU demand in energy for transport, that renewable energy EU policies have to be revised, and that sources different from biomasses have to be endorsed to produce renewable energy. The other way round, in case of a jump in available technologies for raw materials, producing chemicals from biomasses is possible, and oleochemistry could become a key paramount industry sector in implementing biobased chemistry. Therefore it can be expected that oleochemistry will play a major role in arising bioeconomy concepts combining traditional and innovative approaches for a sustainable future.

References

1. Baumann, H., Bühler, M., Fochem, H., Hirsinger, F., Zoebelein, H., and Falbe, J. (1988) *Angew. Chem. Int. Ed. Engl.*, **27**, 41–62.
2. Biermann, U., Bornscheuer, U., Meier, M.A.R., Metzger, J.O., and Schäfer, H.J. (2011) *Angew. Chem. Int. Ed.*, **50**, 3854–3871.
3. Wertz, J.–.L. and Bédué, O. (2013) *Lignocellulosic Biorefineries*, EPFL Press.
4. J.–L. Wertz, O. Bédué and J.P. Mercier (2010) *Cellulose Science and Technology*, EPFL Press.
5. D. Swern (1979) *Bailey's Industrial Oil and Fat Products*, John Wiley & Sons, Inc.
6. G. B. Martinenghi (1963) *Tecnologia Chimica industriale degli oli, grassi e derivati*, Hoepli.
7. COMMISSION REGULATION (EU) No 142/2011 of 25 February 2011 *implementing Regulation (EC) No 1069/2009 of the European Parliament and of the Council laying down health rules as regards animal by-products and derived products not intended for human*

consumption and implementing Council Directive 97/78/EC as regards certain samples and items exempt from veterinary checks at the border under that directive.

8. Sharma, H., Giriprasad, R., and Goswami, M. (2013) *J. Food Process. Technol.*, **4**, 1000252.

9. Baranowsky, K., Beyer, W., Billek, G., Buchold, H., Gertz, C., Grothues, B., Sen Gupta, A.K., Holtmeier, W., Knuth, M., Lau, J., Mukherjee, K.W., Münch, E.-W., Saft, H., Schneider, M., Tiebach, R., Transfeld, P., Unterberg, C., Weber, K., and Zschau, W. (2001) *Eur. J. Lipid Sci. Technol.*, **103**, 505–508.

10. Salimon, J., Salih, N., and Yousif, E. (2010) *Eur. J. Lipid Sci. Technol.*, **112**, 519–530.

11. Meier, M.A.R., Metzger, J.O., and Schubert, U.S. (2007) *Chem. Soc. Rev.*, **36**, 1788–1802.

12. Hill, K. (2000) *Pure Appl. Chem.*, **72**, 1255–1264.

13. Manawwer, A., Deewan, A., Eram, S., Fahmina, Z., and Sharif, A. (2014) *Arabian J. Chem.*, 7, 469–479.

14. Salimon, J., Salih, N., and Yousif, E. (2012) *Arabian J. Chem.*, **5**, 135–145.

15. Schneider, M.P. (2006) *J. Sci. Food Agric.*, **86**, 1769–1780.

16. Harword, H.J. (1962) *Chem. Rev.*, **62**, 99–154.

17. Isbell, T.A. (2011) *Grasas Aceites*, **62**, 8–20.

18. Lie Ken Jie, M.S.F., Pasha, M.K., and Syed-Rahmatullah, M.S.K. (1997) *Nat. Prod. Rep.*, **14**, 163–189.

19. Nora, A. and Koenen, G. (2010) *Metallic Soaps in Ullmann's Encyclopedia of Industrial Chemistry*, Wiley-VCH Verlag GmbH.

20. Baerlocher http://www.baerlocher.com/de/home/ (accessed 30 September 2015).

21. Sogis http://www.sogis.com/ (accessed 30 September 2015).

22. Pagliaro, M., Ciriminna, R., Kimura, H., Rossi, M., and Della Pina, C. (2007) *Angew. Chem. Int. Ed.*, **46**, 4434–4440.

23. Yang, F., Hanna, M.A., and Sun, R. (2012) *Biotechnol. Biofuels*, **5**, 13.

24. EU Regulation (2009) Regulation 1069/2009/EC of the European Parliament and the Council of 21 October 2009.

25. Oil World (2013) ISTA Mielke GmbH Oil World Annual 2013.

26. DIRECTIVE 2009/28/EC OF THE EUROPEAN PARLIAMENT AND OF THE COUNCIL of 23 April 2009 *on the promotion of the use of energy from renewable sources and amending and subsequently repealing Directives 2001/77/EC and 2003/30/EC.*

27. Mosquera, J. (2014) Presentation "Bioeconomy overview" at the Fine, Specialty and Consumer Chemicals Industry Sector (CEFIC) Managers Meeting, April 25, 2014.

28. Moncef Hadhri (2015) Cefic Economic Affairs, Industrial Policy Department, CEFIC, "Chemical Trend Reports", April 14, 2015.

29. Metzger, J.O. and Bornscheuer, U. (2006) *Appl. Microbiol. Biotechnol.*, **71**, 13–22.

30. Abdelmoez, W. and Mustafa, A. (2014) *J. Oleo Sci.*, **63**, 545–554.

31. Espinosa, L.M. and Meier, M.A.R. (2011) *Eur. Polym. J.*, **47**, 837–852.

32. Sunde, K., Brekke, A., and Solberg, B. (2011) *Energies*, **4**, 845–877.

33. Knothe, G. (2010) *Prog. Energy Combust. Sci.*, **36**, 364–373.

34. COM (2010) 2020 – EUROPE 2020: A Strategy for Smart, Sustainable and Inclusive Growth.

35. COM (2012) 60 – Innovating for Sustainable Growth: A Bioeconomy for Europe INSE

36. COM (2011) 112 – A Roadmap for Moving to a Competitive Low Carbon Economy in 2050.

37. DIRECTIVE 2008/98/EC OF THE EUROPEAN PARLIAMENT AND OF THE COUNCIL of 19 November 2008 on waste.

38. COM (2014) 398 – Towards a Circular Economy: A Zero Waste Programme for Europe.

39. European Commission (2014) EU Transport in Figures – Statistical Pocketbook 2014.

20
Arkema's Integrated Plant-Based Factories
Jean-Luc Dubois

20.1
Introduction

The industry has been using biomass well before the word "biorefinery" was invented. Biorefinery refers to a collection of processes for the conversion of biomass-derived starting materials, in a way that reminds the petroleum refineries. However, sometimes there are not much refining (distillation/separation) activities but rather chemical synthesis. Another important common feature is that the economic sustainability is reached in creating value from every drop of starting material and making several products, from various processes to make low-value/ high-volume and high-value/low-volume products. Similarly to oil refineries which are producing low-value fuels and high-value lubricants, solvents, and petrochemicals, a "biorefinery" produces energy and high-value products.

It is rather difficult to introduce completely new chemicals on the market, and history teaches us that it usually takes more than 20 years to penetrate the market. Therefore, alternatives are trying to make an existing fossil-based molecule from renewable materials (also called *drop-in*), benefiting from an existing market (customers and volumes) and bypassing regulations such as REACH since the molecule made is the same.

20.2
Arkema's Plant-Based Factories

Renewable products generate about 700 M€ of turnover (sales) in Arkema (as of 2014). Castor oil is strategic for Arkema in that regard. Arkema produces monomers for long-chain polyamides (Polyamide 11 (PA11) and Polyamide 10.10, which are 100% renewable) from castor oil in two plants. Castor oil has unique features that make it a starting material of choice: it is the sole industrial oil which contains ricinoleic acid (12-hydroxy-9-octadecenoic acid, also called *12-hydroxyoleic acid*) in large concentration (more than 85%). It is quite uncommon to find a natural oil (Genetically Modified Organism (GMO)-free)

Chemicals and Fuels from Bio-Based Building Blocks, First Edition.
Edited by Fabrizio Cavani, Stefania Albonetti, Francesco Basile, and Alessandro Gandini.
© 2016 Wiley-VCH Verlag GmbH & Co. KGaA. Published 2016 by Wiley-VCH Verlag GmbH & Co. KGaA.

with such a high concentration in a single fatty acid, which in addition shows unique reactivity and properties, which either help to isolate it or facilitate downstream processing. Industrial use of castor oil is not competing with food consumption since the oil is non-edible (although it has been used as laxative). Arkema is also consuming linseed oil and soybean oil to produce epoxidized oils which are used as plasticizers in polymers. Terpenes and limonenes, and virtually anything which has a C=C double bond, can be epoxidized within the same plant. Fatty acids obtained after hydrolysis and/or saponification of vegetable oils and animal fats are converted to fatty nitriles and further to fatty amines, which are used as surfactants and additives. Lignocellulosic biomass has its own share in Arkema's products portfolio, since locally sourced pinewood is converted to activated carbon by thermochemical conversion.

Selected vegetable oils have a high H/C and low O/C ratio. So they have a high energy content, and that's what makes them attractive to produce biodiesel, for example. The fatty acids have a linear straight chain structure which makes them ideal starting materials for long-chain amino acids or diacids. Some long-chain fatty acids bear unsaturations such as oleic acid (also described as C18:1, ∂-9, which indicates that it has 18 carbons and 1 double bond located at the ninth carbon counting from the acid group), linoleic acid (C18:2, ∂-9, 12), linolenic acid (C18:3, ∂-9, 12, 15), erucic acid (C22:1, ∂-13), and so on. When looking at epoxidation (to make epoxidized oils or polyols) or at alkyd resins, the number of unsaturation in the oil will be valuable. In the former case, number of unsaturations determines the number of epoxy rings/hydroxy that can be formed. In the latter case, it impacts the drying properties brought by cross-linking of polymeric chains in the latter case. The location of the double bond is also important: for example, azelaic acid ($HOOC-(CH_2)_7-COOH$) is produced by oxidative cleavage (with Ozone, Hydrogen peroxide or other strong Oxidants) of oleic acid. Therefore high-value products can be made if the right fatty acid is available. Unfortunately, they are always found in combination with other fatty acids, which differ solely by the number of carbon atoms or the number of double bonds. Separation is therefore costly and is always a source of problems. The raw materials, processes, and chemistries have then to be selected to facilitate the purification of the final products.

20.2.1
Marseille Saint-Menet (France)

The plant was built in 1955 in the area of Marseille (south of France) (Table 20.1). The choice of the location was dictated by two main factors: the need to import the starting material (castor seeds) – including from former French colonies – and the fact that seed mills were already in place on the Marseille harbor. In addition, the PA11 (Rilsan®) was initially thought to be used for textiles, and the Rhône river connecting the Lyon area (textile) and Marseille was a driver for the logistics issues.

The plant is processing castor oil which is now imported, although castor has all the attributes to be produced in Europe, and was produced at industrial scale in

Table 20.1 Castor oil to aminoundecanoic acid plant – Marseille Saint-Menet (France) plant data.

Start-up date	1955
Total plant capacity	52 000 tons of oil
Employees	319 + 24 (management)
Land	13 ha
Main source of energy	Natural gas
Classification	Seveso II
Product brand names	Monomer of Rilsan® Polyamide 11 (glycerine, heptanaldehyde, heptanoic acid, heptanol, Esterol, undecenoic acid, and so on), OLERIS® (coproducts)

Ukraine during the Soviet period. The oil is converted to its methyl ester through a transesterification process (a biodiesel-like process) which leads to the coproduction of glycerine (about 10% of the oil input). The methyl ester of ricinoleic acid is then thermally cleaved to methyl 10-undecenoate (a C11 omega unsaturated fatty ester) and heptanaldehyde. The other fatty esters do not react and are recovered downstream and commercialized under the name of esterol (e.g., as solvent and concrete demolding agent).

The unique combination of a OH group and a vicinal double bond in the ricinoleic acid chain forming a chain $-CHOH-CH_2-CH=CH-$ makes the cleavage highly selective.

Heptanaldehyde is recovered and isolated to be commercialized for fine chemicals synthesis, for example, in the flavor and fragrances markets. It is also oxidized into heptanoic acid which finds applications in the aviation synthetic lubricant markets, among others; and it is reduced into heptanol.

In the next step, methyl undecenoate is hydrolyzed to undecylenic acid, and methanol is recovered and recycled at the first step. Undecylenic acid (also commercialized) is further reacted with hydrogen bromide which is produced on-site to produce the anti-Markovnikov 11-bromoundecanoic acid. Last, this intermediate is reacted with ammonia to release the 11-aminoundecanoic acid which is the monomer of the Rilsan PA11 produced by Arkema. Although the monomer is produced in Marseille, the polymer is produced in several other locations, including Serquigny in Normandy. PA11 is a technical polymer with unrivaled properties which finds applications in numerous markets including the automotive industry, oil and gas, leisure, and so on, where customers are looking for performance.

20.2.2
Hengshui (China)

The plant was built more recently but is using an old technology (Table 20.2). While the Marseille plant is using a thermal cleavage process, the Hengshui plant is using an alkaline cleavage process. Here again the specificity of the ricinoleic is used to cleave the fatty acid and only this one. The castor oil is first hydrolyzed, and high-quality glycerine is recovered and commercialized. The fatty acids are

Table 20.2 Castor oil to sebacic acid plant – Hengshui (China) plant data.

Start-up date	2008
Total plant capacity	40 000 tons of monomer (sebacic acid), 28 000 tons (2-octanol), 80 000 tons (sodium sulfate)
Employees	585
Land	22 ha
Main source of energy	Coal
Product brand names	OLERIS® (coproducts), HYPROLON® (polyamide)

Table 20.3 Epoxidation plant – Blooming Prairie (US) plant data.

Start-up date	1970 (as Viking Chemical)
Total plant capacity	27 000 tons (total capacity)
Employees	41 + 5 (management)
Land	4 ha
Main source of energy	Natural gas
Product brand names	Vikoflex®

then reacted in sodium hydroxide at high temperature. Several reactions are taking place simultaneously: dehydrogenation (hydrogen transfer) of the OH group, isomerization of the double bond, hydration of the double bond, retroaldolization, and finally oxidation of the aldehyde group to lead to sodium azelate. The coproduct of the cleavage is formally 2-octanone, which benefits from the hydrogen transfer reactions and leads to 2-octanol. Sodium azelate is recovered and reverted to azelaic acid with sulfuric acid coproducing sodium sulfate. The other fatty acids are recovered after the cleavage reaction and commercialized. In another plant, sebacic acid is also converted to the corresponding diamine, which if combined with sebacic acid leads to polyamide-10.10 (Hiprolon®). Sebacic acid is also used to produce polyesters and esters used as solvents and plasticizers and if combined with other diamines leads to other polyamides.

20.2.3
Blooming Prairie (United States)

The site is using two processes for epoxidation of double bonds, with peracetic and performic acid. It has the required flexibility to adapt to large demand of epoxidation chemistry of double bonds such as linseed oil and soybean oil but also terpenes, limonenes, and alpha olefins, for example. The plant was formerly a creamery (from 1931) which was reconverted to a chemical plant in 1970 and is located next to a rail track for easy transportation of products (Table 20.3). Specialty epoxides serve wide market applications such as lubricant additives, flavor and fragrances, urethanes, specialty coatings, pharmaceuticals and agricultural chemicals, and so on.

Table 20.4 Feuchy (France) plant data.

Start-up date	1950 (formally a fertilizer plant since 1928)
Total plant capacity	32 000 tons (total capacity)
Employees	120 + 10 (management)
Land	29 ha
Main source of energy	Natural gas and electricity
Classification	Seveso II
Product brand name	CECA – Cecabase®

Table 20.5 Activated carbons from locally sourced pinewood – Parentis (France) plant data.

Start-up date	1941 (started in 1920 as a brick factory)
Total plant capacity	150 000 tons (locally sourced maritime pinewood total capacity)
Employees	n.c.
Land	25 ha
Product brand name	Acticarbon® commercialized by Arkema's business unit CECA

20.2.4
Feuchy (France)

Fatty acids are first converted to nitriles by reaction with ammonia in the presence of a catalyst. Then the nitriles are selectively hydrogenated to amines which are used as surfactants, asphalt and fertilizer additives, additives for crude oil and natural gas production, and so on. More than 400 products have been registered, and the site is using a combination of 20 different reactions, operating up to 315 °C and 30 bars.

Since the plant is using ammonia and other chemicals, it is listed as Seveso II (Table 20.4).

20.2.5
Parentis (France)

Activated carbons are used as filtration agents in food and pharmaceutical industries but also as supports for catalysts. The main properties expected from them are linked to the high surface area generated per gram of carbon, to the porosity network and to the adsorption efficiency to retain impurities, for example. The site is processing locally sourced pinewood which has been grown since the nineteenth century in a sustainable way.

The plant (Table 20.5) uses thermal (steam) and chemical activation processes of the produced charcoal. Sawdust (chemical activation) and crushed wood (steam activation) are used to make two different kinds of products. In the chemical activation process, phosphoric acid acts as a catalyst and is recovered at the end of the reaction.

20.3
Cross-Metathesis of Vegetable Oil Plant

20.3.1
From Rapeseed Oil to a Synthetic Palm Kernel or Coconut Oil

Arkema has been investigating metathesis chemistry for the production of new monomers (bifunctional molecules) for several years already. Based on this experience, the following part has been constructed as an example of a simple economic analysis to determine if a biorefinery, converting vegetable oils to platform molecules, would be worth to build.

Europe is currently importing palm kernel and coconut oils for industrial applications (e.g., in surfactants) [1]. A 2012 data indicates that the major importing countries are Netherlands and Germany, well ahead of Belgium, Spain, France, Italy, and the United Kingdom [2]. These oils are hydrolyzed to fatty acids and hydrogenated to fatty alcohols. There is no competition with locally produced oils since only tropical oils contain the medium-chain-length (mostly 10 and 12 carbons) fatty acids. Since the industry in Indonesia, Malaysia, and the Philippines are heavily investing in downstream processes, they also compete with European plants in the downstream processing. Metathesis is a chemical reaction (Nobel Prize-winning chemistry in 2005) catalyzed by metals (such as ruthenium complexes in the case of vegetable oil) which exchanges (scramble) the groups on each side of double bonds. Then a $R_1HC{=}CHR_1$ reacting with a $R_2HC{=}CHR_2$ would give $2R_2HC{=}CHR_1$. This is an equilibrium-limited reaction, so to shift the reaction near completion, it is necessary to have a reactant in excess or to withdraw continuously a product. Catalysts used in these reactions (on vegetable oils) are not yet very stable and have limited productivity. In Table 20.6, a composition of rapeseed oil, which is commonly grown in Europe and also known as *canola* in the United States, shows a large amount of unsaturated C18 fatty acid chain length, while palm kernel and coconut oils are rich in C10, C12, and C14 chain length. It is not possible to directly substitute one by the other. After a cross-metathesis reaction with 1-butene, assumed to be carried out at about 80% conversion, the metathesized oil has now chain length distribution much closer to the tropical oils. If needed, the double bonds could be easily hydrogenated to mimic even more the medium-chain-length fatty acids. Metathesis appears then as a way to produce in Europe an industrial oil from locally sourced biomass.

20.3.2
Preliminary Economic Analysis

In order to make a first estimate of the economic model, let's assume that we can reach 80% conversion and a product distribution giving a C10/(C10 + C12) ratio of 0.4. Note that to mimic better the tropical oils it would be better to have a lower ratio, but this could be achieved at the expense of higher conversion and higher catalyst consumption. In addition, in recent years the price of the C10 fatty acid

Table 20.6 Fatty acid distribution.

Composition	Rapeseed/canola oil	Coconut oil	Palm kernel oil	Metathesized oil
Fatty acid		mol%		
C6:0	0.0	0.5	0.9	0.0
C8:0	0.0	5.3	10.0	0.0
C10:0/C10.1	0.0/0.0	3.8/0.0	9.0/0.0	0.0/24.4
C12:0/C12:1	0.0/0.0	53.0/0.0	49.9/0.0	0.0/37.5
C13:0/C13:1	0.0/0.0	0.0/0.0	0.0/0.0	0.0/3.2
C14:0/C14:1	0.0/0.0	14.7/0.0	14.5/0.0	0.0/2.0
C15:0/C15:2	0.0/0.0	0.0/0.0	0.0/0.0	0.0/4.8
C16:0/C16:1/C16:3	5.0/0.0/0.0	7.6/0.4/0.0	7.6/0.4/0.0	5.0/1.0/0.6
C18:0	2.0	1.5	1.5	2.0
C18:1	55.5	10.9	4.7	11.1
C18:2	22.1	1.9	1.5	4.4
C18:3	10.6	0.0	0.0	3.0
C20:0/C20:1	0.0/2.7	0.3/0.0	0.0/0.0	0.0/0.5
C22:0/C22:1	0.0/2.1	0.0/0.0	0.0/0.0	0.0/0.4

Note*: Usually what is reported is weight percentages, but for easiness of the analysis mol% has been preferred.
Note*: Metathesized rapeseed oil is assumed to be obtained at 80% conversion and a C10/(C10 + C12) ratio of 0.4. This ratio is higher than in coconut and palm kernel and was selected on purpose since C10 fatty acid has potentially a higher value than C12 fatty acid.

(decanoic acid) has been increasing due to a higher interest for this fatty acid. This ratio is achieved assuming a recycle of the butene self-metathesis product 3-hexene in order also to reduce the butene consumption. The catalyst efficiency is assumed to give a turnover number (or number of reactions per number of mole of catalyst) of 100 000. This kind of efficiency has been reported in patent literature [3]. It also represents an aggressive assumption of about 5 ppm mol (micromole of catalyst per mole of double bond).

The plant size is assumed to be of 300 000 tons per year capacity (similar to a large biodiesel plant). From the rapeseed oil composition reported in the table above, we can calculate a mass balance. For 1000 tons of oil, the plant will produce 777 tons of metathesized oil and 365 tons of olefins, ranging from C4 to C12, assuming that there are no further cross-metathesis reactions between reaction products (there should be also some C18 olefins and diesters, but here for simplicity we assume that they are also converted to shorter molecules). Note that this assumption is partly wrong, since we know that in an equilibrium-limited reaction all products and reactants continue to react, generating new products. In a real process it will be very important to control the operating conditions in order to avoid these consecutive reactions, for example, in using sufficient quantities of 1-butene or in limiting the conversion.

In these conditions, the plant would consume annually 42 500 tons (assuming that any excess used would be recycled and assuming that 1-butene self-metathesis product is recycled and cofed with 1-butene so that no butene is lost

Table 20.7 Simple calculation on cost of production.

	Unit price $/ton ($/kg cat) ($/employee)		M$ per year	$ per ton of metathesized oil
Oil (1 000 tons/year)	300	800	240	1 030
Butene (1 000 tons/year)	42.5	1 400	59.5	255
Catalyst (kg per year)	6 270	5 000	31.4	135
Utilities	—	—	33.1	142
Total variable cost	—	—	—	**1 562**
CapEx (M$) – 10 year depreciation	*100*	—	10.0	*42.9*
Labor	*27*	*60 000*	1.62	*7.0*
Total fixed cost	—	—	—	**49.9**
Metathesized oil (1 000 tons/yr)	233	—	—	*1 001*
Olefins (1 000 tons/year) (credit)	109	1 300	−142	−611
Total cost of production	—	—	376	**1 612**

Note: Mass balances are based on assumptions, not on real experimental data. Catalyst price is an assumption so that the total catalyst cost remains below 10% of total plant cost. (Above 5% it should be considered as a reactant rather than as a catalyst only. CAPEX-Capital Expenditure.)

in this secondary product) of 1-butene and 6270 kg of catalyst (Table 20.7). It is necessary to operate with some excess of butene to avoid the formation of waxes (oligomers of vegetable oils), although they might have a good value.

The plant capital cost is assumed at 100 M$ (all data have been calculated in US$, since raw materials and benchmarked oils but also coproduct prices are usually in US$). There is no detailed flowsheet nor equipments listed, but this is a fair estimate of what the plant could cost from expert judgment. There is a large uncertainty on this number, but it has a small weight on the total cost of production (it represents only about 5% of the metathesized oil production cost). Similarly, the plant is supposed to be operated by 27 people, at an average operator cost of 60 000 $ per year. The labor cost weights even less although there is also a large uncertainty on the number of employees needed. The major contributors to the cost are the rapeseed oil and the 1-butene consumed. The valorization of the olefins produced (1-decene and others) is an important driver to bring the cost of production down since it is equivalent to 38% of the total plant cost. Some cost contributors have been omitted since they do not significantly affect the result. Based on these assumptions, the cost of production of the metathesized oil is 1001 $ per ton, thanks to the significant contribution to the coproduced olefins which are valorized to an equivalent of 611 $ per ton of metathesized oil.

However, this table is oversimplistic since it considers data at a single date, in a configuration that may never be reproduced (and some minor costs might have been omitted). Historical data for all raw materials and coproducts can be reprocessed in a retrospective analysis (assuming a plant built to start production in 2000). The cost of production of the metathesized oil can then be compared with the market price for palm kernel and coconut oils (Figure 20.1). A cross-metathesis plant that would be using rapeseed oil is most probably located in Europe, while palm kernel and coconut oils are produced in Southeast Asia; a shipping cost from

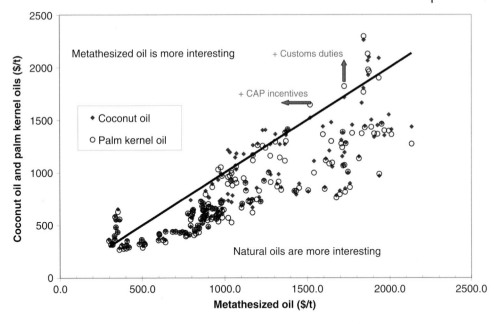

Figure 20.1 Comparison of tropical oils (palm kernel and coconut oils) with metathesized oil cost of production, based on historical data from the 2000 to 2014 period. Note: Historical prices for tropical oils were taken as CIF Rotterdam (meaning delivered in Rotterdam), while rapeseed oil is Free On Board (FOB) (meaning on board of a ship in Rotterdam).

Asia to Europe should be added to their cost. This was taken into account in the prices for tropical oil which are listed as CIF Rotterdam, meaning that they are delivered in Rotterdam. Similarly customs duties – and that's where legislator can play a role to facilitate the introduction of new technologies in Europe – should also be added for a fair comparison. But unfortunately there are no duties on coconut from the Philippines, for example. From this figure, we can see that the metathesized oil is not too far to be competitive with the imported oils on historical basis. Unfortunately (or fortunately) a profit should be added to the plant operation cost, so further improvement of the economics is still needed. Some Common Agricultural Policy (CAP) incentives for production for industrial product could also improve the economics.

20.3.3
Risk Analysis

Another model can therefore be constructed to take into account the uncertainties that we have around the process and the starting materials and coproduct values in the future. Historical data for the period 2000–2014 have been collected from various sources such as Index Mundi [4] (for vegetable oils and crude oil), ICIS pricing [5], CMAI (for 1-butene) [6], and Customs (for 1-decene) [7]. When data where missing, correlations have been established, such as 1-butene versus

Table 20.8 Correlation matrix for raw materials, coproducts, and benchmarks.

First line, product prices vs. time in the 2000–2014 period; second line, statistical distribution of prices in the period; following lines, correlations matrix between reactants and products, including crude oil (Brent).

Raffinate-2 (since 1-butene is extracted from Raffinate-2), and validated on the few data available. Historical data have been plotted in the first line of Table 20.8. In this time period, we have seen large variations in the prices of energy and crude oil but also in prices for vegetable oils. Data have also been processed to show the statistical distribution of prices.

Correlations between all the starting materials and products or benchmarks have also been plotted. Not surprisingly there is a fairly good correlation between

Table 20.9 Uncertainty on main parameters assuming a normal (Gaussian) distribution.

Plant capacity	Rapeseed oil	1-Butene	Catalyst		CapEx depreciation	Olefins
300 kt, rapeseed oil	$/ton	$/ton	$/kg	kg/1000 ton oil	$/ton (plant capacity)	$/ton
Standard deviation, σ	100	100	500	1	5	100
Mean	850	1252	5000	21	33.3	1280

1-butene and crude oil prices and between palm kernel oil and coconut oil. The distribution of data points around the correlations illustrates the dependencies that need to be simulated in the next step. A Monte Carlo model was created including the parameters listed in Table 20.9 to simulate all possible cases. For example, it means that there are 95% chances to have rapeseed oil between 650 and 1050 $ per ton. Based on the correlation matrix (Table 20.8), a dependency was created between rapeseed oil and 1-butene and between rapeseed oil and 1-decene (although rather loose in the latter case). The correlation coefficient was adjusted to find the same type of correlation distribution.

From the simulation model with 10 000 iterations (that number can be extended to increase the accuracy), a mean cost of production of 1039 $ per ton and a standard deviation of 115 $ per ton were found (Figure 20.2), meaning that there are 90% chances to find the production cost between 849 and 1228 $ per ton and 95% chances to have a product cost below 1228 $ per ton.

All these models and simulations illustrate that the most impacting parameters are the purchasing cost of rapeseed oil and 1-butene, but also that the value generated from coproducts cannot be neglected. In the present case, it is a significant driver to bring the cost of production down. This is typical of the "biorefinery" model described initially where value generation from coproducts is a key factor for success. The Tornado plot (Figure 20.3) also illustrates the individual impact of each parameter, at its extremes (mean $\pm 2\sigma$). The sensitivity to the catalyst was assessed both on its price and on its efficiency (assuming a variation on catalyst consumption). Capital cost is only the fifth parameter, meaning that there is probably more benefit to gain in increasing the capital cost if cheaper rapeseed oil (or a cheaper oil) can be processed and/or if a lower amount of catalyst can be consumed.

20.3.4
Lessons for Research and Development Program

The preliminary analysis illustrates where the efforts of research and development (R&D) programs should be focused:

1) Seek for cheaper oils. But taking into account that different fatty acid profile would affect the product distribution (mass balances), palm oil or animal fats

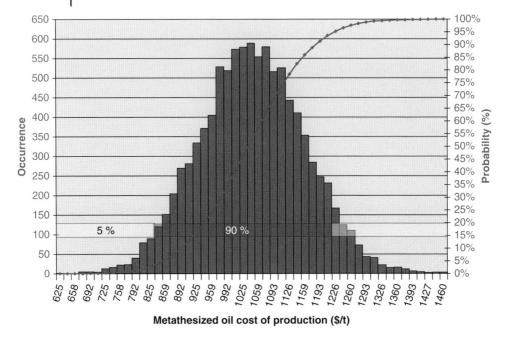

Figure 20.2 Distribution of cost of production from the Monte Carlo simulation, with standard deviations listed in Table 20.9.

are cheaper, but cannot give the expected similarity to coconut oil since they are very rich in saturated chains. Other oils could give also other fatty acid profiles more in line with market expectations.

2) Look at energy optimization. In the model the utilities cost is assumed to be 10% of the variable cost (raw materials and catalyst). If the process can consume less energy, there is a potential source of improvement.

3) Valorize the coproducts. That's should be the leitmotiv for any biorefinery.

4) Seek alternatives to 1-butene. Ethylene could be used, but it is known to inhibit strongly the catalysts, thereby increasing catalyst consumption, and it would significantly affect the product distribution (chain length of the fatty acids). Propylene or 2-butene could be suitable for the reaction but would lead to uneven numbers in fatty acids (C11, C13, etc.) which are not common in the market. So it means that market applications would have to be developed.

5) Look for catalysts that would then be much less inhibited by ethylene (or 1-butene), even at the expense of their cost, to reduce overall catalyst cost contribution.

6) There is probably more value to capture in the downstream processing, and the same analysis should be done taking into account fatty acids and fatty alcohols. And that's where the complexity of the biorefinery model starts to expend.

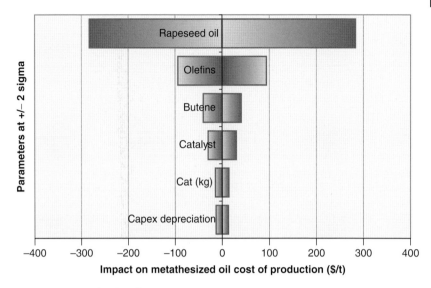

Figure 20.3 Tornado plot of the main parameters.

7) Additional value from the unsaturated versus saturated medium-chain-length fatty acids should be captured. The oils produced have similar chain length than coconut and palm kernel oils, but they have additional functionalities (unsaturations) which should have a value that would need to be investigated.

20.3.5
Lessons for Legislators

The price of biomass in Europe is generally higher than in other parts of the world. This is due to many reasons, including geoclimatic conditions which are limiting the yields. Subsidies for biofuels and not for chemicals are also impacting prices. Customs duties can improve the competition with imported oils. Any CAP mechanism that can help European farmers reduce their cost of production for industrial applications would be beneficial. A global view on the value chain from the farm to the biorefinery needs to be taken, in order to identify all products and coproducts that can be generated, as well as to improve the global economics.

20.4
Summary and Conclusions

There are two main types of biorefineries:

1) Single source of raw material, represented by our castor oil and pinewood plants. They have multiple products linked with the stoichiometry of the

reactions. They need to find the best values for the coproducts that are always produced in the same quantities. They address several markets which have to grow at the same time to satisfy all coproducts.

2) Multiple sources of starting materials, represented by our epoxidized oil and fatty amines/nitrile plants. In this case, multiples sources of biomass and non-biomass products are possible. The plants have to be flexible and can address several markets which can grow independently. The plants can easily accommodate new products and new raw materials.

Arkema has built and operates several plant-based factories (or biorefineries). They have a long history, which means that they survived past economic crisis, such as the oil shocks during the 1970s and 1980s, and more recently the ups and downs of the crude oil in 2008. The common feature is that they produce products which have technical advantage against fossil-based materials. It just happened to be "green" or renewable, since it was the best way to make those products. The plants have a complex network of products and coproducts which contribute to the economic sustainability, but the current equilibrium was not found overnight, which has been built over decades.

Any new "biorefinery" should also look at the valorization of every bit of material available to reach economic sustainability. When a plant is built, it is to last over several years (or decades) so the economics should not be done on a single set of data but taking into account the uncertainties. Upstream and downstream integration (and increased complexity of the plant) is often necessary to reach the economy of scale.

Acknowledgments

The refinement of this preliminary economic analysis is now part of the cosmos project which has been granted in the Horizon 2020 research and innovation program under grant agreement No 635405.

References

1. Faostat http://faostat.fao.org/site/535/default.aspx#ancor (accessed 30 September 2015).
2. Fediol http://www.fediol.eu/ (accessed 30 September 2015).
3. Cohen, S., Luetkens, M.L., Balakrishnan, C., and Snyder, R. Methods of Refining Natural Oil Feedstock. US Patent 8,957,268 Feb 17, 2015.
4. Index Mundi Coconut oil (Philippines/Indonesia), bulk, c.i.f. Rotterdam; Palm kernel Oil (Malaysia), c.i.f. Rotterdam; Rapeseed Oil; Crude, fob Rotterdam; Crude Oil (petroleum), Dated Brent, light blend 38 API, fob U.K., US Dollars per Barrel. http://www.indexmundi.com/commodities (accessed 30 September 2015).
5. ICIS http://www.icispricing.com/ (accessed 30 September 2015).
6. IHS http://www.cmaiglobal.com/ (accessed 30 September 2015).
7. Zauba https://www.zauba.com/customs-import-duty/(as example) (accessed 30 September 2015).

21
Colocation as Model for Production of Bio-Based Chemicals from Starch

Ruben Jolie, Jean-Claude de Troostembergh, Aristos Aristidou, Massimo Bregola, and Eric Black

21.1
Introduction

Today's society is highly dependent on finite fossil resources for the production of energy, fuels, and chemicals. The still growing demand for these resources and their gradual depletion and the awareness of the environmental impact of gas emissions from large-scale use of fossil fuels are strong drivers to explore renewable resources for sustainable production of energy, fuels, and chemicals. One of the main and most versatile renewable resources is organic biomass [1, 2]. Biomass is typically composed of a complex mixture of various macro- and micromolecules. In the past decades, new concepts and technologies have continuously been emerging to fractionate and convert biomass into a diverse spectrum of valuable products, originally for food and feed but also for industrial applications replacing oil-based fuels and chemicals. In analogy with the mineral oil refineries in the petrochemical industry, manufacturing sites applying these concepts have been termed *biorefineries*. Many types of biorefineries exist already at commercial scale and can be classified based on the raw materials that are processed. The most common categories are starch (e.g., corn/maize, wheat, potato), sugar (e.g., cane, beet), fats and oils (e.g., oilseeds, animal fats), and lignocellulose (e.g., wood, grass, residues, and waste) biorefineries [2, 3]. The current chapter will focus on the starch-based biorefineries that process maize (corn) and wheat into an array of intermediate and end products to serve food, feed, and industrial markets, hence playing a central, established but still expanding role in today's bioeconomy.

The use of starch from crops like maize and wheat for nonfood applications is not free of controversy. The growth in nonfood use of food crops (primarily as biofuels) is considered to be one of the (many) interrelated factors that contribute to the complex problem of food insecurity around the world today, besides other obstacles like production shortfalls, supply disruptions (e.g., due to political instability or weather), food waste, government policies that inhibit trade and

Chemicals and Fuels from Bio-Based Building Blocks, First Edition.
Edited by Fabrizio Cavani, Stefania Albonetti, Francesco Basile, and Alessandro Gandini.
© 2016 Wiley-VCH Verlag GmbH & Co. KGaA. Published 2016 by Wiley-VCH Verlag GmbH & Co. KGaA.

negatively affect farmers, the impact of agriculture on the environment, growing resistance to the use of agricultural technology, and price volatility [4–8]. While nourishing the world's growing population remains the most important challenge, it should yet be recognized that a *balanced* utilization of renewable biomass, such as agricultural crops and lignocelluloses, for industrial scopes (fuels and/or chemicals) can play an essential role in mitigating the depletion and environmental effects of natural nonrenewable resources [9]. A prerequisite for this is to have competitive bio-based value chains. Moreover, the right *balance* will likely undergo a stepwise evolution. Eventually it should be the ambition to deploy only nonfood (e.g., lignocellulose) biomass for nonfood applications and to reserve as much as possible the edible (e.g., starch) fractions of all biomass for human consumption, but technology has not progressed sufficiently to already make this move economical today [10]. Bio-based products can only be economically successful in today's and tomorrow's markets if they can compete in terms of price and performance with their petrochemical counterparts. This requires optimized bioconversion technologies (to convert the biomass into a useful reaction substrate and to convert this substrate into a bio-based end product) but also large amounts of readily available price-competitive raw materials and efficient supply chains. Today, for lignocelluloses, all of these elements are still in their development stage, and it will take several more years of additional development to become truly economical and widely deployed (see, for instance, in [3, 11–13]). Starch, on the other hand, can already rely on decades of industrial practice and optimization at commercial scale. It is readily available on the market in amounts that exceed the demand for food applications and has traditionally served nonfood applications such as paper and corrugated materials [14]. With lignocellulose being the future, starch has an important role to play as renewable resource in the bio-based economy of today because it enables the development and deployment of the biotechnologies downstream of the substrate production, thus generating proofs of concept that represent essential fuel for the engine of bioeconomy investments.

For biomass to become a true, sustainable alternative to fossil resources, it is essential that bio-based products can be manufactured economically, regardless of the specific raw material used. Further innovation and optimization in all steps of both existing and new value chains are needed and should be pursued without changing direction each time the mineral oil price changes. Maximal valorization of all parts of the biomass as food, feed, or industrial products according to the biorefinery concept is one important enabler. Share and leverage existing assets and specific expertise of different partners along the value chain by deeply integrating and/or colocating subsequent bioprocessing activities are an additional element that can further improve the economics of green technologies. The current chapter will extensively discuss these concepts from a scientific as well as economic point of view, taking the starch biorefineries as highly relevant and successful case study.

21.2
Wet Milling of Cereal Grains: At the Heart of the Starch Biorefinery

Since ever, starch is a very important source of food, feed, and industrial products. Technologies related to starch production and utilization have been developed fast during the last century, and they are largely reported in literature [15]. Most of the starch produced globally, either for sale as starch or for conversion to other products, is derived by wet milling of maize (corn), wheat, and potatoes. The isolation of starch from the original raw materials follows a process based on a variety of physical and chemical separation steps. While originally aiming specifically at the isolation of starch, the present wet milling processes – as true biorefineries – succeed in the fractionation of a wide range of coproduct streams enriched in proteins, fibers, and/or fats, seeking optimum use and maximum value from each constituent of the raw material. The specific processing steps as well as the specific side streams depend on the type of raw material. In the context of this chapter, focus is put on the wet milling of the cereal grains maize (corn) and wheat.

21.2.1
Wet Milling of Maize (Corn)

The wet milling of maize, as illustrated in Figure 21.1, starts with soaking the cleaned grains in a dilute sulfur dioxide solution to soften the kernels, a process called *steeping* [16, 17]. During the soaking, soluble nutrients are absorbed in the

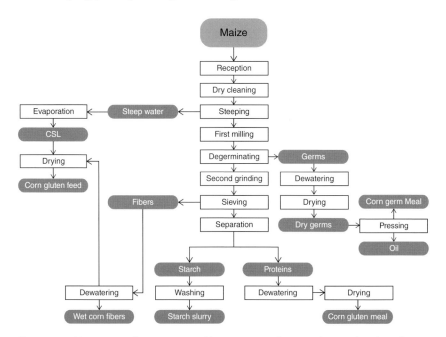

Figure 21.1 Maize wet milling process yielding a variety of intermediate and end products.

steeping water. This water is often evaporated to concentrate the nutrients, forming a product called *corn steep liquor* (*CSL*) or *heavy steep water*. The soaked kernel is coarsely milled to allow removal of the germs in cyclone separators. From the dried germs, the *oil* is recovered by pressing, leaving the *corn germ meal*. The oil extraction can be done either on-site or in a separate crush plant. The remainder of the kernels is subjected to a fine grinding which releases starch and gluten protein from the fiber. Fibers (*corn bran*) are caught on screens and often combined with coproduct streams like CSL and/or corn germ meal to produce *corn* (*gluten*) *feed*, which can be dried (for preservation) or not. Combined, these streams represent about 20–25% of the processed corn. The starch–protein suspension passes through a centrifuge, separating the dense proteins from the starch. While the first are dried and mostly used in animal feed (*corn gluten meal*), the latter is further washed for removal of traces of protein to obtain a >99% pure starch slurry.

21.2.2
Wet Milling of Wheat

Wet milling of wheat, graphically represented in Figure 21.2, shows similarities but is clearly distinct from the maize process [18, 19]. The wheat grains are first dry-milled to obtain *wheat flour*, *wheat germs*, and *wheat bran*. The flour is mixed with water to yield a dough, on which the wet milling or wet separation process is applied. Different industrial processes exist to physically separate the various components present in the dough, but in principle they all rely on the differences in water solubility, density, and particle size of the composing macromolecules.

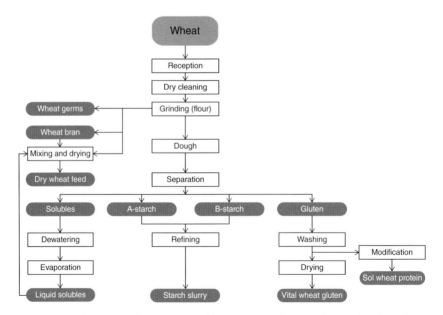

Figure 21.2 Wheat wet milling process yielding a variety of intermediate and end products.

Modern-day wet milling methods typically use a combination of screens, washers, (decanter-type) centrifuges, and/or hydrocyclones for separation of starch and protein from the water dispersion. Besides *A-starch* (i.e., first-grade, large starch granules) and wheat *gluten*, a number of coproduct streams are generated, often referred to as *B-starch* (i.e., second-grade, small starch granules) and *wheat solubles* (fine fibers, pentosans, water solubles, and process water). The latter is sometimes called *wheat C-starch*, because it contains residual starchy material that did not end up in the A- or B-starch streams. The solubles are often concentrated and mixed with the bran to form a *(dry) wheat feed*. The wheat gluten can be dried right away to *vital wheat gluten* or, alternatively, first enzymatically modified to *soluble wheat protein* before drying.

21.2.3
Downstream Processing of Starch Slurry

The fractionation of the starchy raw materials into a *starch slurry* and their other constituents is only the first step of biorefining at or near the cereal wet mill. Today, all starch producers also own, to a different extent, the downstream processing of the starch slurry, which is schematically summarized in Figure 21.3. On the one hand, starch slurries can be dried as such or subjected to thermal or chemical starch modifications to obtain a variety of *native and modified starches* for food as well as nonfood applications [20]. On the other hand, starches are converted

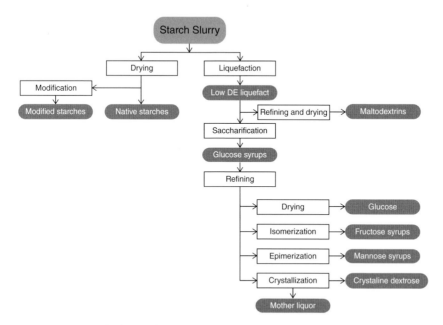

Figure 21.3 General scheme of the downstream processing of a starch slurry, mostly executed at the wet mill site (DE = dextrose equivalent).

chemically and/or enzymatically by liquefaction and saccharification in a portfolio of liquid syrups with varying dextrose equivalent (DE) and sugar composition. These include glucose, fructose, mannose, and maltose, among others [21, 22].

Depending on the end application, different refining steps are performed on the syrups to reach the required purity level. Given the nonsterile nature of part of the process chain, additional and costly processing steps need also to be executed to increase the shelf life of the commercial products. Common is concentration to high dry substance by evaporation, which also favors shipment costs. Alternatively, crystallization of glucose syrup yields *crystalline dextrose* (monohydrate or anhydrous), leaving a *mother liquor* also called *dextrose greens*.

21.2.4
Cereal Grain Wet Mill Products as Raw Materials for Bio-Based Chemicals

All the above elucidates the broad portfolio of intermediate and end products that originate from of a cereal grain wet mill and glucose refinery, going well beyond the classical examples of starch slurries and glucose syrups. An important potential application of wet mill products is in the production of bio-based chemicals, both bulk and fine chemicals, using further bioprocessing technologies [22, 23]. The best-known case is probably the use of starch hydrolyzates as carbon source in microbial fermentation processes, but there is more: also coproducts of the starch production can find applications as nutrients in fermentation processes, and other (bio)chemistries than fermentation exist to convert starch into value-added molecules (see also Section 21.4). To run these bio-based processes, a company could build a stand-alone greenfield plant and have its raw materials shipped there. However, to improve overall bioprocess economics, an alternative, smarter approach could be considered.

21.3
The Model of Colocation

21.3.1
Creating Value through Economies of Scale

A common method to assess the attractiveness of investing into a manufacturing plant, for instance, for a bio-based product, is to use a *discounted cash flow* (DCF) analysis to determine the net present value (NPV) of future cash flows created by investing capital into a project [24, 25]. This methodology calculates the time value of money by discounting gains in future years to reflect their value in today's currency. This deduction is exponential over time. Therefore, the cash generated in later years is more heavily discounted than the cash created in earlier years. The initial capital used to fund the business operations is not discounted, as it is deployed today. The NPV of the project is the summation of the future

cash flows created *minus* the capital invested. If this calculation results in a positive value, then the determination is that spending capital on the project today
will generate more value in the future. However, if the result of the calculation is a
negative value, then the determination is that the investment will result in a loss of
value. The more risk involved in building a business, the higher the expected rate
of return is for an investor. Within a DCF analysis, higher risk leads to a higher
discount of future cash flows generated by the business, thereby requiring the
business to produce greater cash flows to justify the risk of the initial investment.

Based on the construct of the DCF analysis, the business attractiveness is
improved if (i) the future cash flow projections increase, (ii) the up-front capital
demands of the business decrease, and/or (iii) the risk is lowered, thereby
lowering the discount rate.

Increasing the cash flow can be accomplished by reducing the cost of goods sold
(COGS) of the product, thus allowing higher margins of the business. The COGS
is the direct cost incurred by a company to produce the goods to be sold. It is
the result of all fixed costs and all variable costs involved in the direct production process. Fixed costs (e.g., plant, equipment) are independent of production
output, while variable costs (e.g., labor, materials) do vary with output; they rise
as production increases (Figure 21.4). The unit cost is the total COGS divided by
the total units produced during a time period. Generally, one way to lower the
achievable overall unit cost is by increasing the production rate through the production plant, because fixed costs are then shared over a larger number of goods
resulting in a lower per-unit fixed cost. The cost advantage arising from increased
production output is often referred to as *economies of scale*. Economies of scale do
not only affect fixed costs but may also reduce variable costs per unit because of
operational efficiency gains, synergies, and improvements in the company's purchasing power. This would be graphically reflected in Figure 21.4 by a decrease in
the slope of the total cost line.

21.3.2
The Benefits Gained from Colocation at Starch Biorefineries

Investments in businesses processing biomass, such as starchy cereals, into different end products can be judged by using the DCF analysis outlined earlier. Based

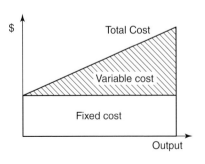

Figure 21.4 Total, variable, and fixed costs in
function of production output.

on the drivers in this model, an attractive way to improve overall economics hence allowing better cost positions for the various product offerings of businesses all along the bio-based value chain is the colocation at a raw material processing site (e.g., a wet mill) forming a biorefinery campus, because it contributes to (i) higher cash flows, (ii) lower capital demands, and (iii) lower risk for all partners involved. This hub-and-spoke method has long been used in the manufacturing of petrochemicals. It has created large highly efficient manufacturing operations and a model of manufacturing that the bioeconomy can utilize to create competitive solutions for customers. The various benefits of how colocated biobusinesses can operate with superior economics to stand-alone greenfield plants in light of the DCF analysis will be examined here.

21.3.2.1 Increased Cash Flow

Colocation can boost the (future) cash flow and improve the cost position of bio-based companies by lowering the COGS in various ways.

Probably the most obvious benefit of neighboring processing activities sits in the variable cost of the raw materials. As the latter constitutes a significant part of the final unit cost of a bio-based product, lowering it will have a substantial effect on the overall profitability of a biobusiness. Colocation at a biorefinery campus achieves this by reducing or eliminating the transportation cost since the (co)product(s) of one process can be supplied directly by pipeline for use as raw materials for another process. A typical example is the supply of glucose syrups from a cereal wet mill and refinery as carbon source for a fermentation process. Additional advantages of the proximity of the feedstock production site (e.g., the wet mill) relate to the avoidance of the variable costs of unit operations for making materials shippable and storable (e.g., purification, drying, evaporation, etc.) and the occurrence of a wider variety of types and purities of raw materials than there is on the market. Close interaction between the various partners can allow selection and further optimization of the most suitable and hence cost-effective feedstock for the given application.

Another way to lower costs of inputs by colocation relies on increasing the size of manufacturing operations. Economies of scale have an effect on the cost of both raw materials and utilities. In the case of processing cereals for starch, the larger the wet mill, the lower the fixed cost contribution is to the unit price of the derived products. This benefit may be maximized by grouping bio-based businesses together to create more demand in a centralized biorefinery. In fact, by colocation, economies of scale are created without all individual partners necessarily being *at scale*. They are reduced or eliminated if the bio-based business builds on an independent site, utilizes different raw material feedstocks, or is dependent on different energy source to operate the production assets.

Once operations are colocated, the manufacturing operations may have the additional opportunity to outsource tasks to on-site providers. The purchasing power of companies is increased by bundling the site's demand for products and services. This grouping allows the participating companies to lower their collective cost for services like preventative maintenance, in-process chemicals, and site security.

21.3.2.2 Reduced Capital Demand

The capital cost of a stand-alone greenfield commercial-scale bioprocessing plant typically ranges between US$100 and US$300 million [26]. This cost includes buildings and equipment used in the production, quality assurance, and storage of product. Also included in this figure is the support infrastructure – utilities for power and steam, wastewater treatment, cooling tower water, warehousing, administrative buildings, fencing, interplant roads, and rail tracks. A company will typically spend 60–70% of the capital of a project on the equipment and infrastructure that is directly used in the production of each unit and 30–40% on the infrastructure that supports the overall operation [27].

The attractiveness of an investment will improve if the capital to build or the working capital needs are decreased. Colocating bio-based manufacturing facilities that utilize similar assets allows for the sharing of these physical assets, thus reducing the investment required to manufacture a product. This typically occurs outside the boundary limits (OSBLs) of the manufacturing plant, where operations like wastewater treatment and utilities (electricity, steam, compressed air, water) are found. The equipment inside the boundary limits (ISBLs), which is directly utilized in the production of the bio-based product, typically offers limited opportunity to share. The shared equipment in the OSBL is typically owned and operated by the biorefinery feedstock provider (e.g., the cereal wet mill and glucose refinery). It may include boilers, aerobic and anaerobic wastewater treatment digesters, cooling towers, and spare parts warehousing, among others. The feedstock provider acts as the core of the biorefinery producing the desired raw material needed to maximize the effectiveness of the bioprocesses. These products are supplied to colocated partners via pipeline, reducing or eliminating the need for working capital utilized for raw material storage. In addition the feedstock provider has preinvested into the OSBL equipment necessary to support colocated partners. The benefit is that these investments are avoided by the colocated partner and are shifted from capital investments into ongoing operating costs. This reduction in up-front capital can increase the NPV of a project.

21.3.2.3 Reduced Manufacturing Risk

The biorefinery concept can reduce the overall risk of colocated bio-based companies. Reduced risk is an increased probability that the manufacturing plant will perform as stated. The two previously discussed benefits of economies of scale and reduced up-front capital improve competitiveness, which can reduce commercialization risk. The manufacturing risk can be reduced by the redundancy and excess capacity of OSBL equipment created by a colocation model.

The overall reliability of a bio-based process is the summation of the reliability of all critical processing equipment. Critical processing equipment is the equipment necessary for operations so that, if unavailable, the manufacturing plant is unable to produce at its nameplate capacity. Direct production equipment (e.g., fermenters, separators, evaporators, etc.) is often owned and operated by the bio-based company, but OSBL equipment can be shared with other colocated partners. This creates an opportunity to access redundant and excess equipment. The ability to share this excess capacity can allow the colocated partner(s) to jointly improve their overall reliability. This is often not an option with a stand-alone facility, as companies cannot afford to build in redundancies for all unit operations. Redundant unit operations and excess capacity lower manufacturing risk, which can be a justification for lowering the *discount rate* variable within the DCF analysis. The discount rate takes into account the time value of money and the risk or uncertainty of future cash flows; the greater the uncertainty of future cash flows, the higher the discount rate. A reduction will reflect as an increase in the NPV of a contemplated bio-based commercial plant.

Utilizing a DCF method, it may be stated that a colocation model for bioprocesses can improve the competitiveness and reliability of a commercial-scale bio-based factory by reducing COGS and up-front capital demands while increasing the plant's overall reliability. This combination can be shared and leveraged by multiple colocated operations, with only an incremental investment on the part of the feedstock provider.

21.4
Examples of Starch-Based Chemicals Produced in a Colocation Model

Cereal grain wet mills produce a variety of intermediate and end products that can function as raw materials for the production of bio-based chemicals for food and nonfood applications. It was argued earlier that colocation with the wet mill and glucose refinery brings numerous advantages to improve production economics. In what follows, this will be illustrated by a number of successful commercial-scale examples of bio-based food and nonfood products produced in a colocation model, primarily focusing on fermentation products but also touching on some others. A schematic overview is provided in Figure 21.5. Other commercial-scale examples that were not selected for discussion here include several amino acids (lysine, glutamate, aspartate), enzymes, and organic acids (itaconic acid, levulinic acid, glucaric acid).

21.4.1
Ethanol

The production of ethanol or ethyl alcohol by yeast fermentation of starch-based feedstocks is among man's earliest ventures into value-added bioprocessing. While the historical focus has long been on food applications, most of the ethanol

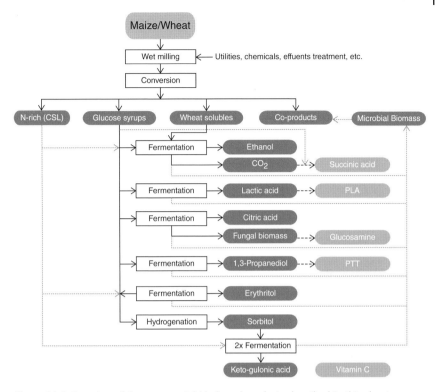

Figure 21.5 Overview of the commercial bio-based products described in this chapter.

produced nowadays is used as biofuel. The common production process for alcohol from cereal starch starts with either dry (whole grain) or wet milling of the grains. In the context of this chapter, two examples of wet mill colocation are considered.

21.4.1.1 Corn Wet Mills for Biofuel Ethanol

Dozens of corn wet mill ethanol colocation facilities exist across the globe, with a particularly high concentration in North America. In that geography, Cargill operates plants in Blair (NE) and Eddyville and Fort Dodge (IA). Other important producers include Archer Daniels Midland (ADM), Tate & Lyle, and Penford Corporation. Typically, the starch slurry coming from the wet mill is liquefied by a high-temperature α-amylase process and saccharified by glucoamylase to a certain DE before transporting it by pipeline to the nearby fermentation plant. After supplementation of the substrate with a nitrogen source, the *Saccharomyces cerevisiae* yeast will ferment glucose into ethanol and CO_2. After fermentation, the resulting "beer" is transferred to distillation columns where the ethanol is separated from the remaining "stillage." The ethanol is dehydrated by means of molecular sieves while the stillage is further processed into dried distillers grains with solubles (DDGS), a high-quality, nutritious livestock feed [28–30]. The CO_2 released during fermentation can be captured and used for multiple applications on- or off-site. These include current uses in carbonated beverages and the manufacture of dry ice, as well as potential future uses as carbon source in fermentation or algae cultivation. In fact, the use of the ethanol coproduct CO_2 in the succinic acid fermentation is applied by BioAmber in its demonstration plant at the agro-industrial complex Les Sohettes in Pomacle-Bazancourt (France).

Colocation of a corn wet mill and an ethanol plant has some specific advantages for both. From the perspective of the wet mill, having an ethanol facility as swing plant next door that can absorb varying volumes of starch and can hence follow the fluctuating demand on starch for other higher-value applications offers a way to stabilize the grind rate on the mill, to maximize the plant capacity utilization, and to respond to changing market prices of the different end products (for instance, due to the seasonal changes in demand for sweeteners). Moreover, the robustness of the ethanol fermentation process allows the addition of wet mill side streams that have no other obvious outlets, such as off-spec hydrolyzate, floats, and broken corn. For the ethanol production, taking the fermentation feedstock directly from the saccharification tanks makes it possible to eliminate costly refining steps (filtration, evaporation) that are typically required for product stability if a starch hydrolyzate would be shipped. In addition, the light steep water from the wet mill process can be used as rich but perishable source of fermentation nutrients, including amino acids, proteins, vitamins, minerals, and cofactors.

21.4.1.2 Wheat Wet Mill for Premium Grain Alcohol

A less widespread but very interesting example of colocation for fermentative production of grain alcohol can be found at some wheat wet mills, including those operated by Cargill in Manchester (United Kingdom) and Sas van Gent (Netherlands) where both premium potable and industrial alcohol are manufactured. Similar processes are being run by Tereos-Syral and Südzucker.

The wet separation process of wheat flour into A-starch, B-starch, and vital wheat gluten generates significant amounts of a coproduct referred to as *wheat solubles* or *C-starch* (see Figure 21.2). This coproduct is not suitable for pure glucose production due to the high fiber and high protein content. However, after subjecting it to an enzyme hydrolysis process similar to the common starch slurry hydrolysis, it represents a rich fermentation feedstock for *S. cerevisiae* yeast for production of ethanol and CO_2. The fermented wheat mash can be turned into high-end alcohol by distillation and rectification steps.

In specific cases in Manchester and Sas van Gent, the ethanol factories were built and operated for several years by the company Royal Nedalco, fermenting a fresh wheat liquefact supplied through a pipeline from the Cargill wheat plant. The distillation residue called *stillage* was returned to Cargill for integration in its animal feed offerings. The supply agreement around quantities and specifications of the biomass streams between the two companies was complemented with arrangements around shared utilities (process water, steam, compressed air), shared wastewater treatment, shared emergency, safety and security plans, as well as land lease. In 2011, Royal Nedalco's alcohol operations were acquired by Cargill. In 2015, an additional alcohol factory, colocated at Cargill's wheat wet mill in Barby (Germany) and demonstrating a similar level of integration between milling and fermentation operations, will come online.

21.4.2
Lactic Acid

Lactic acid or 2-hydroxypropionic acid is a bulk chemical still demonstrating strong market growth for a wide range of applications. Being Generally Recognized as Safe (GRAS) classified, it has a long history as preservative and pH-adjusting agent in the food and beverage sector. Additionally, it is used in the pharmaceutical and chemical industries, for instance, as a solvent and a starting material in the production of lactate esters (ethyl and butyl lactate). The latter have a growing application as biodegradable and nontoxic solvents. Also the polymerization of lactic acid into a biodegradable polymer polylactic acid (PLA), which is used as green alternative to petroleum-derived plastics in food packaging as well as mulch films and garbage bags, has significantly increased global interest in lactic acid manufacturing [31, 32].

Presently, more than 95% of the industrial production of lactic acid is based on microbial fermentation, due to lower production cost and market drivers for bio-based products. It is commercially mature, with many producers globally, including Cargill, Corbion, Galactic, and ADM. Commercial lactic acid fermentations are typically based on glucose derived from starchy materials mostly using *Lactobacillus* species at $30-42\,^\circ$C [33]. The output of fermentation is a crude aqueous lactic acid solution which is concentrated by evaporation. During the bacterial fermentation process, the pH of the culture medium declines owing to the accumulation of lactic acid, thereby affecting growth and productivity of the bacteria. To counteract this acidification, chemicals are added to the culture medium during the process. Consequently, additional operations are required afterward to regenerate the nondissociated acid and to facilitate its extraction from the growth medium.

The bioindustrial campus in Blair, Nebraska (United States), is a compelling example of colocation for the lactic acid value chain. Cargill's corn wet mill supplies by pipeline a glucose syrup as fermentation substrate for the lactic acid fermentation at two different adjacent plants: one of Corbion and one of Cargill, the world's largest. The latter, on its turn, supplies a purified lactic acid stream to the neighboring PLA plant of NatureWorks, a joint venture (JV) between Cargill and PTT Global Chemical, having a nameplate capacity of 140 000 metric tons of biopolymer per year [34]. In the PLA manufacturing process from lactic acid, the acid molecules are chemically converted to the cyclic diester "lactide," of which the ring opens and links together to form long chains of polylactide. The polymerization yields PLA resin pellets that are marketed under the Ingeo® brand name and have the design flexibility to be made into fibers, coatings, films, foams, and molded containers [32, 35].

Until December 2008, the lactic acid fermentation at Cargill used *Lactobacillus*, and the pH was controlled to near neutral by the addition of calcium hydroxide. The lactic acid broth was then acidified by adding sulfuric acid. This process resulted in near stoichiometric usage of lime and sulfuric acid, relative to the lactic acid produced, and subsequent formation and precipitation of gypsum. The gypsum was removed by filtration and the lactic acid concentrated by evaporation. To enable large-scale application of bio-PLA, the reduction of lactic acid production cost through elimination of these by-products has long been a goal of NatureWorks/Cargill, and extensive research was carried out to develop an improved process. In December 2008, a novel yeast-based fermentation technology was introduced [35–37]. It allowed operation of the fermentation at significantly lower pH, thereby significantly reducing the need for lime and sulfuric acid as well as the production of gypsum, while delivering lactic acid production rates and yields similar to those of traditional lactic acid processes. Also, less energy was required to drive the process. Small quantities of gypsum are still being produced, which are used as soil conditioners and so replace mined gypsum. In 2010, Cargill won the "Industrial Biotechnology Award" for distinguished accomplishments in biotechnology, presented by the American Chemical Society. This award honored both the technical innovation that dramatically reduced the chemical costs and environmental impact associated with lactic acid production and Cargill's success in applying the new technology to produce lactic acid at industrial scale.

21.4.3
Citric Acid

Citric acid is a weak organic acid with three carboxyl groups. It is a very-high-volume commodity product with a global annual production surpassing 1.7 million tons and still growing. Thanks to its nontoxic nature, it is used in foods and beverages, in household detergents and cleaners, or in pharmaceutical formulations. Besides, about 10% of the volume is used in industrial applications (textile, cement, metallurgy) [38].

It was originally discovered in citrus fruits, and the first large-scale production of citric acid was by isolation from lemon juice at the end of the nineteenth century. After discovery of efficient citric acid production by certain *Penicillium* and *Aspergillus* strains, Pfizer began industrial-scale fermentative manufacturing shortly after World War I. Nowadays, citric acid is primarily produced by microbial fermentation, mostly using the filamentous fungus *Aspergillus niger* in aerobe bioreactors on starch- or sucrose-based media. The best strains of *A. niger* can produce more than 200 g/l of citric acid, even at low pH (≤ 2.0) [38–41]. Key manufacturers in the citric acid market include ADM, Anhui BBCA Biochemical Co, Cargill, DSM, Gadot Biochemical Industries, Jungbunzlauer, and Tate & Lyle.

Citric acid fermentation processes run by Cargill (United States and South America), ADM, and other companies are based on the biorefinery concept, using primarily corn starch as the carbohydrate feedstock from an adjacent corn wet mill process. As true biorefineries, these processes can generate various valued coproducts in addition to the head product through complementary chemical or enzymatic approaches, which can help maintain the economic viability in a mature market. The fungal biomass associated with the citric acid process is in the 100 000 s of million tons per year and is composed of substantial amounts of chitin and β-glucans [42]. Through elaborate process development, Cargill was able to launch in 2003 a vegetarian glucosamine product, Regenasure®, derived from the citric acid fungal fermentations by an acid hydrolysis process of the chitin portion of the fungal mycelium run on-site.

21.4.4
Sugar Alcohols

Sugar alcohols are a class of polyols that can be derived from sugars. One well-known, commercially applied technology for the production of sugar alcohols is the hydrogenation of sugar aldehydes or ketones at high temperature under hydrogen gas pressure and in the presence of a nickel catalyst [43]. Sorbitol is the sugar alcohol derived from glucose and also the most common polyol. Other polyols derived from starch by hydrogenation include maltitol and mannitol. Various liquid syrups produced at starch biorefineries, including glucose syrup at different DE, mannose syrup, and mother liquor (see Figure 21.3), can be employed as feedstock for the hydrogenation. In starch plants of companies like Cargill, Roquette, Tereos-Syral, ADM, and Corn Products International, hydrogenation plants are often integrated into or colocated with the wet mill and are supplied with raw materials straight from the glucose refinery.

In addition, there also exist polyols that are not obtained by sugar hydrogenation. Erythritol, for example, is produced by fermentative conversion of glucose using an osmophilic, non-GMO yeastlike fungus *Moniliella pollinis* [44]. The process was developed in the 1990s by Nikken/Mitsubishi in Japan and by Cerestar in Belgium and industrialized in Italy. In 1999, Mitsubishi and Cargill established a JV to produce erythritol at Cargill's corn mill in Blair, Nebraska (United States). The corn milling plant supplies glucose syrup and steep water as feedstocks for the fermentation, while the microbial biomass produced during fermentation is mixed in the feed of the corn milling plant. Steam and electricity are also supplied by the biorefinery site. Today, the erythritol plant in Blair is fully owned by Cargill. Erythritol is also manufactured in Europe and China by other companies.

Sugar alcohols can be used either in liquid form or as powder produced by crystallization. Main applications are food, pharmaceuticals, and oral and personal care products. They have a lower caloric content but also a lower sweetness than their sugar analogs and are not metabolized by bacteria in the mouth, thus not contributing to tooth decay [44–46]. Besides, some of them can be used as precursor for value-added chemicals, as will be illustrated in the next section. Recently, certain sugar alcohols are also finding newer applications as intermediates for the production of hydrocarbon fuels through aqueous-phase catalysis [3, 43].

21.4.5
2-Keto-L-gulonic Acid

The sugar alcohol derived from glucose, sorbitol, forms also the starting point for vitamin C production. The current industry standard is a complex, multistep fermentation process, invented by Yin in China [47]. Sorbitol is converted in 2-keto-L-gulonic acid, from which vitamin C (ascorbic acid) can be obtained by chemical ring closure (lactonization). Since its implementation in China and Europe in the 1990s, this bioprocess has displaced the traditional chemical Reichstein process for vitamin C production [48, 49].

Keto-L-gulonic acid production from sorbitol actually entails two fermentation processes. Sorbitol is first converted into sorbose by a microbial oxidation using *Gluconobacter oxydans*. In a second step, sorbose is oxidized into 2-keto-L-gulonic acid by a mixed culture comprising *Ketogulonicigenium vulgare* and *Bacillus megaterium* [48, 49]. Synergic interactions between the two microbial cultures have been extensively studied. Several metabolites produced by one culture are used by the other culture [50].

Keto-L-gulonic acid production is operated by Cargill in Krefeld (Germany) in an integrated way with the sorbitol production, which on its turn is colocated with the corn wet mill. The colocation benefits in this case go beyond the supply of the fermentation carbon source: in both fermentation steps, CSL originating from the corn steeping process is used as nitrogen source. Additionally, the fermentation plant relies on chemicals and utilities from the corn milling plant, including soda for acid neutralization, acid for acidulation, and steam for sterilization, product concentration, and crystallization.

21.4.6
1,3-Propanediol

1,3-Propanediol (PDO) is an important versatile industrial chemical used as building block in the production of various polymers including polyesters (e.g., polytrimethylene terephthalate (PTT), a fiber for carpets), polyethers, and polyurethanes. It is also widely used in food, cosmetics, liquid detergents, and industrial applications like antifreeze [51]. Its market volume is expected to grow from an estimated $310 million in 2014 to more than $600 million by 2021 [52].

The conventional chemical PDO synthesis starts from the petroleum derivatives acrolein (Degussa–DuPont process) or ethylene oxide (Shell process) [51]. More recently, a bioconversion process for PDO has been developed and commercialized using a genetically modified strain of *Escherichia coli*. The PDO pathway which was successfully transferred to *E. coli* originates from *Klebsiella pneumoniae*. This organism naturally converts glycerol into 3-hydroxypropionaldehyde (3-HPA) by a coenzyme B12-dependent dehydratase and subsequently reduces the 3-HPA to 1,3-PDO by an NADH-dependent 1,3-PDO oxidoreductase [51, 53, 54]. The *E. coli* strains were actually engineered to be able to utilize glucose instead of glycerol as the primary fermentation substrate [55]. Technologies to use crude glycerol, the main side stream from biodiesel production at oilseed biorefineries, have been studied widely but are not yet commercially applied.

Since 2006, the DuPont Tate and Lyle Bio Products JV produces Bio-PDO™ in Loudon (Tennessee, United States) [56]. The current plant capacity is £140 million per annum. The facility was located in Loudon to take advantage of the availability of corn glucose, the main feedstock for Bio-PDO™ production, which is produced at the Tate & Lyle wet milling facility on-site. A proprietary strain and fermentation process, jointly developed by both companies, is deployed, requiring also water, vitamins, minerals, and oxygen. After fermentation, the PDO is separated from the aqueous broth through distillation and further purified to obtain a clear, slightly viscous liquid. Different grades are obtained, targeting different applications (food, personal care, and industrial) but all considered readily biodegradable. According to DuPont Tate and Lyle Bio Products, a peer-reviewed life cycle assessment demonstrates that the production of bio-based PDO offers significant environmental benefits including up to 40% less greenhouse gas emissions and 40% less nonrenewable energy used in production versus petroleum-based glycols like propylene glycol. The American Chemical Society awarded the Bio-PDO™ research teams the "2007 Heroes of Chemistry" award for their achievements [56].

21.5
Summary and Conclusions

Biomass represents an attractive sustainable alternative to fossil resources for the production of a portfolio of chemicals. However, the key challenge is to manufacture bio-based products in an economical way to make them truly cost competitive. Efficient bioconversion and extraction technologies play a central role, but also many other elements impact the economics of a biobusiness. According to the DCF analysis, the economics can generally be improved by increasing cash flow (for instance, by lowering the COGS of the product through economies of scale), reducing capital cost, and/or reducing risk. As outlined in this chapter, all of these can be positively influenced by colocating different pieces of the bio-based value chain at the biomass processing site forming a biorefinery campus.

The type of biorefinery based on traditional wet milling of cereal grains to generate starch as well as a range of useful coproducts has proven particularly successful as platform to create at the same location an even wider variety of derived products using fermentation and other conversion technologies for food as well as nonfood applications. Commercial examples, some of which were extensively discussed here, include bioethanol, sugar alcohols, PDO, several organic acids, amino acids, and enzymes. In many cases, the benefits of colocation go well beyond the "over-the-fence" supply of low-cost feedstock: many other assets and know-how present at the wet will can be leveraged, and economies of scale can be created without the different individual operations necessarily being *at scale*.

By contributing to the commercial viability of bio-based value chains in various ways, the model of colocation can be considered an intrinsic enabler of the bio-based economy of the future, also when the transition from food crop biomass to nonfood biomass like lignocelluloses will be established.

References

1. Rostrup-Nielsen, J.R. (2005) *Science*, **308**, 1421–1422.
2. Cherubini, F. (2010) *Energy Convers. Manage.*, **51**, 1412–1421.
3. Maity, S.K. (2015) *Renewable Sustainable Energy Rev.*, **43**, 1427–1445.
4. OECD/FAO (2011) *OECD-FAO Agricultural Outlook 2011*, OECD Publishing, Paris.
5. Gustavsson, J., Cederberg, C., Sonesson, U., van Otterdijk, R., and Meybeck, A. (2011) *Global Food Losses and Food Waste – Extent, Causes and Prevention*, Food and Agriculture Organization of the United Nations (FAO), Rome.
6. Interagency Agricultural Projections Committee (2012) *USDA Agricultural Projections to 2021*, Office of the Chief Economist, World Agricultural Outlook Board, U.S. Department of Agriculture, 102 pp.
7. Bain & Company (2014) *Enabling Trade: From Farm to Fork*, World Economic Forum, Geneva, 47 pp.
8. Cargill Corporate Affairs (2014) *Food Security the Challenge*, Cargill Incorporated, http://www.cargill.com/corporate-responsibility/food-security/the-issues/index.jsp (accessed 30 September 2015).
9. European Commission (2012) Innovating for Sustainable Growth: A Bioeconomy for Europe, COM(2012)60, Brussels, Belgium.
10. Philp, J.C., Ritchie, R.J., and Allan, J.E.M. (2013) *Trends Biotechnol.*, **31**, 219–221.
11. Limayem, A. and Ricke, S.C. (2012) *Prog. Energy Combust. Sci.*, **38**, 449–467.
12. Kim, T.H. and Kim, T.H. (2014) *Energy*, **66**, 13–19.
13. Maity, S.K. (2015) *Renewable Sustainable Energy Rev.*, **43**, 1446–1466.
14. Starch Europe http://www.starch.eu/european-starch-industry (accessed 30 September 2015).
15. BeMiller, J.N. and Whistler, R.L. (2009) *Starch: Chemistry and Technology*, 3rd edn, Academic Press, San Diego, CA, 879 pp.
16. Eckhoff, S.R. and Watson, S.A. (2009) in *Starch: Chemistry and Technology*, 3rd edn (eds J.N. BeMiller and R.L. Whistler), Academic Press, San Diego, CA, pp. 373–439.
17. Corn Refiners Association http://corn.org/process/ (accessed 30 September 2015).
18. Maningat, C.C., Seib, P.A., Bassi, S.D., Woo, K.S., and Lasater, G.D. (2009) in *Starch: Chemistry and Technology*, 3rd edn (eds J.N. BeMiller and R.L. Whistler), Academic Press, San Diego, CA, pp. 441–510.
19. Sayaslan, A. (2004) *LWT Food Sci. Technol.*, **37**, 499–515.
20. Chiu, C. and Solarek, D. (2009) in *Starch: Chemistry and Technology*, 3rd edn (eds J.N. BeMiller and R.L. Whistler), Academic Press, San Diego, CA, pp. 629–655.
21. Hobbs, L. (2009) in *Starch: Chemistry and Technology*, 3rd edn (eds J.N. BeMiller and R.L. Whistler), Academic Press, San Diego, CA, pp. 797–832.
22. de Troostembergh, J.C. (1996) in *Biotechnology*, 2nd edn, vol. 6 (eds H.J. Rehm and G. Reed), Wiley-VCH Verlag GmbH, Weinheim, pp. 31–46.
23. Lichtenthaler, F.W. and Peters, S. (2004) *C.R. Chim.*, **7**, 65–90.
24. Slater, S.F., Reddy, V.K., and Zwirlein, T.J. (1998) *Ind. Market. Manage.*, **27**, 447–458.
25. Sadhukhan, J., Ng, K.S., and Martinez Hernandez, E. (2014) *Biorefineries and Chemical Processes: Design, Integration and Sustainability Analysis,*

Wiley-VCH Verlag GmbH, Weinheim, 674 pp.

26. Lynd, L.R., Wyman, C., Laser, M., Johnson, D., and Landucci, R. (2005) Strategic Biorefinery Analysis: Analysis of Biorefineries NREL/SR-510-35578, 40 pp., http://www.nrel.gov/docs/fy06osti/35578.pdf (accessed 30 September 2015).

27. McAloon, A., Taylor, F., Yee, W., Ibsen, K., and Wooley, R. (2000) Determining the cost of producing Ethanol from Corn Starch and Lignocellulosic Feedstocks NREL/TP-580-28893, 44 pp., http://www.nrel.gov/docs/fy01osti/28893.pdf (accessed 30 September 2015).

28. Bai, F.W., Anderson, W.A., and Moo-Young, M. (2008) *Biotechnol. Adv.*, **26**, 163–172.

29. Arshadi, M. and Grundberg, H. (2011) in *Handbook of Biofuels Production – Processes and Technologies* (eds R. Luque, J. Campelo, and J. Clark), Woodhead Publishing, Cambridge, pp. 199–220.

30. RFA http://www.ethanolrfa.org/ (accessed 30 September 2015).

31. Wee, Y.J., Kim, J.N., and Ryu, H.W. (2006) *Food Technol. Biotechnol.*, **44**, 163–172.

32. Nampoothiri, K.M., Nair, N.R., and John, R.P. (2010) *Bioresour. Technol.*, **101**, 8493–8501.

33. John, R.P., Nampoothiri, K.M., and Pandey, A. (2007) *Appl. Microbiol. Biotechnol.*, **74**, 524–534.

34. NatureWorks LLC http://www.natureworksllc.com/ (accessed 30 September 2015).

35. Vink, E.T.H., Davies, S., and Kolstad, J. (2010) *Ind. Biotechnol.*, **6**, 212–224.

36. McMullin, T. (2009) Biotechnology to the bottom line-low pH lactic acid production at commercial scale SIM Annual Meeting and Exhibition, Society of Industrial Microbiology, Toronto, Canada (presentation).

37. Suominen, P., Aristidou, A., Pentilla, M., Ilmen, M., Ruohonen, L., Koivuranta, K., and Roberg-Perez, K. (2012) Genetically modified yeast of the species and closely relates species, and fermentation processes using same. US Patent 8097448.

38. Berovic, M. and Legisa, M. (2007) in *Biotechnology Annual Review*, vol. 13 (ed. M. Raafat El-Gewely), Elsevier, pp. 303–343.

39. Max, B., Salgado, J.M., Rodriguez, N., Cortes, S., Converti, A., and Dominguez, J.M. (2010) *Braz. J. Microbiol.*, **41**, 862–875.

40. Aristidou, A.A., Baghaei-Yazdi, N., Javed, M., and Hartley, B.S. (2012) in *Fermentation Microbiology and Biotechnology*, 3rd edn (eds E.M.T. El-Mansi, C.F.A. Bryce, B. Dahhou, S. Sanchez, A.L. Demain, and A.R. Allman), CRC Press, Boca Raton, FL, pp. 225–262.

41. Dhillon, G.S., Kaur, S., Verma, M., and Brar, S.K. (2012) in *Citric Acid: Synthesis, Properties and Applications* (eds D.A. Vargas and J.V. Medina), Nova Science Publishers, Hauppauge, NY, pp. 71–96.

42. Saad, M.M. (2006) *Res. J. Agric. Biol. Sci.*, **2**, 132–136.

43. Zhang, J., Li, J., Wu, S., and Liu, Y. (2013) *Ind. Eng. Chem. Res.*, **52**, 11799–11815.

44. de Cock, P. (2012) in *Sweeteners and Sugar Alternatives in Food Technology* (eds K. O'Donnell and M. Kearsley), Wiley-Blackwell, Hoboken, NJ, pp. 215–241.

45. Munro, I.C., Berndt, W.O., Borzelleca, J.F., Flamm, G., Lynch, B.S., Kennepohl, E., Bär, E.A., and Modderman, J. (1998) *Food Chem. Toxicol.*, **36**, 1139–1174.

46. Livesey, G. (2003) *Nutr. Res. Rev.*, **16**, 163–191.

47. Yin, G., Lin, W., Qiao, C., and Ye, Q. (2001) *Acta Microbiol. Sin.*, **41**, 709–715.

48. Bremus, C., Herrmann, U., Bringer-Meyer, S., and Sahm, H. (2006) *J. Biotechnol.*, **124**, 196–205.

49. Pappenberger, G. and Hohmann, H.-P. (2014) *Adv. Biochem. Eng./Biotechnol.*, **143**, 143–188.

50. Du, J., Zhou, J., Xue, J., Song, H., and Yuan, Y. (2012) *Metabolomics*, **8**, 960–973.

51. Saxena, R.K., Anand, P., Saran, S., and Isar, J. (2009) *Biotechnol. Adv.*, **27**, 895–913.

52. Markets and Markets http://www.marketsandmarkets.com/PressReleases/

1-3-propanediol-pdo.asp (accessed 30 September 2015).

53. Forage, R.G. and Lin, E.C.C. (1982) *J. Bacteriol.*, **151**, 591–599.

54. Tong, I.T., Liao, H.H., and Cameron, D.C. (1991) *Appl. Environ. Microbiol.*, **57**, 3541–3546.

55. Nakamura, C.E. and Whited, G.M. (2003) *Curr. Opin. Biotechnol.*, **14**, 454–459.

56. DuPont Tate & Lyle http://www .duponttateandlyle.com (accessed 30 September 2015).

22
Technologies, Products, and Economic Viability of a Sugarcane Biorefinery in Brazil

Alan Barbagelata El-Assad, Everton Simões Van-Dal, Mateus Schreiner Garcez Lopes,
Paulo Luiz de Andrade Coutinho, Roberto Werneck do Carmo, and Selma Barbosa Jaconis

22.1
Introduction

The growth of environmental awareness directly affects several sectors of society. Within the government, these effects are mainly evident in the growth of legislation, the creation of clean technology development incentives, and the use of renewable raw materials (RRMs) in industry. Within the corporate sphere, environmental awareness has influenced the development of new technologies, product selection processes, and their future visions. Sustainability is now considered as a part of corporate business strategies, and the use of RRM is one way to guarantee assistance.

Within the scientific sphere, significant effort has been made to produce energy and products from clean technologies with diminished environmental impact. The use of RRMs in product and energy production can address the environmental concerns in the government, corporate, and scientific sectors, which have played an important role in promoting the creation and use of so-called biorefineries.

The activities of these sectors should lead to the establishment of a new production chain, the technologies should be proven, several players throughout the chain must be strengthened, and products/markets must be developed.

Many diverse alternative technologies have been proposed, which suggests that the processes and winning technologies that will eventually become standard have not been defined yet. The innovations in this sector could be rated based on the possible evolution of the productive chain itself, and these changes may be related to biomass production, the pretreatment of biomass for chemical transformations, and the processes that these chains will use to generate chemical products.

Efforts regarding biomass technologies, which are generally based on biotechnology and genetic engineering, are used to increase productivity and can help adapt crops to specific soil conditions and climates. In addition, these efforts aim to improve the cultivation process, including harvest and culture preservation.

The development of raw materials becomes the strategic focus of companies and countries, with higher productivity on marginal lands, lower chemical input, and

Chemicals and Fuels from Bio-Based Building Blocks, First Edition.
Edited by Fabrizio Cavani, Stefania Albonetti, Francesco Basile, and Alessandro Gandini.
© 2016 Wiley-VCH Verlag GmbH & Co. KGaA. Published 2016 by Wiley-VCH Verlag GmbH & Co. KGaA.

drought resistance as key factors that affect the sustainability of biorefineries. Currently, sugarcane is used as a reference material for providing sugar and provides a substantial fraction of the lignocellulose that can be converted into chemicals and energy. In addition, new tools for metabolic engineering are becoming available, and new traits are being developed to increase resistance to insects and drought and to adapt sugarcane to other climates/regions.

The pretreatment stage of technologies is important for establishing this new chain. The pretreatment stage provides basic elements for chemical transformations in the next stage and allows the use of several types of biomass and residues. In addition, this stage will provide some solutions for scale and logistical problems. Thermochemical technologies (pyrolysis/gasification) generate bio-oil/syngas, and the chemicals and biochemicals (hydrolysis) release the sugar held in lignocellulose.

The conversion of lignocellulose into biofuels and biochemicals involves biochemical (fermentation) and/or chemical (catalysis) processes. The process that will be used and the final product are chosen as a function of the raw material.

This study aims to identify the factors that will make biorefineries possible in the future. In addition, this study discusses the aspects that will guide this implementation, including the biomass to be expended, available/developing technologies, and the identification of products with market penetration potential and intrinsic competitiveness. Overall, this discussion is based on Braskem's roadmap of chemicals from RRMs [1]. This tool was developed by the company to monitor, evaluate, and identify opportunities in the area.

This paper includes six sections, including this section. In the following section, the actors that are involved in constructing this new productive chain are discussed, some of the types of biomasses that have been considered are listed (with special emphasis on sugarcane), and the technologies and products that have previously been considered are also discussed. In Section 22.3, we present details regarding a currently operating sugarcane biorefinery facility in Brazil that is used to produce alcohol, electricity, and sugar. Section 22.4 describes the Braskem methodologies by assembling a roadmap of the evolution of chemicals from RRMs and the evaluation process used for analyzing alternative biorefinery models. In Section 22.5, we demonstrate how biochemical production could increase the competitiveness of biofuels in the biorefinery industry. Finally, the conclusions are presented in Section 22.5.

22.2
Biorefineries: Building the Basis of a New Chemical Industry

According to the NREL [2], a biorefinery includes an installation that integrates processes for converting biomass into electric energy, fuels, and chemicals. According to the International Energy Agency Bioenergy Task 42 [3], a biorefinery sustainably converts biomass into a large range of products.

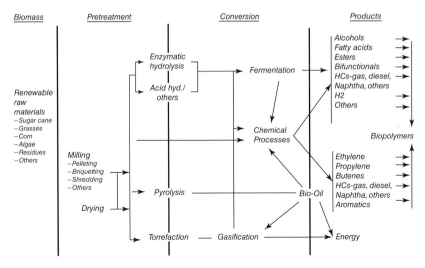

Figure 22.1 Schematic representation of a biorefinery.

This definition covers a large range of technologies that are used for producing energy. In addition, these technologies produce the building blocks that are already, or will be, the feedstocks for many fuels and chemical products. According to the World Economic Forum [4], biorefineries are capable of transforming biological material into fuels, energy, chemicals, and food. Currently, standard biomass and conversion processes have not been defined.

The availability of raw materials and points of view regarding which technologies should be used will likely differ around the world.

Figure 22.1 illustrates several possible approaches for biomass technologies (physical processing, thermochemistry, chemistry, biotechnology, etc.) and products (fuels, chemicals, polymers, etc.).

Currently, a variety of players participate in biomass technologies, including major energy companies (Total, Petrobras, BP, Shell, Exxon, etc.), major chemical companies (BASF, Bayer, Dow, Braskem, DSM, Mitsubishi, Mitsui, Lanxess, etc.), technology startups (Genomatica, Amyris, Verdezyne, Gevo, Global Bioenergies, Metabolic Explorer, Metabolix, Virent, Cobalt, etc.), biotechnology companies (Dupont/Genencor, Novozymes, and Codexis), venture capitalists (Khosla Ventures, TPG, Stark, Alloy Ventures, etc.), and technology institutions and universities around the world (Fraunhofer Institute, MIT, Korea Research Institute of Chemical Technology (KRICT), University of California, Wageningen University, NREL, etc.). Governments (the United States, Europe, China, Brazil, Korea, etc.) and companies at the end of the production chain (Coca-Cola, P&G, Nike, Ford, etc.) indirectly encourage research by supplying resources for technology development [1].

The business model and participants (shareholding interest, technology licensing, supply of resources aiming future benefits, etc.) in this chain are not

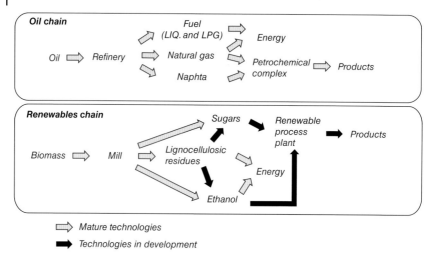

Figure 22.2 How will the chemical chain work for renewable raw materials?

well defined. Will we use a chain similar to the oil supply chain in the future (Figure 22.2)? Will there be room for producers that do not integrated raw materials?

Currently, the consolidated production model of biofuels/food focuses on sugar/ethanol production and on some electric energy (Brazilian case). However, will there be room for nonintegrated companies that receive basic raw materials (sugar, oil, etc.) from current producers for commodity production? In addition, will space be available for biomass treatment stations and, if so, where will they be located? Ideally, the creation of these stations would be related to the production/release of second-generation (2G) sugars and based on considerations of possible problems related to logistics/scale. First, this new chain model should consider complete integration regarding biomass and the commodity chemical product during energy and biofuels production. Currently, the model used by Dupont, which receives biomass and produces biofuels and chemicals, serves as a model biorefinery of the Brazilian alcohol/sugar industry [1]. Sugar sales for ethanol production are not economically viable. Differences in the production of other basic chemicals should not be expected. However, it is not possible to highlight how these challenges will differ as the production chain improves (Figure 22.3), which also impacts the final model to be adopted.

In addition to these issues, it is important to consider that the success of 2G sugar/ethanol production demonstrates the viability of biomass pretreatment technologies. This success enables the use of lignocellulosic sugars, which would allow aspects of the supply chain (quantity, logistics, etc.) and avoid questions regarding competition against food. The first commercial plants using this technology began operating this year. Based on the available information from licensors and the literature, it is possible to verify that the production costs of released sugars and the ethanol produced from sugarcane bagasse would be very

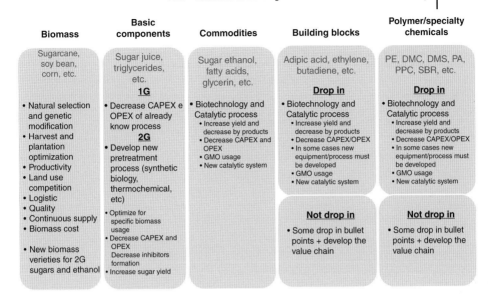

Biomass	Basic components	Commodities	Building blocks	Polymer/specialty chemicals
Sugarcane, soy bean, corn, etc.	Sugar juice, triglycerides, etc.	Sugar ethanol, fatty acids, glycerin, etc.	Adipic acid, ethylene, butadiene, etc.	PE, DMC, DMS, PA, PPC, SBR, etc.
	1G		**Drop in**	**Drop in**

• Natural selection and genetic modification
 • Harvest and plantation optimization
 • Productivity
 • Land use competition
 • Logistic
 • Quality
 • Continuous supply
 • Biomass cost
 	
 • New biomass verieties for 2G sugars and ethanol
	• Decrease CAPEX e OPEX of already know process
	 2G
	 • Develop new pretreatment process (synthetic biology, thermochemical, etc)
	
	 • Optimize for specific biomass usage
	 • Decrease CAPEX and OPEX Decrease inhibitors formation
	 • Increase sugar yield
		• Biotechnology and Catalytic process
		   • Increase yield and decrease by products
		   • Decrease CAPEX and OPEX
		   • GMO usage
		   • New catalytic system
			• Biotechnology and Catalytic process
			   • Increase yield and decrease by products
			   • Decrease CAPEX/OPEX
			   • In some cases new equipment/process must be developed
			   • GMO usage
			   • New catalytic system
				• Biotechnology and Catalytic process
				   • Increase yield and decrease by products
				   • Decrease CAPEX/OPEX
				   • In some cases new equipment/process must be developed
				   • GMO usage
				   • New catalytic system

			Not drop in • Some drop in bullet points + develop the value chain	**Not drop in** • Some drop in bullet points + develop the value chain

Figure 22.3 Challenges in each step of the chemical production chain for renewable raw materials.

similar to the cost on the market [1]. These plants still encounter processing problems that need to be solved but should achieve greater capacity in the next 5 years. The learning curve and the introduction of new specific biomasses to the sector (sugarcane energy, for instance) should reduce the costs of sugar and alcohol production relative to first-generation plants.

22.2.1
Biomasses/Residues and Their Costs

Annual worldwide biomass production has reached 180 billion tons [2]. This biomass includes various types of crop biomass (sugar and starch crops, herbaceous and woody biomass, oilseeds, etc.). As shown in Figure 22.4, 75% of these biomass sources consist of carbohydrates (in cellulose, hemicellulose, starch, and sucrose forms), and the other 20% consist of lignin. The great availability of cultures and lignocellulosic-based residues can explain the large interest in the development of 2G sugars that could constitute the base of this new industry.

The transition from the use of raw petrochemical materials to biomass is hampered by several factors regarding the definition of its technical and economic viability. These factors include the following: compositions that vary within the harvest; a viable production scale; productivity; the yield per hectare, quality and densification of crops; distinct shapes for different biomasses and their residues; competition against food; logistics for obtainment, transportation,

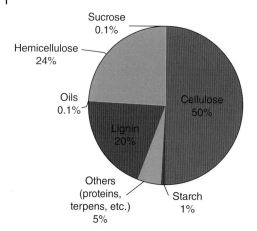

Figure 22.4 Constituent parts of the biomass produced in the world [2].

transshipment, and distribution; and interdependent scales and seasonality that result in the idleness of expensive structures [5]. Thus, although several factors allow RRMs to influence the chemical industry, their participation and the definition of winning products and principal actors is not determined by their cost, availability, or related logistical factors.

Availability and costs allow different agricultural residues to be seen as potential biomasses for the generation of renewable chemicals. LMC [6] and Embrapa [5] analyzed the potentials of several residues/energetic cultures from a variety of regions around the world, including grain straws and energy grasses from Canada and the United States, palm biomass from Southeast Asia, residual straw and energy grasses from Europe, grasses and sugarcane bagasse from Brazil, and so on. These prices range from US$ 34/t to US$ 354/t depending on the biomass, the local offer conditions and the opportunity cost for the producer.

Biotechnology and genetic engineering have been used to increase availability and reduce production costs. Studies in these areas have targeted the search for greater productivity per hectare, climate adaptation, soil conditions, and better pest resistance. In Brazil, Embrapa has worked on genetic selection and improvement for the last 30 years, which has increased productivity in several crops, including soy, potato, and beans. Regarding sugarcane, Brazil currently has several genetic improvement programs, including Rede Interuniversitária de Desenvolvimento do Setor Sucroalcooleiro (RIDESA), Centro de Tecnologia Canavieira (CTC) Campinas Agronomic Institute (IAC), and CanaVialis (founded in 2004, works together with Allelyx, a company dedicated to the development of noncommercial transgenic varieties) [1].

The search for the ideal raw material has become a strategic focus of companies and countries. Currently, sugarcane is used as a reference material for making sugar available while allowing a significant fraction of the lignocellulosic biomass (bagasse and straw) to be converted to 2G sugars (pentoses and hexoses) that can be transformed into biochemicals, biopolymers, and biofuels. For sugar, food production is associated with the production of biomass that is suitable for chemical production.

Competition against food production is addressed in this scenario because there is a significant amount of lignocellulosic residue for the installation of biochemical, biofuel, and renewable energy production units in the same area used for food production.

Despite the economic importance of sugarcane, it accounts for relatively little land use compared with grain production. In 2006, the worldwide sugarcane cultivation area was nearly 20.4 million hectares, and those of soy, corn, rice, and wheat were 93, 144.4, 154.3, and 216.1 million hectares [7].

The availability of land and the climate of Brazil make it suitable for becoming a sector leader in sugarcane production in the future. Several startups and companies in the biochemical/biofuel area already exist (Amyris, Solazyme, Abengoa, etc.) and have looked for locations in Brazil. Today, Brazil is the first world producer of sugar and the second producer of ethanol from sugarcane [8].

Bagasse, which results from extracting sugarcane juice, is mainly used as fuel in boilers to generate electricity and steam, which makes the industry energetically self-sufficient and is responsible for supplying approximately 7% of the consumed electricity in the country [8]. The use of this bagasse in sugar production through polysaccharide hydrolysis of the cell wall has been extensively studied. These sugars, which are obtained from the deconstruction of the cell wall, may be used in chemical or biochemical routes to produce ethanol or several other compounds for the chemical and pharmaceutical industries. Straw, which was previously underutilized, is becoming a new biomass alternative for the production of chemicals and energy. With the advent of harvest mechanization, new regulations were created to limit fires and provided opportunities for the development of equipment that will provide transportation to low-cost plants. Currently, bagasse and straw prices depend on the use of opportunity costs, including the use of sugarcane as a fuel source in sugarcane plants or for generating energy for the supply chain.

According to Danelon *et al.*, the bagasse/straw price/cost of opportunity is a function of the existence or absence of electricity cogeneration at the producer. When using cogeneration, the function of the settled process' efficiency is considered. In this case, prices could range from US$ 12/t to US$ 68/t [9]. Recently, GranBio announced that it would acquire sugarcane straw for US$ 25/t [10]. Considering that GranBio would be assuming the costs of harvesting the material, it is possible to assume that the total cost of the straw for GranBio would reach between US$ 40 and US$ 50/t.

22.2.2
Competing Technologies/Products in Development: The Importance of Biotechnology

First-generation sugars from the sugarcane, corn, beet, and cassava agroindustries can be directly converted to chemicals by using biotechnological (Genomatica,

Solazyme, Amyris, Myriant, Bioamber, Verdezyne, Green Biologics, Reverdia, Global Bioenergy, etc.) or catalytic (Rennovia, Virent, Avantium, Segetis, etc.) processes [1]. In fermentation processes, most innovation efforts have focused on the discovery or creation of microorganisms capable of producing molecules of interest. The reduction in the number of production steps and, consequently, the investment and operational costs of an industrial plant play decisive roles in the economic viability of the process. For catalytic processes, the growing importance of multifunctional catalysts also enables the reduction of chemical production steps when using RRMs.

Lignocellulosic biomass can be converted into fuel or chemical products through chemical/biochemical or thermochemical processes. Using a biochemical process, a pretreatment is required to release the sugars and the lignin that will later be converted into chemicals of interest. This pretreatment may include a biotechnological step with the use of enzymes (Beta renewables, Abengoa, Dupont, Inbicon, POET, etc.) or may involve the use of specific process/catalyst conditions (Renmatix, Midori, Zeachem, etc.) [1]. Next, the C_5 and C_6 sugars that are released in the pretreatment can be converted into fuel or chemical products by using biotechnological processes. However, these technologies should be reassessed because the conditions of biomass pretreatment may result in the formation of undesired compounds that affect fermentation yield or catalytic reactions. Thermochemical pathways usually presuppose syngas production from biomass, which is then converted into the desired chemical through biological (Coskata, Lanzatech, etc.) or catalytic (Range Fuels, Choren, ECN, Enerkem, Ineos, Chemrec, etc.) pathways [1].

Based on internally developed technology (see Section 22.4.2), we identified that biomass pretreatment by chemical/biotechnological processes would have several advantages over thermochemical/gasification processes. Thermochemical/gasification processes present challenges because of the high investment required for their implementation and because of their lower yields relative to chemical/biotechnological methods. The high investment per ton of product for thermochemical processes results from the problems associated with preparing the biomass for feeding the gasifier, the need to clean the obtained syngas when recovering the energy generated in the process, the final purification, and the gas conversion process itself. The number of steps involved and the drastic conditions required result in a high capital expenditures (CAPEX). On the other hand, preexisting knowledge in this area (coal gasification and the production of chemicals from it) could involve less technological risk regarding development. Currently, this technology can only be applied by using biomass with a variable composition and negative cost, such as urban waste.

In 2013, an overview of biofuel projects, labeled according to the production pathway, was presented [3] that included 83 projects developed by 54 companies or research organizations. The biochemical pathway represented 65% of the projects and appeared more advanced when the high number of pilot plants and operational demonstrations were considered. Generally, a delay in the delivery of the commercial plants was observed.

Today, only six plants are operational or are functioning in the initial stages of operation. All of these plants are based on biotechnology (POET, Dupont, Abengoa, Beta renewables, GranBio, and Raizen). Companies based on thermochemical processes, such as Range Fuels and Choren, have problems, and Coskata replaced biomass with shale gas as a raw material due to its availability and cost [1].

Priority is being given to companies and academies that use biotechnology in pretreatment processes. Biomass conversion requires more attention by those responsible for deciding where and how to invest in an area. The ability to engineer biological systems has significantly improved due to scientific breakthroughs in emerging disciplines, such as metabolic engineering and synthetic biology. Synthetic biology is the "design and construction of new biological systems that do not exist in nature through the assembly of well-characterized, standardized, and reusable components" [11]. *Metabolic engineering* is broadly defined as "the development of methods and concepts for analysis of metabolic networks, typically with the objective of finding targets for engineering of cell factories" [11]. Both concepts are generally called *engineering of biological systems* and draw on a wide range of disciplines and methodologies to design new biological functions, assemble biological systems, and increase their performance.

A schematic representation of the interconnectivity of different biological engineering-based technologies used to create a biorefinery is shown in Figure 22.5. The biorefinery value chain begins with different raw materials, including 2G crops, chemical-producing crops, or biofuel-producing algae. Recent advances in metabolic engineering have made it possible to increase the yield/productivity of desired compounds and have introduced novel biosynthetic pathways for a variety of species to enhance nutritional or commercial value. To improve metabolic engineering capabilities, new transformation techniques were developed to allow for gene specific silencing strategies or the stacking of multiple

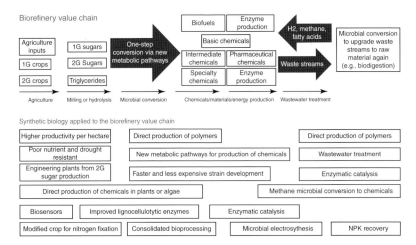

Figure 22.5 Biorefinery and synthetic biology [12].

genes within the same region of the chromosome. The "omics" era provided a new resource for determining uncharacterized biosynthetic pathways and enabling novel metabolic engineering approaches. These resources currently allow the advanced metabolic engineering of plant production systems and the synthesis of increasingly complex products in engineered microbial hosts. The status of current metabolic engineering efforts is highlighted for the *in vitro* production of paclitaxel and the *in vivo* production of β-carotene in Golden Rice and other food crops [13]. Additionally, using high-yield crops and lignocellulosic sugars could significantly affect the scale of production and costs of sugar, respectively. Hence, raw materials could be designed/selected to meet resources constrains.

Sugars can be used to produce a wide range of biofuels, chemicals, foods, pharmaceuticals, feed, and enzymes, as outlined in Table 22.1. During the last decade, the engineering of biological system has made use of novel pathways to optimize metabolic performance. More than fifty new metabolic routes have been created and expressed in different hosts, from algae to plants [23]. Bioengineering has resulted not only in the production of new chemicals but also in the use of new carbon sources, such as 2G sugars, methane, and CO_2 [23]. Thus, engineered biological systems could play a key role in value chain consolidation by creating more efficient and sustainable routes for producing food, biofuels, energy, chemicals, and materials, especially in the context of biorefineries [23].

Enzymes can be used for lignocellulose hydrolysis, downstream enzymatic catalysis, or sold as end products, for example, for detergent or laundry applications (Figure 22.5). One interesting case is the chemical manufacturing route to sitagliptin, a dipeptidyl peptidase-4 inhibitor treatment for type II diabetes

Table 22.1 Examples of basic and specialty chemicals produced in a one-step conversion from renewable sources to the final chemical via new synthetic biology routes.

Compound	Institution/Company	Main applications	Reference
Vitamin C	Genencor and Eastman	Food	[14]
Isoprene	Genencor and Goodyear Braskem, Amyris, and Michelin Ajinomoto and Bridgestone	Rubber	[15]
Isobutylene	Global Bioenergies	Polymers and chemicals	[16]
Adipic acid	Verdezyne	Polymers and esters	[17]
1,4-Butanediol	Genomatica	Polymers and chemicals	[18]
FAEEs	LS9	Diesel	[19]
Farnesane	Amyris	Diesel	[20]
Propylene	Global Bioenergies	Polypropylene	[16]
Methionine	Metabolic Explorer	Food and feed	[21]
1,3-Propanediol	Genencor	Polymers	[22]

that was developed by Merck and Codexis. By designing and generating new enzyme variants, Codexis identified a novel enzyme that provided detectable initial activity. This enzyme was improved by more than 25 000-fold to generate a highly active, stable, and enantioselective enzyme from a starting enzyme that did not previously exist in the natural world. The application of synthetic biology and enzymes in organic synthesis is currently marketed under the trade name Januvia® [24].

The main by-products of those processes, for example, liquid waste streams, microbial residues, and CO_2, can be used in other processes to conserve energy and chemicals. For example, vinasse could be used as a raw material for hydrogen [12], short-chain fatty acid [12], or bioplastic [25] production. Furthermore, methane or H_2 from biodigestion could be used to produce electricity/energy and stimulate 2G production by reducing the amount of lignocellulose burned as fuel. Alternatively, methane could be used in the production of chemicals and biofuels using microbial catalysis, especially if more efficient methane biological metabolism is developed [26]. Calysta and Intrexon's industrial products focused on the conversion of methane to chemicals when using microbial catalysis. The resulting chemicals included lactic acid, farnesene, and isobutanol, which were already produced at industrial scale by using sugar as a raw material.

By taking advantage of available inexpensive and renewable electricity, CO_2 can be converted to chemicals via microbial electrosynthesis (Figure 22.5). Microbial electrosynthesis bypasses photosynthesis by using microorganisms that can directly extract energy from electricity or H_2 and produce chemical compounds [23, 27]. Potential advantages of microbial electrosynthesis include the 100-fold higher efficiency of photovoltaics in harvesting solar energy, which eliminates the need for arable land and avoids the potential environmental degradation associated with intensive agriculture [27]. During microbial electrosynthesis, electrons are supplied via an electric current and are used by microorganisms to reduce carbon dioxide and yield industrially relevant products. The production of acetic acid [28], isobutanol, and 3-methyl-1-butanol [29] using microbial electrosynthesis has been demonstrated. Electrosynthesis could significantly affect the overall yield of chemicals from biomass while decreasing the carbon footprint toward zero-carbon emissions [23].

Recent developments in the engineering of biological systems can bridge gaps to create biorefineries for producing chemicals, biofuels, energy, and electricity [23]. However, sustainable production is based on the development of adjacent technologies related to energy efficiency, wastewater treatment, soil conservation, biomass logistics, the discovery of new drugs and improved materials, among other factors [23]. For example, synthetic biology could offer a more efficient use of natural resources by creating symbiotic microbes that could decrease water [28] and nitrogen [30] requirements in plants.

Biorefineries can use various feedstocks that can be processed by using multiple conversion technologies and are used in companies that produce a diverse portfolio of products and services (e.g., chemicals production, wastewater treatment, and lignocellulose hydrolysis). All of these processes will benefit from raw material logistics, common facilities, utilities, wastewater treatment, cooling towers, and other shared equipment, including bioreactors. Biorefineries will benefit from economy of scale (reduction of fixed cost and CAPEX), bioenergy, and bioelectricity [12].

With advances in biological engineering, the importance of product dimensions has increased in the sector structure, and some variables should be highlighted. Biotechnology provides opportunity for the appearance of new molecules with low costs and high functionality.

Bioproducts can include final or intermediate chemicals "drop in" or "not drop in" commodities or specialties. The "drop in" products are those that directly correspond with a petrochemical counterpart (the green polyethylene in comparison with the petrochemical polyethylene, for example). Products that are "not drop in" constitute new molecules and require some type of market or technological development before being used by the final user and its complementary agents. Commodity products are those with large production volumes and low added value and are sold under well-known specifications. In this case, competitiveness is based solely on the product cost to the user. Specialties normally involve lower product volumes and higher added values, and their competitiveness is defined by a better cost-to-benefit ratio for the final user.

"Drop in" products have several inherent advantages beyond the fact that they are intermediate or final products, commodities, or specialties. Because "drop in" products do not require any additional market/technological development in the present production chain, their diffusion is mostly related with its production cost. "Not drop in" products require a more careful analysis. For intermediate products, it is necessary to understand how the chain will evolve to the corresponding final product. For example, levulinic acid can be obtained directly from lignocellulosic biomass at a low cost and can be used as an intermediate for the production of adhesives, solvents, plastics, and polymers but has not been marketed yet because it is not possible to determine how the chain will evolve to include this chemical in the final products. To be redirected to the commodities market, "not drop in" products must be competitively priced against the products that they aim to replace. Even then, market penetration is sometimes difficult. Few companies are willing to take risks in developing molecules that require significant development.

Bioproducts directed to specialty market compete based on their cost-to-benefit ratio for clients, which require that the producers have the ability to sell the product. This opens the field for bioproducts because the fact that they are not "drop in" products is characteristic of this sector that searches for constant differentiation and aims to deliver products with high performance. Furthermore, this fact may explain the product repositioning adopted by some PLA and PHA producers. PURAC developed a PLA with better thermal properties and redirected it toward the engineering polymers market. METABOLIX, after the failure of its joint venture with ADM, repositioned PHA to look for new applications in the plasticizer market.

22.3
Sugarcane-Based Biorefineries in Brazil: Status

The development of new biorefineries in Brazil will benefit from the efforts for continuous improvement that have been applied in the country during the last decades to all stages of sugar and alcohol production from sugarcane. The improvements that have been achieved in sugarcane cultivation and harvesting, the technologies used for "fertirrigation" with vinasse, and the high level of energy integration illustrated by the green power generated from bagasse and straw that is delivered to the grid can all be readily applied to any downstream process that uses any of the traditional products from sugar mills, including sucrose (either as sugar or as sugarcane juice or syrup), ethanol, and lignocellulosic materials, such as bagasse and straw (tops and leaves).

22.3.1
An Industry with a History of Evolution

Sugarcane was initially recognized only as a food crop and has been used for manufacturing sugar since colonial times in the "engenhos." In addition, sugarcane was used to produce beverages such as "cachaça," which is produced by fermenting the sugar and distilling the resulting liquid. At the beginning of the twentieth century, ethanol from sugarcane was used in industry manufacture chemicals such as acetic acid and ethylene. However, the sugarcane industry for chemical production was eventually displaced by the petrochemical industry. Manual harvesting was typical, and because the sharp edges of leaves pose a serious risk to workers, the dry leaves used to be burned before harvesting.

The oil crisis in the 1970s, which hit Brazil very hard because of its dependency on foreign oil, marks a turning point for the sugarcane industry as ethanol-powered cars entered the market and the addition of ethanol to gasoline became the norm. Backed by significant research efforts [31], the sugarcane industry responded by increasing its capacity and productivity, effectively becoming a sugar and ethanol industry. By-products such as bagasse (the solids left after milling the sugarcane) and vinasse (the liquid residue from distillation) remained unsolved environmental problems. Gradually, vinasse was applied in "fertirrigation," where the concentrated nutrients were sent back to the field. The development of furnaces capable of burning bagasse resulted in another step change, allowing the mills to generate all the power and steam they needed and to export the excess power to the grid. As mills began building high-efficiency boilers, the green power from bagasse grew in importance. Today, approximately 6% of the electrical power generation in Brazil comes from sugarcane mills [32, 33].

Both the area dedicated to sugarcane and the weight of sugarcane processed in the mills have increased steadily [34]. Technology improved dramatically in the years after the oil crisis, with significant reductions in the fermentation time and increased total yields, before reaching a maturity plateau in the 1990s [35, 36].

The sugarcane industry in Brazil still faces many challenges. For example, burning straw in the field is now prohibited, which forces the mills to process straw with bagasse to either generate power or serve as a lignocellulosic raw material for 2G plants. Mechanical harvesting has been instrumental for increasing productivity; however, increased root damage has become an issue. Current harvesting machines, which typically leave the straw in the field to transport only the sucrose-rich stalks, may need to be replaced with lighter, less-damaging equipment that can be used to take straw to the mill.

Another relevant change for biorefineries is the corporate profile. Family-owned, isolated mills that were run very conservatively have given way to consolidated groups that are run by companies such as Raizen (Shell and Cosan), Guarani (Tereos and Petrobras), BP, Odebrecht Agroindustrial, and Bunge.

Sugarcane presents a challenge when compared with grains such as corn, which can be stored. Sugarcane juice can only be processed during the season, and mills can stop if heavy rains fall during the harvest. Using grains such as sorghum or corn during the off-season can result in higher utilization in mills. Two mills, both located in the State of Mato Grosso, are known to use corn during the off-season and are now referred to as flex mills by the local press [37]. At least one more flex mill will open in the State of Goiás in 2016 [38].

22.3.2
Existing Biorefineries in Brazil

22.3.2.1 Bioethylene and Biopolyethylene

Biopolyethylene (bio-PE) is a biopolymer with the same composition as conventional polyethylene, which is manufactured from bioethanol. A chemical plant converts ethanol into ethylene by dehydration, and conventional polymer plants use bioethylene as a raw material for bio-PE. When Braskem decided to build its pioneering plant, the company decided to maximize its use of existing assets and logistics rather than build a new, dedicated biorefinery, which would be a capital intensive and risky investment. The company procures fuel-grade bioethanol, which is widely available in Brazil, and sends it to the Green Ethylene plant in Triunfo, Brazil. The green, polymer-grade ethylene from the plant is sent to one of the four polymer plants the company owns in the same area to produce bio-PE or to a neighboring plant that produces elastomers with a 50% biocontent. This approach resulted in a "virtual" biorefinery in the sense that the biotechnology processes are not located at the same site as the final product transformation units. This biorefinery remains the largest single-site producer of bioplastic in the world, producing more than 20 grades of polyethylene resins.

22.3.2.2 Terpenoids

The production of farnesene and other similar products was initiated by Amyris in 2012 and is an example of integrated production site. A biotech-based farnesene plant was installed next to Usina Paraíso, a sugar mill at Brotas in the state of

São Paulo. The mill, a traditional producer of ethanol and sugar from sugarcane, provides the Amyris plant with sugarcane juice and utilities [39].

Farnesene is a branched, unsaturated hydrocarbon that is produced from sugars that are fermented with proprietary yeasts modified by synthetic biology. Farnesene is commercialized under the trade name Biofene® and can be further processed into a number of other products. For example, farnesene can be hydrogenated to its corresponding paraffin, farnesane, to produce renewable fuels. The biorefinery includes an extensive portfolio of products besides those mentioned above. The plant can also be used to produce fragrances and flavors by using different strains of yeast in the same equipment that is used to produce farnesene. Amyris has already produced patchouli oil fragrances at the Brotas plant [40].

A line of solvents (marketed as Myralene), sugarcane diesel fuel, and cosmetics based on squalene are also included in the company portfolio [41]. The production of lubricants from farnesene illustrates how the biorefinery concept can be extended. Amyris formed a joint venture with the Brazilian fuel distributor Cosan, named Novii, to generate a line of lubricants with renewable content [42].

22.3.2.3   Other Biorefineries

Brazilian biorefineries also produce butanol, fatty oils, and 2G ethanol.

The Solazyme plant adjacent to Bunge's Moema Mill uses microalgae to transform sugar into tailor-made vegetable oils. As of 2014, the plant was able to produce high-lauric oil (containing 45% 12-carbon fatty acid radicals), mid-oleic oil (containing 60% oleic acid), and high-oleic acid (containing >85% oleic acid) [43].

Industrial, large-scale, 2G ethanol plants became a reality in 2014. The GranBio Bioflex 1 plant in Northeast Brazil uses sugarcane straw from a neighboring crop to produce 82 million liters per year of fuel-grade ethanol in a dedicated 2G plant. Raizen used a different approach in their Costa Pinto mill by adding a 2G pretreatment plant to the existing mill that is capable of processing both sucrose from sugarcane and the sugars from 2G pretreatment. Raizen can now produce ethanol that is partially 2G.

HC Sucroquímica, located in the state of Rio de Janeiro, uses the fermentation process known as *ABE* to ferment sugar into a mixture of acetone, butanol, and ethanol. This plant, located next to Paraíso mill, has been operational since 2008 [44].

22.4
A Method for Technical Economic Evaluation

22.4.1
Braskem's Roadmap for Chemicals from RRMs

The production and commercialization of products from RRMs has created a complex and turbulent organizational scenario. The Technology Roadmapping (TRM) method is very useful for understanding this dynamic scenario and may be useful for decision making in the renewable space. TRM(s) are considered flexible techniques that can be used as visualization and communication tools and represent a living document and learning process. This tool provides a method for identifying, evaluating, and selecting technology alternatives that can be used to meet important problems and specific needs relative to the present and future trends [45, 46].

Braskem uses the TRM to structure existing information regarding the development and production of chemicals from RRMs [1, 47] and intends to connect drivers/markets (purposes – know-why), products (delivery – know-what), and technologies (resources – know-how) over time (time – know-when). This structure allows for a greater understanding of this new industry environment and helps identify opportunities in the area. Braskem's TRM analyzes information from approximately 107 sources of information, tracking approximately 120 different products and more than 900 players that are pursuing new renewable technologies. This TRM is based on a relational database, is structured in Microsoft Dynamics CRM with all relevant information, and serves as a systematic update for renewable scenarios. These tools (the TRM and the database) are shared with the company's employees in Braskem's intranet to disseminate knowledge and consider different levels of secrecy when desired.

In the chart shown in Figure 22.6, the X-axis corresponds to the time horizon and is divided into five periods. The monitoring process and the database allow the information to be framed in the correct period. The current period contains products that are already being produced on a commercial scale. The short-term vision contains information about announced construction plans. The medium-term vision contains information that is reserved for information forecasting from patents with a long-term vision that is based on technical articles. The Y-axis is

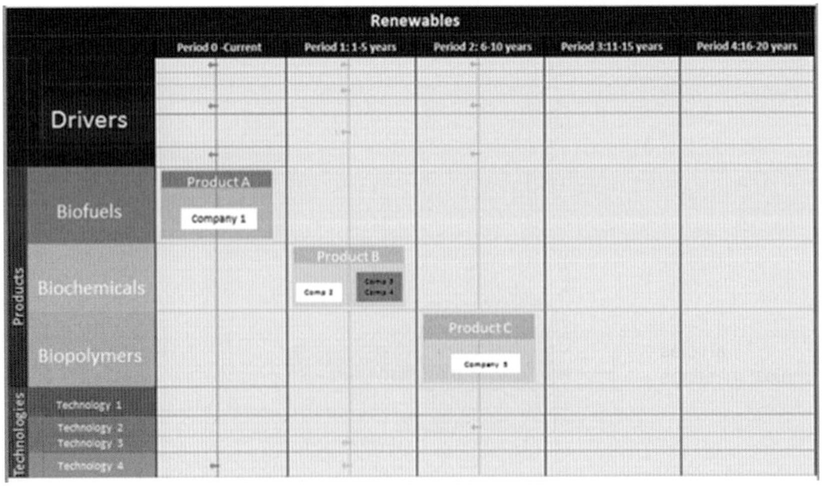

Figure 22.6 Braskem's technology roadmapping for chemicals produced from renewable raw materials.

divided into three dimensions or layers, drivers, products, and technologies. In the first layer, "drivers" represent the market and societal factors that should guide the pace of launching the products and processes, such as the bio-based content, low carbon footprint, sustainability strategy, and value chain (among others). In the TRMs second layer, "products" are divided into three categories, biofuels, bio-based chemicals, and biopolymers. The technologies required for developing each product from biomass are identified in the third layer.

Braskem's TRM and the relational database provide access to several pieces of information about companies, universities, venture capitals, partnerships, feedstocks, development stages, plant locations, and capacities. Indeed, this database provides qualified information for different valuable analyses, such as feasibility analyses, potential partner identification, and intellectual property space. Furthermore, the roadmap helps Braskem make strategic decisions regarding trends, challenges, and new opportunities for the company when facing bio-based products. Furthermore, the roadmap orientates the company's strategy based on a greater understanding of the industry environment and on the identification of the strategies of competitors and potential partners.

22.4.2
Braskem's Method for the Evaluating the Chemicals Produced from RRMs

To develop new products from renewable sources, Braskem established an evaluation procedure that includes several technical and economic analyses.

Inputs of new possible products result from many different sources, including market studies, clients' demands, metabolic pathway analysis, and the development of new catalysts. Additionally, numerous inputs have resulted from the Braskem technology roadmap for renewable products, as discussed in Section 22.4.1.

Figure 22.7 shows an overview of the Braskem methodology for prospecting and evaluating renewable chemicals. All of the represented steps occur in parallel and interact with each other.

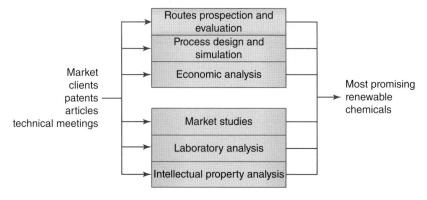

Figure 22.7 Overview of Braskem methodology for renewable chemical evaluation.

Because sugarcane sugar and biomass are widely available in Brazil, they are often considered the main raw materials in company evaluations. A large number of possible routes may lead to the production of the same chemicals, as exemplified for biobutadiene in Figure 22.8. These routes include biological conversion using genetically modified organisms (GMOs) and chemical conversion, which often involves the use of new catalysts. Information available in patents and articles along with internal studies are used to construct this type of diagram.

Each route has its own theoretical maximum yield that strongly depends on its reaction thermodynamics and the produced by-products. Yield plays an important role in the economic viability of a given route for several reasons. In addition to the higher conversion of raw materials in products, higher yields result in smaller quantities of by-product. Thus, high yields generally lead to reduced need for purification, which reduces CAPEX and operational expenditures (OPEX). Therefore, determining the maximum theoretical yield of each route is the first step toward selecting which routes should be evaluated in more detail. To detail the economic potential of selected routes, it is necessary to design an industrial process for each route. A generic model of the industry includes the following sections: the preparation of raw materials, reaction, product purification, utilities, storage tanks, and effluent treatment.

In this stage of process development, several assumptions must be made due to the limited amount of information available and the large uncertainties of some parameters. Often, safety factors are used to over design processes to compensate for possible details that have not been considered at this stage. Therefore, the performance of sensitivity studies is important for verifying the economic robustness of each process and route.

The use of commercial process simulators can be helpful because of their extensive database of thermodynamic properties and preprogrammed unit operations. Thus, mass and energy balances can be obtained for each process. However, new technologies often include new process operations and compounds that are not available in commercial simulators, which require the use of empirical correlations and information available in articles and patents.

To economically evaluate all processes and routes, CAPEX and OPEX must be determined for each case. Thereafter, it is possible to estimate a minimum selling price (MSP) for each scenario.

The MSP must consider variable costs, fixed costs, and return on investment (ROI). Variable costs include the costs of raw materials, utilities, and by-products. Fixed costs include operating labor, maintenance, taxes, insurance, and other costs. The ROI depends on the plant lifetime and on the desired rate of return of the project.

In this design stage, technology uncertainties remain important. Thus, the MSP should be regarded as an indicator for technology comparison and not as a final and accurate value. Figure 22.9 shows an MSP comparison for butanol production considering different levels of integration with the sugarcane mill. The MSP is presented on a percentage basis using fossil-based butanol as a basis for comparison.

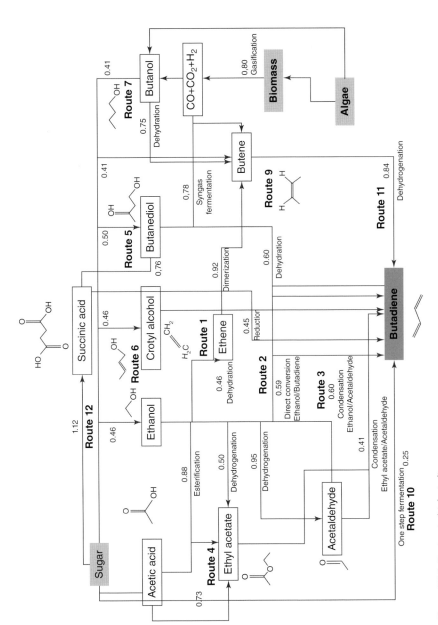

Figure 22.8 Routes to biobutadiene.

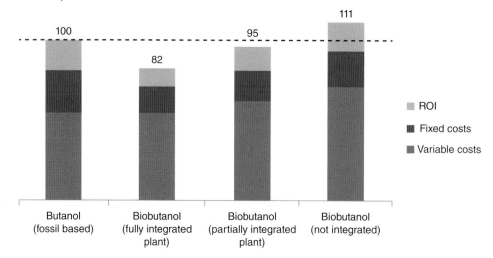

Figure 22.9 Minimum selling price comparison of biobutanol production considering different levels of integration with the sugarcane mill.

For a more thorough economic assessment, the net present value (NPV) and the internal rate of return (IRR) for each case is calculated. For this step, price forecasting for raw materials, utilities, and products for the plant lifetime are essential.

The calculation of economic indicators (MSP, NPV, and IRR) alone is not sufficient for evaluating a process and/or a product. Other factors, such as market size, logistics, and the required level of investment should be considered as well.

Market analysis fundamentally aims to characterize the market size, main global players, competing products, and competing technologies. In addition, market analysis aims to provide information regarding market acceptance. For example, market analysis should show whether a product is drop in or not and if drastic changes in the value chain (already discussed in Section 22.2.2) are required. Understanding the product price formation and price forecasting is also important.

Contact with clients, suppliers, and other stakeholders of the value chain are important to understand the possible challenges faced when integrating a new product into the market. These challenges are not only related to the replacement of existing products but also include possible issues and synergies with logistics and transportation and the substitution of existing machines. Contact with clients is also important to test and validate the performance of new products and applications. Internal laboratory testing can be used to compare the performances of new products against their existing competitors.

In parallel, an extensive intellectual property analysis is performed to monitor technology development and the market and to properly direct investments. This analysis can be used to avoid efforts aimed at developing metabolic pathways that have already been patented.

For approximately 3 years, Braskem has used this method to evaluate opportunities for chemicals obtained from RRMs. Over 50 different possible products have been evaluated. Less than half of the evaluated products had the potential to compete with fossil-based competitors. Of these products, approximately 70% were based on biotechnological processes and 30% employ specific catalysts. Thermochemical processes via syngas were not considered competitive because of their high CAPEX and lower yields.

Some of the winning products can already be found in the market or will be available in the next few years, including 1,3-propanediol (Dupont); 1,4-butanediol (Genomatica); and succinic acid (Myriant, Bioamber, etc.). All of these products were produced from direct fermentation of sugar. Other products include mixtures of diols produced by glycerin or sugar hydrogenolysis (Global Biochem, ADM Oleom) and epichlorohydrin produced from glycerin (Vinithay, joint venture between Solvay and PTT) in Thailand. The production of farnesene from the direct fermentation of sugars (Amyris), isobutanol (Gevo), and butanol (Cobalt, Green Biologics, etc.) is also worth mentioning. Although these products were developed for fuel applications, they are only competitive if used as chemicals.

22.5
The Sugarcane Biorefinery of the Future: Model Comparison

In this section, different biorefinery designs are analyzed, from conventional Ethanol production (Section 22.5.1) to integrated plants that produce chemicals, ethanol, and raw sugar from both sugarcane biomass and juice (Section 22.5.4). The methodology presented in Section 22.4.2 was applied to each design to estimate the CAPEX and OPEX for each case. The output of this analysis is an early-stage economic evaluation of each design under specific premises that are described here. Thus, a comparative analysis of the different approaches can be made. One major goal is to determine the potential of integrated chemical production in a biorefinery and to provide a preliminary indicator of economic feasibility in each case.

The basis of all analyses presented here is a field that delivers 4 million tons of raw sugarcane to the factory each year. All economic evaluations were performed based on ethanol because it is the only common product. Therefore, the prices for other by-products were fixed, and the MSP for ethanol was evaluated for each scenario. The ROI was set at 15% per year over a lifetime of 20 years.

The prices considered for major feedstocks and products were based in average 2014 values and internal analyses and are presented in Table 22.2.

22.5.1
Conventional Sugarcane Ethanol Plant

A typical large-scale conventional sugarcane mill was designed. This mill normally uses cogeneration to generate electrical power, which is partially exported

Table 22.2 Prices of feedstocks and products.

Feedstock/product	Price
Sugarcane	26 US$/t
Bagasse (dry basis)	50 US$/t
Straw (dry basis)	50 US$/t
Raw sugar	304 US$/t
Electricity	68 US$/MWh
N-Butanol	1200 US$/t
Succinic acid	1800 US$/t

to the grid. The typical values vary between 5 and 85 MW and are affected by the plant capacity, level of heat integration, steam generation pressure, and cogeneration technology. An average export of 9.2 MW was assumed in this scenario, and no straw supply was considered. The overall scheme for this case is shown in Scheme 22.1, and the economic results are presented in Figure 22.10.

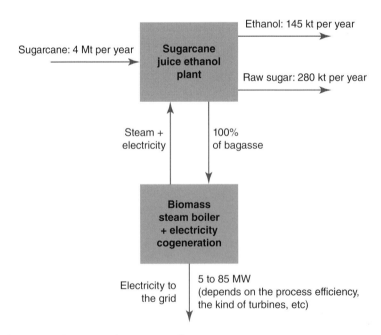

Scheme 22.1 Conventional sugarcane mill.

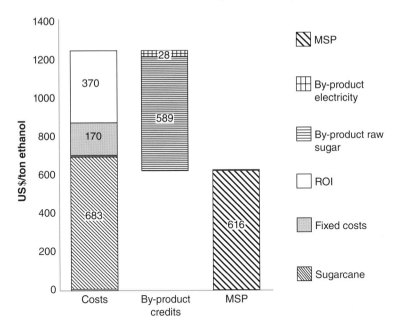

Figure 22.10 Ethanol MSP conventional mill.

22.5.2
Stand-Alone Cellulosic Ethanol Plant

A stand-alone cellulosic ethanol plant was designed, assuming that this facility would be built next to a conventional and existing sugarcane mill. This conventional factory was estimated to offer 30% of its total bagasse by generating less power surplus and by implementing some inexpensive process optimizations. It was also assumed that 50% of the total straw from the field could be collected and delivered to the stand-alone cellulosic ethanol plant.

The cellulosic sugar conversion plant (pretreatment and hydrolysis) was simulated based on several patents (US 5,562,777 [48]; US 5,705,369 [49]; US 5,125,977 [50]; US 5,879,463 [51]; US 2014/0363856 [52]; US2012/0009632 [53]). The other plant sections are similar to conventional plants.

This process produces lignin as a by-product, which is partially consumed in a biomass steam boiler that supplies steam for the plant. The excess lignin is sold back to the stand-alone sugarcane mill to generate extra electricity. The overall scheme for this case is shown in Scheme 22.2, and the economic results are presented in Figure 22.11.

22.5.3
Integrated Conventional and Cellulosic Ethanol Factory

A greenfield ethanol plant was designed by considering the conventional production of raw sugar, biomass pretreatment and hydrolysis, and the fermentation of

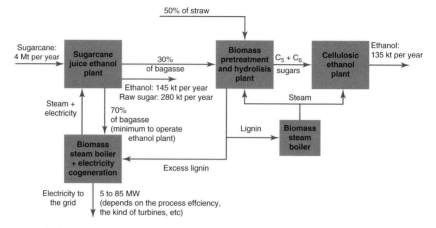

Scheme 22.2 Conventional ethanol and stand-alone cellulosic ethanol plants.

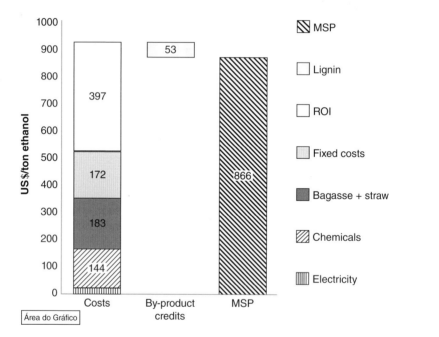

Figure 22.11 Ethanol MSP stand-alone cellulosic plant.

sugarcane juice and all cellulosic sugars to ethanol. To increase ethanol production, the factory should consume minimal bagasse in the boilers. Thus, no electric surplus was considered, and all of the produced lignin was sent to the boilers. It was estimated that 67% of the total bagasse and 50% of the straw would be supplied to the biomass pretreatment section. The cellulosic sugar conversion section (pretreatment and hydrolysis) was simulated using the same technology at that shown in Section 22.5.2. The overall scheme for this case is shown in Scheme 22.3, and the economic results are presented in Figure 22.12.

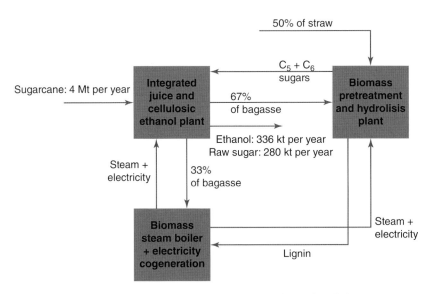

Scheme 22.3 Integrated conventional and stand-alone cellulosic ethanol plant.

22.5.4
Biorefinery Producing Ethanol, Raw Sugar, and Succinic Acid

A greenfield biorefinery complex was designed by considering the conventional production of raw sugar, biomass pretreatment, and hydrolysis, and by using fermentation to convert sugarcane juice and all C_6 sugars to ethanol. The cellulosic C_5 sugars were used to produce succinic acid. As described in Section 22.5.3, part of the bagasse and all of the lignin is burned in the biomass boiler to produce steam for the plants, and the remaining bagasse was sent for biomass pretreatment. It was estimated that 35% of the total bagasse and 50% of the straw would be supplied to the cellulosic sugar conversion plant. The cellulosic sugar conversion section (pretreatment and hydrolysis) was simulated using the same technology as shown in Section 22.5.2. The succinic acid plant was simulated based on several patents (US 2013/0096343 [54]; WO 2010/063762 [55]; WO 2011/082378 [56]; WO 2012/018699 [57]; and WO 2012/138642 [58]). The overall scheme for this case is shown in Scheme 22.4, and the economic result is presented in Figure 22.13.

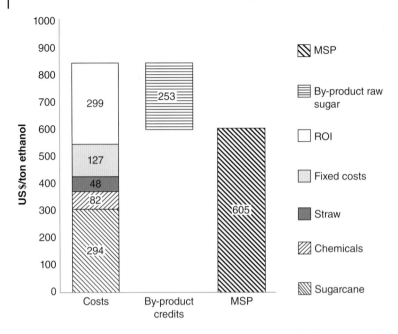

Figure 22.12 Ethanol MSP from an integrated conventional and cellulosic ethanol plant.

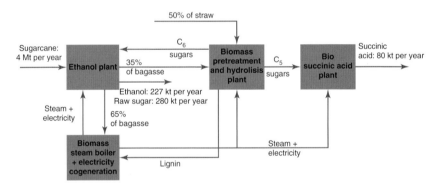

Scheme 22.4 Biorefinery producing ethanol, raw sugar, and succinic acid.

22.5.5
Biorefinery Producing Ethanol, Raw Sugar, Succinic Acid, and Butanol

A greenfield biorefinery complex was designed, taking into consideration the conventional production of raw sugar, biomass pretreatment, and hydrolysis and the fermentation of sugarcane juice to ethanol. The cellulosic five-carbon sugars would be fermented to produce succinic acid, and the cellulosic six carbon sugars would be fermented to produce 1-butanol. As described in Section 22.5.3, part of the bagasse and all of the lignin would be burnt in the biomass boiler to produce steam for the plants, and the remaining bagasse would be sent to biomass

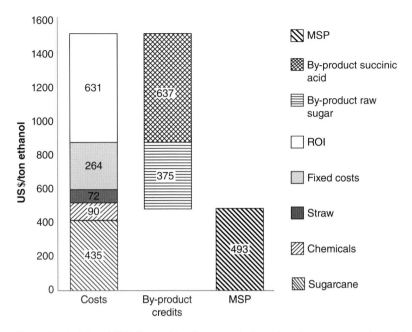

Figure 22.13 Ethanol MSP from a biorefinery producing ethanol, raw sugar, and succinic acid.

pretreatment. Overall, approximately 31% of the total bagasse and 50% of the straw would be supplied to the cellulosic sugars conversion plant. The cellulosic sugar conversion section (pretreatment and hydrolysis) was simulated using the same technology as that of Section 22.5.2, the succinic acid plant was simulated using the same technology as that of Section 22.5.4, and the butanol plant was simulated based on several patents (WO 2008/137402 [59]; WO 2008/124490 [60]; WO2008/121701 [61]; WO2008/080124 [62]; WO2008/052973 [63]). The overall scheme for this case is shown in Scheme 22.5, and the economic results are presented in Figure 22.14.

22.5.6
Model Comparison

By comparing the economic results from the models, it is clear that integration and product diversification play a positive role. A comparison of the minimum ethanol selling prices for the different cases is presented in Figure 22.15. The stand-alone cellulosic ethanol showed the worst economic results, and a fully integrated biorefinery producing two chemicals from biomass, ethanol, and sugar showed the best results. Integration and chemical production may be essential to make a plant competitive and to achieve economic viability. However, considering a biorefinery that produces even more products with the same amount of feedstock (sugarcane and straw) would reduce the scale of all plants and could increase the energy consumption of the complex, which could negatively affect the overall economics.

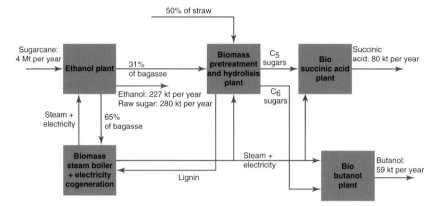

Scheme 22.5 Biorefinery producing ethanol, raw sugar, succinic acid, and butanol.

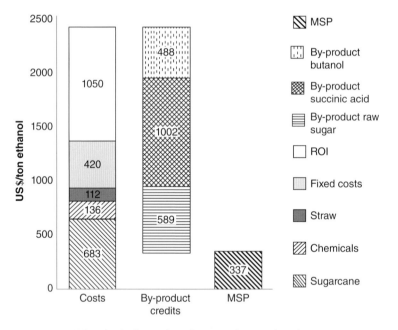

Figure 22.14 Ethanol MSP from a biorefinery producing ethanol, raw sugar, succinic acid, and butanol.

22.6
Conclusions

Countries, companies, and academic researchers are all committed to finding sustainable solutions for the economic, environmental, and social aspects of the industry. Until recently, economic issues limited the implementation of measures that aimed to reduce environmental problems. However, new technological

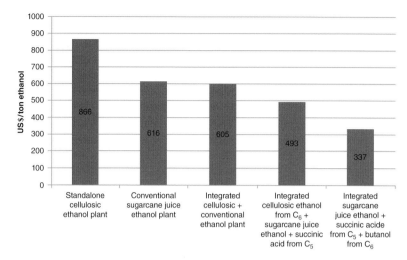

Figure 22.15 Ethanol minimum selling prices comparison.

advances should change this situation. As we have shown, chemicals are already being produced from biomass that are competitive with corresponding oil-derived products. This has encouraged companies from several industry sectors to invest in the field. Consequently, a new production chain is emerging.

The evolution of the sector should lead to increases in the technologies used for producing a wide range of products, from commodity chemicals to pharmaceuticals. The development of raw materials will account for regional aspects, such as climate and water availability. Some raw materials have higher potential, such as sugarcane, while others will be developed for use during the sugarcane off-season or for regions where sugarcane is not grown. The agricultural machinery for biomass collection and processing will evolve with pretreatment technologies. This situation means that decisions must be made at the levels of public policy and corporate strategy that have opposite risks, specifically, decision makers risk betting on several alternatives and investing poorly in their development or choosing a losing option and wasting investments and effort.

Therefore, the first competitiveness attribute is the ability to capture the sector dynamics and guide investments and policies in an uncertain environment. The goal of this attribute is to construct a solid technological base for future competitiveness in renewable chemistry. It appears that industrial biotechnology will inevitably serve as the basis of the industry in the future. However, advances in biotechnology might change this. Genetic enhancements that improve productivity, allow growth in areas with different soil types and climates, and create better pest resistance have reduced biomass production costs. Simultaneously, the development of enzymes as specific catalysts and microorganisms that allow direct production of chemicals from biomass or one of its components in a single step have increased the profitable use of raw materials with lower investments.

Sugarcane stands out as the biomass with the greatest potential. The mastery of this culture, land availability, and proper climate all favor sugarcane production in Brazil. The direct production of sucrose chemicals has already begun, with the Amyris and Bunge–Solazyme plants serving as examples. With the implementation of 2G ethanol production by GranBio and Raizen, new opportunities have become accessible. C_5 and C_6 sugars from lignocellulose will be available and may be used in chemical production, which has been demonstrated to be a more profitable option than ethanol-based production.

The second attribute relates to the raw material structure and chain integration. Requirements exist for productivity, availability, quality, cost and environmental performance, and a supply logistics chain occurs. The need for this structure may indicate a business model that presupposes the integration of biomass with the most viable chemical product.

The third attribute is related to the choice of processes and products. This choice can be closely related to the chemical composition of raw materials and products. For example, products that have higher oxygen content tend to have higher yields and thus lower production costs. Green processes with lower production costs tend to be more competitive than conventional processes, and their products could even be applied in new markets. Chemicals that are not currently produced on an industrial scale can also become economically viable. In both cases, application development is required, either to discover new applications and markets for an existing product or to find uses for a new molecule. We believe that economically attractive production of green chemicals is possible, whether they are "drop in" or not.

References

1. Coutinho, P.L.A., Morita, A.T., Lopes, M.S., and Jaconis, S.B. (2013) Roadmap de Matérias-primas Renováveis, Internal Report Braskem S.A. – Inovação e Tecnologia Corporativa – Prospecção e Avaliação de Tecnologias Renováveis, pp.1-167.
2. NREL (0000) What is a Biorefinery? http://www.nrel.gov/biomass/biorefinery .html (accessed 10 April 2015).
3. IEA Bioenergy (0000) IEA Bioenergy Task42-Biorefineries, http://www.iea-bioenergy.task42-biorefineries.com/en/ ieabiorefinery.htm (accessed 4 April 2015).
4. King, D., Inderwildi, O.R., and Williams, A. (2010) *The Future of Industrial Biorefineries*, World Economic Forum.
5. Gonçalves, S.B., Souza, D.T., Dias, J.M.C., Cançado, L.J., Oliveira, P.A., and Pacheco, T.F. (2014) Estudo de Fontes de Biomassa para produção de açúcares de segunda geração, Relatório Final, Embrapa Agroenergia, contrato de parceria EMBRAPA/BRASKEM.
6. LMC International (2012) Feedstock for Bio-Based Chemicals: Which will be Competitive.
7. Novacana (0000) A Produção de cana de açúcar no Brasil (e no mundo), http:// www.novacana.com/cana/producao-cana-de-acucar-brasil-e-mundo/ (accessed 7 May 2015).
8. Souza, Z. (2014) Bioelectricity Overview, Clean Technology Seminar, Brasilia, http://www.unica.com.br/documentos/ apresentacoes/pag=1 (accessed 7 May 2015).
9. Danelon, A.F., Silva, M.T., and Xavier, C.E.O. (2014) *O bagaço de cana de açúcar como insumo para a geração*

de vapor e eletricidade: uma análise de mercado, Pecege ESALQ-USP.

10. Novacana (2013) Brasil terá 3 usinas de etanol celulósico até 2015, Jornal Valor Econômico, 20 de setembro de 2013, http://www.novacana.com/n/etanol/2-geracao-celulose/brasil-usinas-etanol-celulosico-200913/ (accesses 30 September 2015).

11. Nielsen, J., Fussenegger, M., Keasling, J., Lee, S.Y., Liao, J.C., Prather, K., and Palsson, B. (2014) *Nat. Chem. Biol.*, **5**, 319–406.

12. Sydney, E.B., Larroche, C., Novak, A.C., Nouaille, R., Sarma, S.J., Brar, S.K., Letti, L.A. Jr., Soccol, V.T., and Soccol, C.R. (2014) *Bioresour. Technol.*, **159**, 380–386.

13. Wilson, A.S. and Roberts, S.C. (2014) *Curr. Opin. Biotechnol.*, **26**, 174–182.

14. Tawasha, M.A. (1999) Eastman, Genencor to Market New Process, http://www.icis.com/resources/news/1999/08/12/91877/eastman-genencor-to-market-new-process/ (accessed 25 October 2014).

15. Whited, G.M., Feher, F.J., Benko, D.A., Cervin, M.A., Chotani, G.K., McAuliffe, J.C., LaDuca, R.J., Ben-Shoshan, E.A., and Sanford, K.J. (2010) *Ind. Biotechnol.*, **6**, 152–163.

16. Marliere, P. (2011) US Patent 20110165644 A1, assigned to P. Marliere.

17. Picataggio, S. and Beardslee, T. (2012) US Patent 20120021474 A1, assigned to Verdezyne Inc.

18. Yim, H., Haselbeck, R., Niu, W., Pujol-Baxley, C., Burga, A., Boldt, J., Khandurina, J., Trawick, J.D., Osterhout, R.E., Stephen, R., Estadilla, J., Teisan, S., Schreyer, H.B., Andrae, S., Yang, T.H., Lee, S.Y., Burk, M.J., and Van Dien, S. (2011) *Nat. Chem. Biol.*, **7**, 445–452.

19. Steen, E.J., Kang, Y., Bokinsky, G., Hu, Z., Schirmer, A., McClure, A., Del Cardayre, S.B., and Keasling, J.D. (2010) *Nature*, **463**, 559–562.

20. Renninger, N. and McPhee, D. (2008) WO Patent 2008045555, assigned to Amyris Biotechnologies, N. Renninger and D. McPhee.

21. Boisart, C., Bestel-Corre, G., Barbier, G., and Figge, R. (2012) WO Patent 2011080301 A2, assigned to Metabolic Explorer.

22. Diaz-Torres, M., Dunn-Coleman, N.S., Chase, W.M., and Trimbur, D. (2000) US Patent 6136576 A, assigned to Genencor International Inc.

23. Lopes, M.S. (2015) *J. Ind. Microbiol. Biotechnol.*, **42**, 813–838.

24. United States Environmental Protection Agency (2010) Greener Reaction Conditions Award, http://www2.epa.gov/green-chemistry/2010-greener-reaction-conditions-award (accessed 11 June 2014).

25. Bengtsson, S., Werker, A., Christensson, M., and Welander, T. (2008) *Bioresour. Technol.*, **99**, 509–516.

26. Haynes, C.A. and Gonzalez, R. (2014) *Nat. Chem. Biol.*, **10**, 331–339.

27. Nevin, K.P., Summers, Z.M., Ou, J., Woodard, T.L., Snoeyenbos-West, O.L., and Lovley, D.R. (2011) *Appl. Environ. Microbiol.*, **77**, 2882–2886.

28. iGEM Imperial College London (2011) http://2011.igem.org/Team:Imperial_College_London (accessed 10 November 2014).

29. Li, H., Opgenorth, P.H., Wernick, D.G., Rogers, S., Wu, T.Y., Higashide, W., Malati, P., Huo, Y.X., Cho, K.M., and Liao, J.C. (2012) *Science*, **335** (6076), 1596.

30. iGEM (2010) BCCS—Bristol, http://2010.igem.org/Team:BCCS-Bristol (accessed 10 November 2014).

31. Nogueira, L.A.H. (2008) *Bioetanol de cana-de-açúcar: energia para o desenvolvimento sustentável*, BNDES/CGEE, Rio de Janeiro.

32. Aneel (2014) Brazil National Power Authority (ANEEL) Informações gerenciais, December, 2014, http://www.aneel.gov.br/arquivos/PDF/Z_IG_Dez_2014_v3.pdf (accessed 30 September 2015).

33. Ministry of Mining and Energy and EPE (2014) National Energy Balance, May 2014, https://ben.epe.gov.br/downloads/S%C3%ADntese%20do%20Relat%C3%B3rio%20Final_2014_Web.pdf (accessed 30 September 2015).

34. Unica (2015) http://www.unicadata .com.br/historico-de-producao-e-moagem.php?idMn=31&tipoHistorico=2 (accessed 23 April 2015).

35. Leal, M.R.L.V. (2014) in *Sugarcane Bioethanol*, R&D for Productivity and Sustainability (ed. L.A.B. Cortez), Edgard Blücher, São Paulo, pp. 561–576.

36. Finguerut, J. (2005) Simultaneous production of sugar and alcohol from sugarcane. Proceedings of the XXV International Society of Sugarcane Technologists – ISSCT Congress, Guatemala City, Guatemala.

37. Novacana (2014) Estudo aponta viabilidade de usinas flex com cana-de-açúcar e milho, http://www.novacana.com/n/ etanol/alternativas/estudo-viabilidade-usinas-flex-producao-etanol-cana-milho-240914 (accessed 31 March 2015).

38. Novacana (2015) Usina flex: Cargill e USJ querem produzir mais etanol de milho que de cana, http://www.novacana .com/n/etanol/alternativas/usina-flex-cargill-usj-milho-etanol-cana-090315 (accessed 31 March 2015).

39. Amyris (2015) Amyris to enter the industrial cleaning products market. Press release, January20, https://amyris .com/amyris-to-enter-the-industrial-cleaning-products-market/ (accessed 31 March 2015).

40. JornalCana (2014) Unidade da Amyris em Brotas volta a operar após entressafra de cana, http://www.jornalcana .com.br/unidade-da-amyris-em-brotas-volta-a-operar-apos-entressafra-de-cana (accessed 30 September 2015).

41. Voogt, J. (2014) *Myralene: Amyris Renewable Solvent*, World Biomarkets, Brazil.

42. Novvi (0000) http://novvi.com/ (accessed 31 March 2015).

43. Ravaglia, L. (2014) *Converting Sugar into Oil*, World Biomarkets, Brazil.

44. Natalense, J.C. (2013) Prospecção tecnológica de biobutanol no contexto brasileiro de biocombustíveis. MSc dissertation. USP/ IPEN, São Paulo.

45. Phaal, R., Farrukh, C.J.P., and Probert, D.R. (2004) *Technol. Forecasting Social Change*, **71**, 5–26.

46. Rinne, M. (2004) *Technol. Forecasting Social Change*, **71**, 67–80.

47. Coutinho, P.L.A. and Bomtempo, J.V. (2011) *Quim. Nova*, **34** (5), 910–916.

48. Farone, W.A. and Cuzens, J.E. (1996) US Patent 5562777 A, assigned to Arkenol, Inc.

49. Torget, R.W., Kadam, K.L., Hsu, T.-A., Philippidis, G.P., and Wyman, C.E. (1998) US Patent 5705369 A, assigned to Midwest Research Institute.

50. Grohmann, K. and Torget, R.W. (1992) US Patent 5125977 A, assigned to The United States of America As Represented By The United States Department Of Energy.

51. Antonio, G.P.H. (1999) US Patent 5879463 A, assigned to Dedini S/A.Administracao E Participacoes.

52. Sisson, E.A., Ferrero, S., Torre, P., Ottonello, P., Cherchi, F., Grassano, G., Oriani, L., and Giordano, D. (2014) US Patent 20140363856 A1, assigned to Beta Renewables S.P.A.

53. Retsina, T. and Pylkkanen, V. (2012) US Patent 20120009632 A1, assigned to American Process, Inc.

54. Tietz, W. and Schulze, J. (2013) US Patent 20130096343 A1, assigned to Thyssenkrupp Uhde Gmbh.

55. Krieken, J.V. and Breugel, J.V. (2013) WO Patent 2010063762 A2, assigned to Purac Biochem Bv.

56. Fruchey, O.S., Keen, B.T., Albin, B.A., Clinton, N.A., Dunuwila, D., and Dombek, B.D. (2011) WO Patent 2011119427 A1, assigned to Bioamber S.A.S.

57. Hermann, T., Reinhardt, J., Staples, L., Udani, R., and Yu, X. (2012) WO Patent 2012018699 A2, assigned to Myriant Corporation.

58. Cockrem, M. and Dunuwila, D. (2012) WO Patent 2012138642 A1, assigned to Bioamber International S.A.R.L.

59. Bramucci, M.G., Flint, D. Jr, Miller, E.S., Nagarajan, V., Sedkova, N., Singh, M. and Van, D.T.K. (2008) WO Patent 2008137402 A1, assigned to Du Pont.

60. Yang, S.-T. (2008) WO Patent 2008124490 A1, assigned to University of Ohio State and S.-T. Yang.

61. Aristidou, A., Fosmer, A.M., Dundon, C.A., and Rush, B.J. (2008) WO Patent 2008121701 A1, assigned to Cargill Inc.

62. Gunawardena, U., Meinhold, P., Peters, M.W., Urano, J., and Feldman, R.M.R. (2008) WO Patent 2008080124 A2, assigned to Gevo.

63. Soucaille, P. (2008) WO Patent 2008052973 A2, assigned to Metabolic Explorer Sa and P. Soucaille.

23
Integrated Biorefinery to Renewable-Based Chemicals

Gianni Girotti and Marco Ricci

23.1
Introduction

The biorefinery concept, despite its increasing use, is not easy to define. According to one of the most authoritative definitions, a biorefinery is an industrial facility where biomass is transformed into fuels and/or value-added products. This definition, however, also applies, for example, to oil mills where olives (the biomass) are processed to extract olive oil (the high-value product), possibly along with other, lower-grade oils, also providing a highly energetic residue which is a valuable solid fuel. Nevertheless, oil mills are not usually regarded as biorefineries, and it is largely agreed that food production and transformation do not belong to the biorefinery realm.

Rather, the biorefinery concept was established and developed keeping traditional oil refinery as a model. Originally, the term *refining* referred to all the increasingly complex technological activities aimed at producing a series of intermediates and commercial products from crude mineral oil, mainly to be used as fuels. For most practical purposes, pure compounds were (and still are) not required; rather, the refining products were mixtures with specific properties. Later on, refining would have been increasingly integrated with the rising petrochemical industry, and refineries began to include also the production of basic chemicals such as hydrogen, olefins, methyl-*tert*-butyl ether (MTBE), and aromatics.

Analogously, in a biorefinery the raw material (the biomass, rather than the crude oil of traditional refineries) undergoes a number of separations into valuable fractions that are then processed and transformed to produce fuels and/or energy and also, possibly, chemical intermediates, including polymers or their monomers.

The analogy holds even further. In traditional refineries, in order to synthesize chemical intermediates, crude oil undergoes an extensive cracking that can be accomplished either thermally or catalytically and that affords relatively simple primary building blocks, mainly olefins (ethene, propene, butenes, and butadiene), butanes, and aromatics (benzene, toluene, and xylenes). These small molecules are then reassembled according to the principles of industrial organic chemistry

Chemicals and Fuels from Bio-Based Building Blocks, First Edition.
Edited by Fabrizio Cavani, Stefania Albonetti, Francesco Basile, and Alessandro Gandini.

to produce a plethora of useful products, including organic intermediates, solvents, surfactants, detergents, monomers, plastics, rubber, synthetic fibers, and so on. In biorefinery too there might be the need to crack the biomass (or at least a substantial part of it) into simpler molecules – to be used as such or enabling further transformations – and this can be accomplished mainly by two different processes: saccharification and gasification. Saccharification allows to transform polysaccharides (cellulose and hemicelluloses, which together account for most of the biomass) into monomeric sugars that can feed several fermentation processes to produce a number of different chemical compounds.

Alternatively, gasification affords synthesis gas, or syngas, a valuable mixture of hydrogen and carbon monoxide that was originally produced from coal and that was already at the root of the carbochemistry development occurred before the Second World War, that is, before the start of the modern petrochemical industry [1].

Thus, in the biorefinery industry, saccharification and gasification play, to certain extent, the same pivotal role of the cracking in the traditional refinery.

A last important analogy between bio- and traditional refineries is most worthy to be underlined. Today, the refinery industry offers an impressive display of efficiency. Nevertheless, at its very beginning only a relatively small part of the crude oil was adequately valorized, and the development of technologies enabling to exploit most of the barrel, including its heaviest fractions, lasted for several decades and is still undergoing. Biorefining, in its turn, is currently in its early days and faces, on this side, the same challenges that traditional refining solved many years ago. In order to extract the highest possible value from biorefining, it will be necessary to take advantage from the whole biomass, rather than from just one or few of its fractions, as still often occurs, but most of the technologies for the exploitation of the whole biomass (including lignin, its most recalcitrant fraction) must still be developed or proven at industrial scale. Thus, it can be safely held that biorefinery profitability can only increase in the next few years.

It must be underlined that biorefineries are largely viewed as industrial manifestations of a new, sustainable economy driven, to the largest possible extent, by the 12 principles of the so-called green chemistry [2], the chemists' answer to the growing demand of collective responsibility toward mitigation of the human impact on the environment and toward resource consumption. Thus, biorefineries strongly pursue, for example, the minimization of waste and the adoption of benign synthetic procedures, and are the most obvious place to apply the seventh of the 12 principles: *A raw material or feedstock should be renewable rather than depleting wherever technically and economically practicable* (*depleting resources* are defined as resources that cannot be replenished within a time frame comparable with the average human life span). This exploitation of renewable raw materials greatly increases the sustainability of chemical productions, decreasing at the same time the economic, political, and social problems arising from our current dependence upon fossil resources such as coal, oil, and natural gas.

Coming from this general view to a Versalis strategy for green chemistry, the main target is the development of platform technologies enabling, as much as possible, the full exploitation of the whole biomass entering the biorefinery, as well as preserving the molecular complexity given by nature to such biomass to the extent needed to avoid unnecessary breaking molecules to simpler ones, depending on the specific biorefinery and chosen biomass.

As a matter of fact, Europe's chemical industry has been experiencing an erosion of its capability to compete with stronger players, faced through restructuration/rationalization of both technology and production sites, as well as decisively taking onboard alternative production methods, lightening the burden resulting from the volatility of oil-based feedstock prices, and growing restrictions on emissions. This does mean embracing a stronger orientation toward technologies free from these pressures, as part of a larger economic and social sustainability vision including a bioeconomy based on biomass whose production must be integrated within the territory hosting the new production sites, that is, the biorefineries, where completely new technologies, different from the ones developed along the last century for extracting value from fossils, will then progressively take place.

Based on a deep evaluation of the whole bioeconomy matter, some preconditions, such as biomass availability, financial and human resources for R&D, and demo plants and initial incentive policies, are strongly required, together with avoiding interference with the food/feed agricultural cycle, which would be a drawback in terms of social acceptance. After that and besides continuing to rely on oil-based productions, the Versalis strategy enabling new real biorefinery businesses is first based on new R&D and industrial activities in the field of green chemistry, mainly driven by the following two criteria:

1) Focusing on new chemical specialties for sectors where Versalis, or its mother company ENI, is already present, taking advantage of in-depth knowledge of market trends.
2) Producing from renewables some of the so-called building blocks currently derived from fossil sources.

Two projects have been chosen from the project portfolio in the field of renewables, as examples of Versalis strategy implementation. Both of them already started, related to Versalis current businesses of synthetic elastomers and of styrene-based thermoplastics: (i) the production of natural rubber from guayule plants and (ii) the development of a process to produce butadiene starting from renewables.

Other projects at different states of advancement are (iii) a metathesis-based biorefinery to produce biobased intermediates for lube, cosmetic, and detergency market sectors, in cooperation with Elevance Renewable Science; (iv) the production of high oleic acid oils from second-generation (2G) sugars rather than from dedicated crops; and (v) the 2G sugar platform technology from biomass, assuring the high quality (low inhibitors and metal contents) required by sophisticated fermentations, including those involving several genetically modified microorganisms.

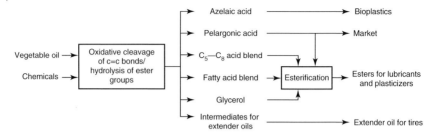

Figure 23.1 Matrica biorefinery based on highly unsaturated vegetable oils. (Adapted from ref. [3].)

A last initiative to mention is *Matrica*, a 50/50 joint venture established in 2011 by Versalis with Novamont, to convert a no longer competitive petrochemical site in Porto Torres, Sardinia (Italy), into a large biorefinery that will produce biomonomers, biolubricants, bioadditives for rubbers, and bioplastics, and that will be integrated with the agricultural supply chain of local crops, developed on purpose by Novamont. In 2014, an industrial plant was started up, able to produce 35 kt per year of products ([3] and Figure 23.1).

All the preceding projects will enable the company to diversify its current businesses. In the following, the already mentioned guayule and biobutadiene projects will be further detailed illustrating how the biorefinery concept is being exploited within Versalis.

23.2
An Alternative Source of Natural Rubber: Toward a Guayule-Based Biorefinery

> *Remember a fella wanted us to put in that rubber bush they call guayule? Get rich, he said.*
>
> John E. Steinbeck, The Grapes of Wrath, Ch. 9.

Natural rubber is a hydrocarbon polymer (*cis*-1,4-polyisoprene) largely found as a water emulsion in the milky latex of hundreds of plant species. Latex is thought to serve as protection in case of wound or as a means to discourage animals from eating the plant. Natural latex mostly occurs in soft, tacky forms impractical for any commercial use and, not surprisingly, lacked significant industrial application before Charles Goodyear and others discovered the vulcanization process in the 1830s and afterward. Upon the addition of sulfur and heat, vulcanized rubber gains strength, shelf life, and other desirable characteristics so that it can be shipped, stored, and transformed into countless industrial products ranging from medical tubes to tires.

The main source of natural rubber is *Hevea brasiliensis*, a tree native of the Amazon basin. South America remained the main source of the limited amounts of latex rubber used during much of the nineteenth century. Although the trade was well protected, in 1876, Henry Wickham smuggled 70 000 *Hevea* seeds from Brazil and delivered them to Kew Gardens, England. Only 2400 of these germinated after which the seedlings were then sent to India, Ceylon (Sri Lanka), Indonesia, Singapore, and Malaysia. Today, Southeastern Asia is, by far, the main producer of natural rubber (more than 90%), whereas South American cultivations have been almost dismissed, mainly due to the occurrence of plant parasites and diseases.

In the meanwhile, along most of the twentieth century, chemists had developed several routes to synthetic rubber. Eventually, the discovery of Ziegler–Natta polymerization allowed to polymerize isoprene with regio- and stereoselectivities that make synthetic *cis*-1,4-polyisoprene virtually undistinguishable from natural rubber, except the cis content which remains not as high as in natural rubber. The latter has not been then abandoned, also because some of its properties, particularly mechanical ones, are actually determined also by its nonrubber components (mainly, small amounts of associated proteins and lipids) and, as a matter of fact, still remain to be matched by synthetic polymers. Thus, in 2013 the total rubber production (27.5 Mt) still included 12.0 Mt (43.6%) of natural rubber [4].

There are, however, a couple of very good reasons for looking for natural rubber sources other than *Hevea*. Latex collection, indeed, is extremely labor-intensive and largely relies upon extremely low salaries and burdensome work conditions. On the other hand, it is still possible that devastating parasite attacks or disease outbursts virtually destroy Southeastern Asian plantations, as already occurred in South America. For these reasons, botanists studied hundreds of other potential rubber plants, and among them, guayule turned out to be the most promising.

Guayule (*Parthenium argentatum*) is a perennial woody shrub, up to 90–100 cm tall, with silvery olive leaves and small white-yellowish flowers. Firstly described in 1859 by Asa Gray, the most influential American supporter of Darwin's theory of evolution, guayule is native to the Chihuahuan arid zone of northern Mexico and southwestern Texas. A first period of strong interest for guayule-derived natural rubber occurred in the United States in the 1920s, after leaf blight decimated the Brazilian rubber tree plantations. Guayule rubber was again regarded as a possible replacement for *Hevea* latex during World War II, when Japan dominated the Pacific Ocean courses and the United States had no access to the Malaysian rubber. The war, however, ended before large-scale farming of the guayule plant began, and the project was scrapped, since it was again cheaper to import rubber from Southeastern Asia [5]. More recently, interest for guayule experienced a new resurgence due to the hypoallergenic properties of its rubber. Indeed, *Hevea* rubber contains proteins that can cause severe allergic reactions in a few people. With the AIDS crisis of the 1980s, the surge in rubber glove usage revealed that about 10% of health care workers are allergic to latex. On the contrary, protein content of guayule-derived natural rubber is quite low so that it does not cause latex allergies. For this reason Versalis, together with Yulex Corporation (San Diego, CA), is running a complex project aiming at establishing guayule cultivation in southern Italy/Europe and at using the resulting biomass to produce hypoallergenic natural rubber.

It is interesting to note that this is not the first attempt to grow guayule in southern Italy, where climate is not so different from that of guayule's native zones. At the end of 1937, the Società Agricola Industriale Gomma Anonima (SAIGA; agricultural–industrial society for rubber) was established to grow guayule and to produce natural rubber. Even in Italy, however, like in the States, experimentation ended just after the War, when natural rubber became again available and cheap, and upon the discovery and the impressive development of synthetic rubber [6].

As already stated, however, in order to make economic the production of natural rubber from guayule, it will probably be necessary to take advantage from its whole biomass, rather than just from the latex. In this respect, guayule is particularly interesting since, along with the latex, it also produces a valuable resin that mostly occurs in its stems, in amounts comparable with those of the rubber: up to 10% of the whole dry biomass. It has been already shown that guayule resin can find different applications in several fields including, for example, adhesives, coatings, and wood protection toward termite attack [7]. Furthermore, some use of the resin's individual components could turn out to be even more promising. Researches about guayule metabolites already started in 1911 [8]. Today, we know [9] that the most important components of the guayule resin are two sesquiterpenes and two triterpenes. The former are two esters of the same sesquiterpene alcohol, parteniol, and are termed *guayulin A* and *guayulin B* (Figure 23.2). Guayulin A accounts for 8–10% of the plant resin, while guayulin B is much less abundant: 1–3% [9]. Usually, however, the major component of guayule resin is a triterpene, argentatin A, that can account for 8 up to 17% of the resin weight. The related argentatin B is also present in significant amounts: 4–7%.

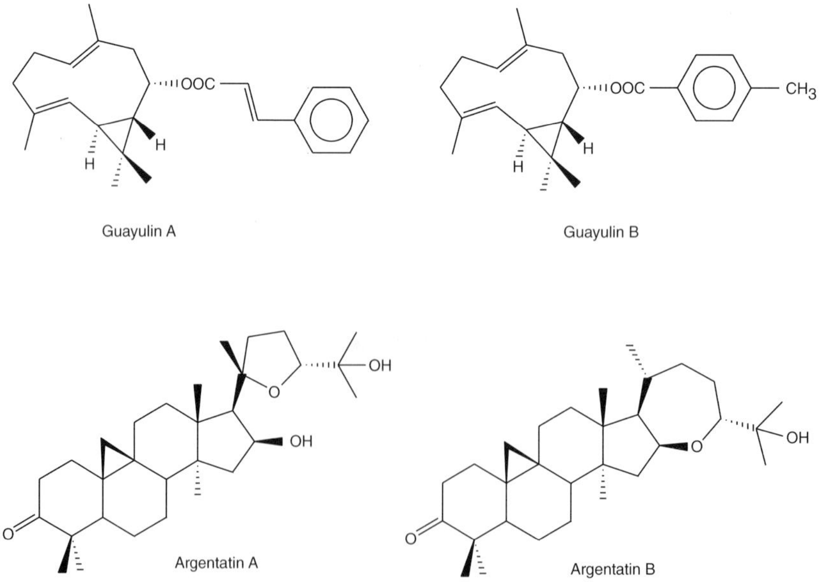

Guayulin A

Guayulin B

Argentatin A

Argentatin B

Figure 23.2 Major sesqui- and triterpenes found in the guayule resin.

Several of these compounds, or of their derivatives, can be of significant practical and economic interest. For instance, similarities have been already recognized between guayulins and, for example, pheromones or even with antineoplastic drugs [10]. In their turn, argentatins, as well as other triterpenes, already turned out to have some anti-inflammatory activity and also some cytotoxicity against human cancer cell lines [11].

Along with the resins, guayule also produces an essential oil, mostly segregated into its leaves where it accounts for about 1% of their fresh weight [12]. Essential oils are valuable mixtures of terpenes (in the guayule case, mostly few mono- and sesquiterpenes) and find widespread use, for example, in the perfumery and cosmetic industries. Finally, after recovery of the rubber, of the resin, and of the essential oil, the residue (bagasse) has a typical lignocellulosic composition and can be saccharified [13] to get 2G sugars that can be used to feed different fermentations aimed at producing a number of useful intermediates and products.

A simplified, integrated scheme for the guayule whole plant valorization is shown in Figure 23.3.

The first project step will target nontire market sectors, especially the ones highly pricing the peculiar features of guayule natural rubber; the engineering for a first extraction plant located in Italy is already in progress. Then, provided needed improvement in rubber extraction technology from guayule shrub, the second step will target larger tire/nontire market sectors, where Versalis already plays a primary role as synthetic elastomer supplier.

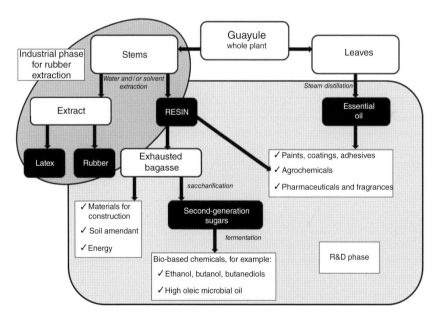

Figure 23.3 An integrated scheme for the guayule whole plant valorization. Primary products of a guayule-based biorefinery are shown in the black boxes.

23.3
Toward Renewable Butadiene

For a company, such as Versalis, deeply involved in the elastomer business, it is extremely appealing to use 2G sugars, possibly produced in a guayule-based biorefinery, as raw materials for butadiene production.

1,3-Butadiene is one of the most important chemicals in the modern industrial chemistry since it is the main raw material for the production of synthetic rubber that, in its turn, is a strategic material for our society due to its key role in the tire production and, consequently, on the entire transportation system. In 2013, the global production of synthetic rubber exceeded 15 Mt [4].

Currently, butadiene is a coproduct of the naphtha cracking process. Due to the current tendency to shift the cracking feed from naphtha to the cheaper ethane, with a marked decrease of the C4 fraction, a butadiene shortage, or at least a marked increase of its price, should be expected. Even more important and due to forecasted lower energy consumption because of better engine efficiency and building insulation, the oil growth rate is decreasing, while chemical consumption does not. This would then result into a shortage of liquid feedstock for petrochemistry, even in the absence of any shift from naphtha to ethane as cracking feed. For this reason, several efforts are currently devoted to develop on-purpose processes for butadiene, thus decoupling its production from cracking processes, and many of these efforts are actually devoted to develop butadiene production from renewables. Thus, in the last few years, a dozen or so of meaningful routes for producing butadiene from renewable feedstock has been envisaged, and several companies and partnerships are actively involved in developing the most promising of these routes and processes.

Virtually all the envisaged routes are hybrid biotechnological/chemical processes with fermentation as the first step, followed by one or more chemical, catalytic transformations.

Fermentations can be either quite traditional (ethanol fermentation, acetone-butanol-ethanol fermentation, etc.) or more innovative (e.g., fermentations to butanediols). Few efforts are also devoted to develop a direct fermentation to butadiene, avoiding any chemical step. This is a tough goal, also because no microorganisms able to do this job are currently known. Thus, although the direct production by fermentation might be the ultimate process for producing butadiene, it is considered a very long-term project with an uncomfortable time to market, and hybrid biotechnological/chemical processes are usually considered more convenient.

Usually, fermentations are fed with sugar or with a mixture of sugars, possibly including 2G ones. In one case, however, the fermentation is fed with biosyngas or even with flue gases.

The most promising approaches are summarized in Figure 23.4.

Referring to this figure, route 1 is actively pursued by the French consortium BioButterfly, led by Michelin. The first two steps (biomass to sugars and sugars to ethanol) are quite conventional even if it is expected to exploit not

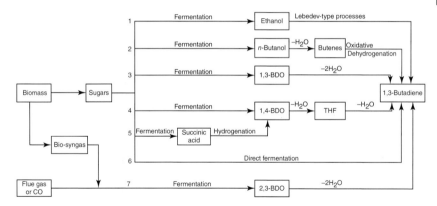

Figure 23.4 Main approaches to renewable butadiene (BDO: butanediol).

only first generation but also 2G sugars. Not so for the third step (ethanol to butadiene) that should take advantage from the Lebedev process (one step) or the Ostromilensky one (two steps) to transform ethanol into butadiene. Both the processes were already run on industrial scale starting from 1928 up to the 1960s. Their mechanisms are still a matter of discussion, but they possibly involve ethanol dehydrogenation to acetaldehyde, which then undergoes the usual aldol condensation/dehydration sequence affording cis and trans crotonaldehydes. These aldehydes react with ethanol in a Meerwein–Ponndorf–Verley reduction resulting in the formation of acetaldehyde and crotyl alcohol that, in its turn, is finally dehydrated to 1,3-butadiene [14, 15]:

$$2C_2H_5OH \rightarrow 2CH_3CHO + 2H_2 \tag{23.1}$$

$$2CH_3CHO \rightarrow CH_3-CH(OH)-CH_2-CHO \rightarrow CH_3-CH{=}CH-CHO + H_2O \tag{23.2}$$

$$CH_3-CH{=}CH-CHO + C_2H_5OH \rightarrow CH_3-CH{=}CH-CH_2OH + CH_3CHO \tag{23.3}$$

$$CH_3-CH{=}CH-CH_2OH \rightarrow CH_2{=}CH-CH{=}CH_2 + H_2O \tag{23.4}$$

Since both the processes had relatively low yields (typically 50–65%), new and more efficient catalysts are required with both redox (including dehydrogenation) and dehydration capabilities, but surprisingly, even the simple reproduction of old catalytic data turns out to be very challenging.

Routes 2 and 5 involve quite conventional fermentations and chemistry, apparently with few or no question marks. Both these routes, however, involve multiple steps and several unit operations, thus suggesting that high investments should be expected. Referring to route 5, it should be noted that sugar fermentation may provide a cheaper and more efficient way to get the intermediate 1,4-butanediol (see in the following). In any case, it is likely that the catalysts for tetrahydrofuran dehydration must be improved, particularly as far as their life is concerned.

Last, not least, several routes (3, 4, and 7) have been defined involving fermentation to a linear butanediol and then its catalytic double dehydration to butadiene. It is worthy to note that route 7 is the only approach to renewable butadiene that does not use sugars as raw materials. Several microorganisms, indeed, are able to grow on gases. This capability is ancient, predating the appearance of photosynthesis, and probably developed to take advantage from gas emissions from hydrothermal vents. It is even possible that gases were the main, or perhaps the only, carbon and energy source for the first life forms. Today, gases both from hydrothermal vents and from several industrial emissions (e.g., by steel manufacturing) have quite similar compositions including carbon monoxide and dioxide and some hydrogen, hydrogen sulfide, and methane. Thus, some flue gases, as well as syngas, can be exploited as both nutrient and energy sources to feed fermentations and to produce a number of chemicals. LanzaTech (United States and New Zealand) is a clear leader in this field and is working to produce butadiene in a two-step process via catalytic dehydration of CO-derived 2,3-butanediol. The route, however, is very challenging. Several difficulties are to be expected, including the fact that catalytic dehydration of 2,3-butanediol mostly affords methyl ethyl ketone rather than the expected butadiene [15], either via a classic pinacol rearrangement (i.e., hydride shift), or because, in the first dehydration step, formation of the internal olefin (actually, an enol: 2-buten-2-ol, probably as a mixture of cis and trans isomers) is thermodynamically preferred with respect to the formation of the terminal one (Zaitsev rule). Should this be the case, the enol would undergo an extremely rapid keto–enol tautomerism, thus affording methyl ethyl ketone:

$$CH_3-CH(OH)-CH(OH)-CH_3 \rightarrow CH_3-C(OH) = CH-CH_3$$
$$\rightarrow CH_3-CO-CH_2-CH_3 \tag{23.5}$$

However, few catalysts, including thorium oxide [16] and alumina [17], are able to direct the dehydration toward the formation of the terminal olefins. In this case, 3-buten-2-ol, the non-Zaitsev product, would form and, upon a second dehydration, would afford butadiene.

Versalis, in close collaboration with Genomatica, also is engaged in the butanediol family of processes to renewable butadiene. Based upon its experience in the metabolic engineering, Genomatica is defining a fermentation to transform sugars, including 2G ones, into butanediols. Versalis, in its turn, would leverage on its catalytic skills to reach efficient dehydration of butanediols to 1,3-butadiene. The required biochemical pathway has been defined and already demonstrated on laboratory scale, even on 2G sugars. At the same time, suitable dehydration catalysts have been identified. Thus, all the main steps of the process have been already defined, and the continuous production of renewable butadiene should follow soon.

References

1. Ricci, M. and Perego, C. (2012) in *Biorefinery: From Biomass to Chemicals and Fuels* (eds M. Aresta, A. Dibenedetto, and F. Dumeignil), De Gruyter, Berlin, pp. 319–332.
2. Anastas, P.T. and Warner, J.C. (1998) *Green Chemistry: Theory and Practice*, Oxford University Press, New York.
3. Matrica http://www.matrica.it (accessed 14 January 2015).
4. Lembaga Getah Malaysia http://www.lgm.gov.my/nrstat/nrstats.pdf (accessed 27 January 2015).
5. Finlay, M.R. (2009) *Growing American Rubber. Strategic Plants and the Politics of National Security*, Rutgers University Press, New Brunswick, NJ.
6. Cianci, A. (2007) *SAIGA, Il progetto autarchico della gomma naturale*, Thyrus, Arrone (in Italian).
7. Nakayama, F.S. (2005) *Ind. Crops Prod.*, **22**, 3–13.
8. Alexander, P. (1911) *Chem. Ber.*, **44**, 2320–2328.
9. Schloman, W.W. Jr., Hively, R.A., Krishen, A., and Andrews, A.M. (1983) *J. Agric. Food. Chem.*, **31**, 873–876.
10. Zoeller, J.H. Jr., Wagner, J.P., and Sulikowski, G.A. (1994) *J. Agric. Food. Chem.*, **42**, 1647–1649.
11. Flores-Rosete, G. and Martínez-Vázquez, M. (2008) *Nat. Prod. Commun.*, **3**, 413–422.
12. Haagen-Smith, A.J. and Siu, R. (1944) *J. Am. Chem. Soc.*, **66**, 2068–2074.
13. See, e.g., Chundawat, S.P.S., Chang, L., Gunawan, C., Balan, V., McMahan, C., and Dale, B.E. (2012) *Ind. Crops Prod.*, **37**, 486–492.
14. See, e.g., Jones, M.D., Keir, C.G., Di Iulio, C., Robertson, R.A.M., Williams, C.V., and Apperley, D.C. (2011) *Catal. Sci. Technol.*, **1**, 267–272.
15. Makshina, E.V., Dusselier, M., Janssens, W., Degrève, J., Jacobs, P.A., and Sels, B.F. (2014) *Chem. Soc. Rev.*, **43**, 7917–7953.
16. Lundeen, A.J. and Van Hoozer, R. (1963) *J. Am. Chem. Soc.*, **85**, 2180–2182; *J. Org. Chem.* (1967) **32**, 3386-3389.
17. Davis, B.H. (1982) *J. Org. Chem.*, **47**, 900–902.

24
Chemistry and Chemicals from Renewables Resources within Solvay

Sanjay Charati, Corine Cochennec, Manilal Dahanayake, Patrick Gilbeau, Marie-Pierre Labeau, Philippe Lapersonne, Philippe Marion, Sergio Martins, François Monnet, Ronaldo Nascimento, and Franco Speroni

24.1
Introduction

For Solvay, a leader in the chemical industry, sustainability is a priority. Aligned with this overarching trend, we are strongly committed in developing chemistry, products, and solutions that deliver optimized performance to the consumer. Among them, chemicals derived from renewable raw materials offer sometimes the best compromise in delivering innovative, sustainable solutions to the chemical industry. Within Solvay, we have developed considerable experience and expertise in the manufacture of differentiated specialty chemicals from renewable resource. In this chapter, we highlight a few of these products and their chemistry, performance, and applications. The diversity of the raw materials used in developing the underlined sustainable chemistry and products and their corresponding performance and benefits to the consumer and the environment will further emphasize the advantage for the use of renewables in the manufacture of performance chemicals. Most of the benefits derived are linked to the very nature of the raw material itself that otherwise could not have been reached with fossil hydrocarbons. Being able to leverage the unique structures that these raw materials offer is the basis for the design of highly specialized innovative products.

Aligned with the strategy of sustainability and responsible care, at Solvay we will not confuse the tool (e.g., using renewable raw materials) and the goal (i.e., being sustainable). Thus Solvay's strategy is always to analyze our products, raw materials, and processes with respect to their complete life cycle and have a balanced approach in using specific raw materials and processes that will ultimately benefit the environment and the consumers.

Chemicals and Fuels from Bio-Based Building Blocks, First Edition.
Edited by Fabrizio Cavani, Stefania Albonetti, Francesco Basile, and Alessandro Gandini.
© 2016 Wiley-VCH Verlag GmbH & Co. KGaA. Published 2016 by Wiley-VCH Verlag GmbH & Co. KGaA.

24.2
Chemistry from Triglycerides

24.2.1
Epichlorohydrin

Epichlorohydrin, a chemical compound of the epoxide family, is an essential feedstock material for the production of epoxy resins. The resins are used for various applications, including corrosion–protection coatings and in the electronics, automotive, aerospace, and power-generating windmill industries. Epichlorohydrin is also used in the production of elastomers, glycidyl ethers, polyols, glycidyl derivatives, polyamide-epichlorohydrin resins, and polyamine-epichlorohydrin resins for water purification and in the reinforcement of paper [1]. Conventionally, epichlorohydrin is produced in a multistep process starting with the reaction of fossil-origin propylene with chlorine at high temperature. The purified allyl chloride is hypochlorinated in an isomeric mixture of dichloro-propanol which is finally dehydrochlorinated with a base to give epichlorohydrin [1–3] (Scheme 24.1).

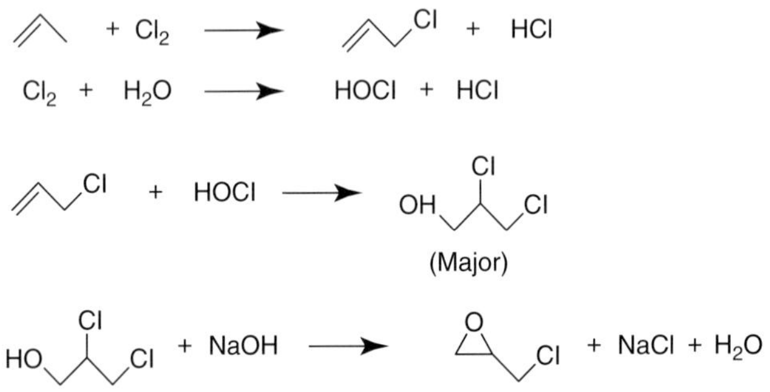

Scheme 24.1 Traditional epichlorohydrin production from propylene.

The isomeric mixture of dichloropropanol is constituted of 2,3-dichloro-1-propanol (about 66%) and 1,3-dichloro-2-propanol (about 33%). The predominance of 2,3-dichloro-1-propanol is a drawback of this process, because the rate of epichlorohydrin formation from this isomer is significantly slower than from 1,3-dichloro-2-propanol and this has negative consequences on the formation of undesired by-products [4]. The theoretical chlorine atom economy of this process is low given that four atoms are used to end up with only one fixed in the epichlorohydrin. The yield of epichlorohydrin on chlorine or propylene is limited to about 70% essentially because chlorinated organic by-products are largely formed during the chlorination and hypochlorination steps [3]. The hydrogen chloride coproduct is only partly recoverable, and usually the part formed during the hypochlorination is produced as a dilute hydrochloric acid solution containing dichloropropanol. This solution is not purified before the dehydrochlorination compelling to double the quantity of required base, to both quench the acid and carry out the dehydrochlorination, and therefore the quantity of waste produced chloride salt. The dilute salt solution is always almost released in the environment, in a river, or in the sea, after treatment.

The Epicerol® bio-based technology processes renewable glycerol to produce epichlorohydrin (Epicerol® is the registrated trademark for the 100% bio-based epichlorohydrin produced by the Solvay process). The renewable glycerol is usually obtained as a by-product in four different processes: soap manufacture, fatty acid production, fatty ester production, and microbial fermentation. It can also be prepared by hydrolysis and hydrogenolysis of carbohydrates [5]. The renewable glycerol has become an attractive cheap raw material during the last decade because of the booming development of fatty acid methyl esters (FAMEs) as biofuels [6]. Around 1000 kg of FAME production involves approximately 100 kg of glycerol as a by-product. Although the fundamental principles of producing epichlorohydrin from glycerol are known from over a century [7], to our knowledge no perennial industrialization of the process had been done before the 2000s. Epichlorohydrin is produced in only two steps. The first step involves the reaction between glycerol and hydrogen chloride using a carboxylic acid as a catalyst, which produces an isomeric mixture of dichloropropanol. The second step is a dehydrochlorination reaction of dichloropropanol with a base (Scheme 24.2).

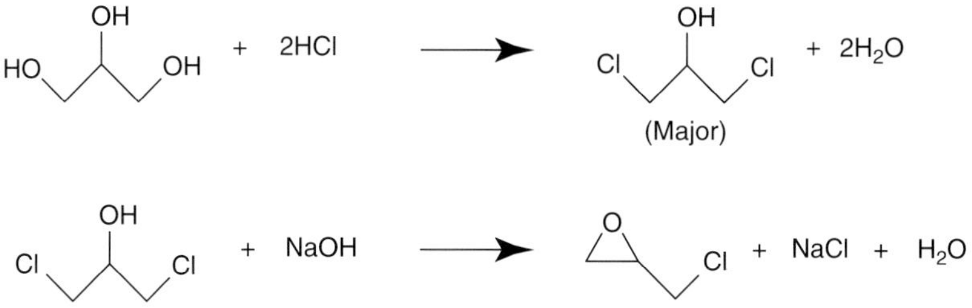

Scheme 24.2 Epichlorohydrin production from glycerol.

The isomeric mixture of dichloropropanol is here largely constituted of the more reactive 1,3-dichloro-2-propanol with a minor proportion of 2,3-dichloro-1-propanol (only several %) allowing a better optimization of the dehydrochlorination step. Epicerol® technology is composed of a modular process and has several environmental advantages against the traditional propylene process. The theoretical chlorine atom economy of the process is far better given that only two atoms are used to end up with one fixed in the epichlorohydrin. The improved selectivity and the consumption of a chlorinating agent in lower quantities result in low production of the chlorinated organic and inorganic compound by-products. A lower consumption of energy and water reduces the emissions to the environment. The process produces only one equivalent of waste chloride. The brine effluent quantity can be reduced as the intermediate dichloropropanol is produced only together with a bistoichiometric water quantity. Moreover, the brine can optionally be recycled to a chlor-alkali electrolysis plant after an adequate purification treatment [8]. Based on the assessment of all the steps from resource extraction to the finished product, Solvay performed a comparative life cycle analysis (LCA), in order to benchmark the Epicerol® bio-based process with a state-of-the-art propylene-based process. This cradle-to-gate environmental assessment was audited and validated by DEKRA, a German leading certification body in environmental management. The main benefits are a 61% reduction of the *global warming potential* (defined as the sum of greenhouse gas (GHG) emissions and biogenic CO_2 capture) and a 57% reduction of nonrenewable energy consumption (related to the efficiency of the processes) (Figure 24.1).

Integrating 1 ton of Epicerol® (instead of classical epichlorohydrin from nonrenewable resource) in a product makes the carbon footprint drop down by 2.56 ton CO_2 equivalent. A first PCT patent for this original Solvay technology has been published in 2005 [9], followed by more than 30 PCT published patents covering the specific chemical and technical aspects of the reaction parameters, process, products, and applications. This Solvay technology has been developed industrially in 2007 in an initially 10 000 ton per year epichlorohydrin plant integrated in a common epichlorohydrin production line within an existing classical propylene-based plant at Tavaux in France. The pure bio-based Epicerol® production is being exploited in a 100 000 ton per year plant in the Thai affiliate Vinythai Public Company Limited at Map Ta Phut in Thailand since 2012. There is also a project of a new 100 000 ton per year plant at Taixing in China.

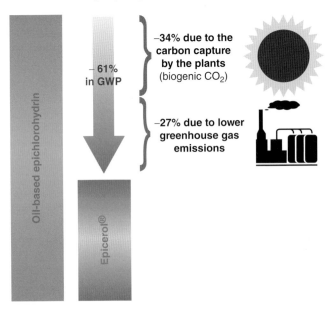

Figure 24.1 Integrating 1 MT of Epicerol® (instead of classical epichlorohydrin from non-renewable resource) in a product makes the carbon footprint drop down by 2.56 MT CO_2 equivalent.

24.2.2
Augeo™ Family: Glycerol as a Platform for Green Solvents

Solvay's Augeo™ family is an innovative portfolio of solvents targeted toward different worldwide markets. Its development was based both on technical performance and on the sustainability axes, which consider economic, social, and environmental factors. All of these molecules are synthesized from a renewable raw material: glycerol, generated by the biodiesel production.

The chemistry of this family is based on the reaction between glycerol and carbonylated compounds (e.g., ketones and aldehydes), whose main products are ketals – molecules that display enhanced performance in terms of solvency power and evaporation, representing excellent alternatives as solvents in different applications and markets (Figure 24.2).

Besides the physicochemical properties, this family intrinsically presents a very low HSE profile, exemplified by the following characteristics:

- Low toxicity to humans and to the environment
- Slight odor

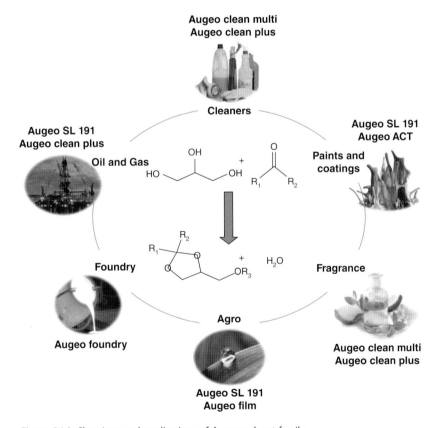

Figure 24.2 Chemistry and applications of Augeo solvent family.

- High flash point
- High boiling point
- Low vapor pressure which allow several products of the Augeo family to be granted the classification of "non-VOC"
- Low carbon footprint, especially when compared with solvents of petrochemical origin.

Due to this combination of physicochemical and HSE aspects, the Augeo family possesses enough versatility to cover a broad range of applications in different market segments, contributing to exclusive performances and properties. Some examples of benefits imparted by the use of Augeo family products in formulations from several different fields are listed below:

- *Paints and Coatings:* Contribution for excellent coating film formation due to high solvency power and dry time adjustment. When used as coalescents in waterborne formulations, in addition to the coalescence performance, they significantly help in reducing the VOC, toxicity, and odor, improving safety during the application.
- *Cleaners for Household and Industrial and Institutional (I&I):* Due to their high capacity as coupling agents to a broad range of surfactants, they allow improvements in the cleaning performance, lessening the impact on the environment and making formulations more friendly.
- *Fragrances:* Augeo family products are suitable to be used in this field with exclusive performance, owing to their slight odor and low toxicity, working in tandem with their high solvency power for a broad range of fragrances.
- *Oil and Gas:* Physicochemical properties that allow their use in formulations geared toward different steps of the process – upstream, transportation, and downstream.
- *Agro:* Solubilization of actives and enhancements in HSE profile.
- *Leathers:* The slight odor and good skin penetration properties result in excellent carrier agents, used in the leather finishing process, generating noticeable improvements in HSE at the workplace.
- *Electronics:* Excellent properties, especially for stripping processes, attributable to the combination of good solvency power and low evaporation rate.

24.2.3
Polyamide 6.10

24.2.3.1 Generalities
Known since the dawn of nylon history [10] and traditionally used in niche applications (e.g., toothbrushes, monofilaments, filters), in recent years Polyamide (PA) 6.10 has got renewed interest as an engineering polymer, due to its peculiar properties and partially biosourced and renewable origin.

This polyamide is industrially obtained by mass polycondensation from hexamethylenediamine (HMD) and sebacic acid (SA). While HMD is a well-known monomer of fossil origin, SA is obtained from the chemical modification of castor oil. Its production process is based on castor oil alkaline fusion followed by dehydrogenation, retro-aldol scission, and final disproportionation at $T = 200-300\,°C$ to give SA and 2-octanol (Scheme 24.3) [11, 12].

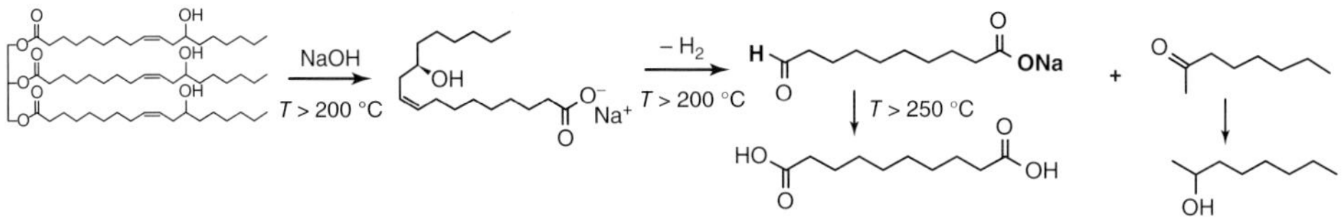

Scheme 24.3 Schematic synthesis path of sebacic acid from castor oil.

The polycondensation of PA 6.10 usually takes place by a classical route starting from the salification of HMD and SA in water, followed by mass polycondensation and extrusion [13, 14].

The solubility of the 6.10 salt in water is close to 60% at 90 °C; therefore industrial polymerization processes and yields obtainable are comparable with batch polymerization of the more common PA 6.6 from the corresponding salt. The PA 6.10 polymer obtained so far is often postpolycondensed in the solid state to reach high molecular weight (Mw). This additional process is required, for example, for the production of extrusion grades for pipes and tubes.

Due to its SA fraction, PA 6.10 counts for the 62 wt% of bio-based content in the polymer chain (according to ASTM D6866-10 and based on ^{14}C content detection). Compared to other polyamides, the production of PA 6.10 is eco-efficient and shows a favorable carbon footprint of around 4.6 kg CO_2 equivalent [15]. Its production process requires at least 20% less nonrenewable resources per weight unit than other conventional aliphatic polyamides having comparable properties, such as PA 12 and PA 6. Its production chain from renewable resources reduces GHG emissions by 50% when compared with such other polyamides derived from nonrenewable sources [16].

24.2.3.2 Polymer Intrinsic Properties

PA 6.10 is a semicrystalline polymer, with a neat melting point ranging between 215 and 220 °C [17–20]. Its main physicochemical properties, in comparison with other aliphatic nylons, are directly related to the ratio between nonpolar, aliphatic $(-CH_2-)_n$ chains, and polar amide bonds $(-CONH-)$, as shown in Table 24.1.

Clearly, PA 6.10 key properties are positioned somehow in the middle versus those of the two major families of aliphatic nylons (short-chain PA 6.6 and 6 on one side and long-chain PA 11 and 12 on the other) and have some peculiar features. For instance, PA 6.10 shows quite a high melting point [18, 19] – close to that of PA 6 – but at the meantime it approaches PA 11 and PA 12 as far as water absorption and density are concerned. Another feature of PA 6.10 is high compatibility or full melt miscibility with PA 6, 6.6, 11, and 12 and interaction with the latter

Table 24.1 Physical properties: comparison of PA 6.10 versus other aliphatic nylons, as a function of increasing CH_2/amide ratios.

PA type	CH_2/amide ratio	Melting point (°C)	Density (g cm^{-3})	Water absorption 23 °C/50% RH (%)	Water absorption at saturation (%)
4.6	4	295	1.18	3.7	13.0
6.6	5	260	1.14	2.5	8.5
6.10	**7**	**218**	**1.09**	**1.6**	**3.3**
6.12	8	210	1.07	1.4	2.8
10.10	9	205	1.05	1.1	1.9
6	5	222	1.13	2.7	9.0
11	10	182	1.04	0.8	1.9
12	11	176	1.01	0.6	1.5

as a nucleating agent as reported by literature [21–24]. Therefore, PA 6.10 may constitute an interesting technical alternative versus other aliphatic polyamides, or a useful complement to them, in order to create materials in which a specific balance of properties is required.

24.2.3.3 Main Technological Properties and Applications

Low water absorption, crystallinity, relatively high melting point, and relatively high CH_2/CONH ratio are the fundamental physicochemical features determining most of the peculiar technological properties of PA 6.10. The combination and balance among such properties make PA 6.10 a new material particularly adapted for specific applications, particularly in automotive and in industrial applications, as an alternative or complement to PA 6, 11, and 12:

Dimensional stability versus moisture and temperature: Due to low water absorption (Table 24.1), PA 6.10 shows higher dimensional stability and limited variation of mechanical properties in presence of moisture versus PA 6 and 6.6. Modulus and strength are only partly affected by moisture, so that properties in conditioned state approach those of PA 11 and 12 (Table 24.2). On the other side, crystallinity and relatively high melting point are compatible with good retention of mechanical properties at high temperature. Such a set of properties makes PA 6.10 a good matrix for unreinforced and glass-reinforced materials where stiffness and strength are required in a broad range of temperature and environmental conditions (e.g., for plumbing applications in contact with water, battery gaskets, quick connectors, smart device covers, joints, etc.).

Environmental stress cracking resistance (ESCR), impact strength, and barrier: ESCR to deicing salts [25] is an important requirement for automotive parts exposed to salt splash (see, e.g., SAE J844 norm). Whereas PA 6 and PA 6.6 are very sensitive to ESCR phenomena in the presence of $ZnCl_2$ or $CaCl_2$ [26–34], PA 6.10 shows high resistance within a wide range of T, t, and σ conditions

Table 24.2 Technological properties of PA6.10 typologies–glass fiber (GF) reinforced and plasticized.

Property	Unit	PA 6.10 GF30	PA 6.10 GF30 high impact	PA 12 GF30	PA 6.10 plasticized	PA 6.10 plasticized high impact	PA 12 plasticized
Melting point DSC (°C)	°C	217	217	176	215	215	175
Density	g cm^{-3}	1.3	1.25	1.23	1.05	1.04	1.02
Water uptake (24 h, 23 °C/saturation)	%	0.4/2.4	0.35/1.9	0.5/1.3	0.65/1.9	0.45/1.75	0.5/1.5
Tensile strength at break[a]	MPa	153/112	124/91	115/95	47/42	40/34	41/37
Tensile strength at yield	MPa	—	—	—	40	33	32
Burst pressure of 6 mm × 8 mm pipes[b]	bar	—	—	—	98/34	82/29	81/28
Elongation at break[a]	%	4.3/7.5	5.2/8.1	6.4/7.7	215/283	185/238	230/269
Tensile modulus[a]	MPa	8900/6100	7400/5200	7100/5200	890/570	830/540	625/530
Flexural modulus[a]	MPa	8100/6200	6900/5300	6600/5600	720/570	685/530	600/610
Charpy notched +23 °C[a]	kJ m^{-2}	12	27	23	24/55	87/117	47/49
Charpy notched −30 °C[a]	kJ m^{-2}	12/16	27/31	23/24	3.9/3.5	20/21	5.3/4.6
Charpy unnotched +23 °C	kJ m^{-2}	87	92	88	NB	NB	NB
Charpy unnotched −40 °C	kJ m^{-2}	76	103	90	NB	NB	NB
Shore D hardness ISO 868	—	84	78	79	68	63	65
Flammability UL 94 (0.8 mm)	—	HB	HB	HB	HB	HB	HB

a) EH0 (dry as molded)/EH50 ISO 1110.
b) Fuel pipes, 6 mm internal and 8 mm external diameter, 1 mm thick.

(Figure 24.3). In most cases, PA 6.10 approaches the performances of the commonly used PA 12 for the production of most critical parts (e.g., fuel pipes, cooling ducts, air brake pipes) where ESCR is not negotiable.

However, requirements for fuel pipe applications overcome ESCR compliance, including flexibility, toughness at low temperature, and barrier to fuel vapors as well.

Furthermore, PA 6.10 can be modified with either plasticizers or toughening agents, in order to obtain materials showing improved flexibility and mechanical properties. Typical applications are air and fuel pipes requiring high burst resistance, strength at low temperature (Table 24.2, Figure 24.3), and ease to be conformed in particular shapes.

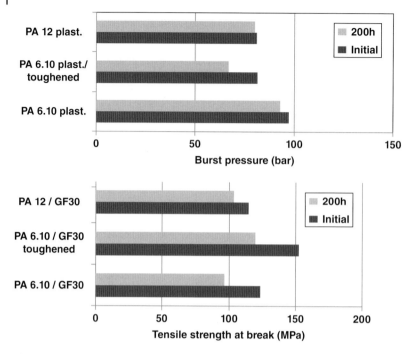

Figure 24.3 Aging in ZnCl$_2$/water 50/50 wt% solution at 80 °C for 200 h: burst pressure of 6 mm × 8 mm plasticized PA 6.10 tubes and tensile strength of 30GF-reinforced PA 6.10 versus PA 12.

A further advantage of PA 6.10 is low permeation to oxygen, carbon dioxide, refrigerant fluids, and polar fluid vapors, such as alcoholated gasoline (Figure 24.4).

As a whole, such properties are relevant, for example, for applications in fuel ducts and in quick connectors for fuel circuits in automotive applications, as well as for air conditioning systems and packaging too.

Among recent applications of PA 6.10, it is worth mentioning the production of automotive cooling system elements (radiator caps, expansion tanks, ducts, etc.) requiring ESCR to calcium chloride. This is a demanding requirement by car manufacturers for vehicles to be used in several cold countries (e.g., Asia, Russia, North Europe) where massive use of deicing salts on roads is the rule.

In such conditions, traditional PA 66 GF30 – normally used for the production of cooling circuit elements – is not compliant, for example, if subjected to pulsating pressure cycles in the presence of CaCl$_2$ solution spray.

GF-reinforced PA 6.10 or PA 6.10/PA 6.6 blends show outstanding ESCR performance versus CaCl$_2$, associated with lower moisture absorption and hence improved hydrolysis resistance in water/glycol mixtures.

Such properties make glass-reinforced PA 6.10 an ideal polymer for a number of further growing applications where chemical resistance, water and hydrolysis resistance, barrier, and mechanical and thermal behavior are required.

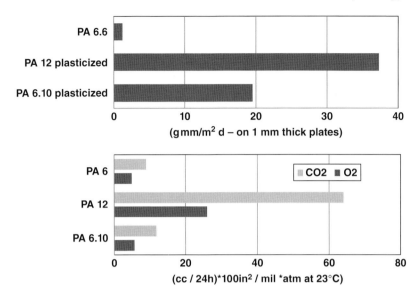

Figure 24.4 Permeability at 40 °C versus E10 fuel (10% ethanol, 45% toluene, 45% isooctane) and at 23 °C versus CO_2 and O_2.

24.2.4
A New Generation of "Sustainable" Viscoelastic Surfactants Based on Renewable Oleochemicals

24.2.4.1 Introduction

Surfactants are the most versatile chemicals in the industry. They are widely used in washing powder, personal care products, and domestic cleaning products including those involved in industrial cleaning. Moreover, they have been extensively utilized in the formulation of paints, coatings, polymer and plastics, agrochemical adjuvants, food production, explosives, leather, textiles, pharmaceuticals, and oil and gas recovery.

Over the last decade "sustainability" has become a priority for surfactant manufactures. This is driven by the consumer demand for "greener" surfactants from renewable raw materials with favorable ecotoxicological properties and minimal environmental impact. In order to meet this demand, oleochemicals are replacing synthetic petrochemicals as the raw material of choice and represent a fast-growing market for the manufacture of surfactants.

24.2.4.2 Micellar Structure and Rheology

At concentrations above their critical micelle concentration (cmc), surfactant molecules may aggregate in the bulk of the solvent into various microstructures such as spheres, cylinders, lamellar forms (flat sheets), and "wormlike" extended micelles. These complex associations depend on the geometry of the surfactant molecules and their molecular environment. The formation of these types of aggregates has been correlated to a packing parameter [35, 36]:

$V_H/(lc - a_0)$ where V_H is the volume occupied by the hydrophobic group, lc is the length of the hydrophobe, and a_0 is the cross-sectional area of the head group. When the value of $V_H/(lc - a_0)$ is between 0 and one-third, then spherical micelles are formed; between one-third and one-half, cylindrical micelles are formed; and between one-half and 1, lamellar micelles are formed in aqueous media. When the value is >1, then reverse micelles are formed in nonpolar media.

Aqueous solutions of spherical micelles at concentrations close to cmc are Newtonian fluids and have viscosities that do not vary very far from that of water. As the surfactant concentration is increased, at least >30 times the cmc, the micelles may become asymmetrical (cylindrical) and when in solution may show moderate increase in viscosity. When the concentration is increased even more, the cylindrical micelles may pack together into a hexagonal arrangement to form long hexagonal liquid crystals, with an accompanying sharp increase in the solution viscosity. Further increases in the concentration to >50 times the cmc may result in the formation of lamellar liquid crystals from hexagonal liquid crystals. This would result in some reduction in the viscosity of the solution (Figure 24.5) [36].

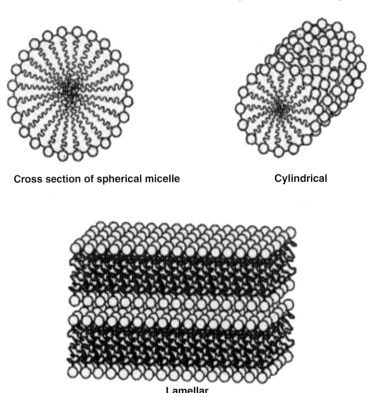

Cross section of spherical micelle   **Cylindrical**

Lamellar

Figure 24.5 Different configurations of micelles.

Figure 24.6 Schematic illustration of the wormlike micelle network.

The rheological properties of cylindrical and lamellar systems have been extensively studied but utilized mostly in personal care formulations such as shampoos, body lotions, body soaps, and skin moisturizers. These formulations require high concentrations, 20–60% of surfactants to form these lamellar and cylindrical micelles. However, in industrial applications, use of such high concentration of surfactants is cost prohibitive. Furthermore, disposal at these high concentrations is a major environmental hazard, and therefore their use has been limited mostly to personal care applications [37–39].

Recent advances in surfactant technology have resulted in the development of certain specialty oleochemical-based amphoteric/zwitterionic [40] and cationic [41] surfactants that could transition, at very low concentrations, from spherical to long "wormlike" lamellar micelles Figure 24.6. The molecular packing parameters for such formations of wormlike micelles approximate to 1/2. The entanglement of these "wormlike" micelles results in a drastic increase in elasticity and viscosity of the fluid, with the elastic forces (G′) exceeding the viscous forces (G″) over a wide frequency. These surfactants are referred to as viscoelastic surfactants (VESs). Under shear, these "wormlike" micelles can break and reform, once the shear is released. As such these VESs are also referred to as *living polymers* to distinguish them from high Mw polymers, which break down under shear, with a loss in viscosity that is irreversible.

This article will focus upon the fastest-growing class of oleo-based amphoteric/zwitterionic surfactants that are specially designed to provide optimal VES properties at extremely low concentrations [40, 42] (Scheme 24.4).

Chemistry and synthesis route for oleo-based amphoteric VES [40, 42].

24.2.4.3 VES Fluids in the Recovery of Oil and Gas

Over the last decade, VES fluids have seen the fastest growth in the recovery of oil and gas from subterranean formations [40, 42–44].

In recovering petroleum from subterranean formations, it is common practice to fracture the rock containing the petroleum in order to create flow channels.

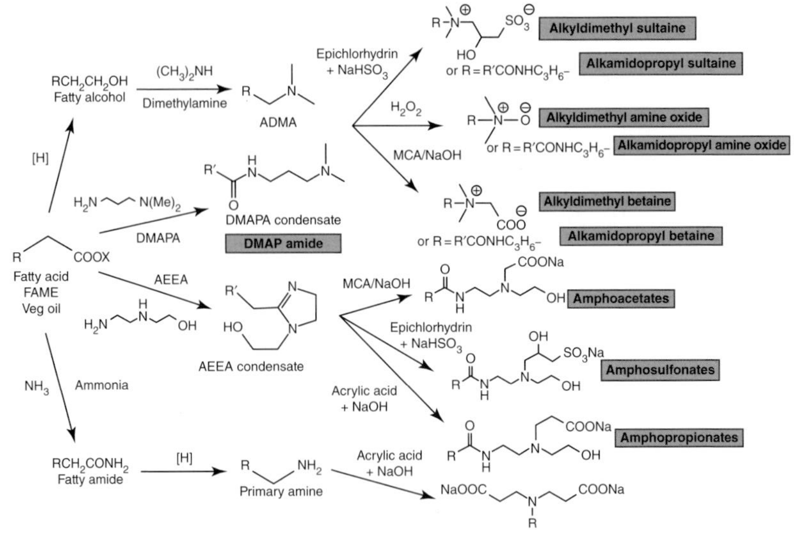

Alkyldimethyl sultaine

Alkamidopropyl sultaine

Alkyldimethyl amine oxide

Alkamidopropyl amine oxide

Alkyldimethyl betaine

Alkamidopropyl betaine

DMAP amide

Amphoacetates

Amphosulfonates

Amphopropionates

RCH₂CH₂OH Fatty alcohol — (CH₃)₂NH Dimethylamine → ADMA

Epichlorhydrin + NaHSO₃

H₂O₂

MCA/NaOH

[H]

H₂N–N(Me)₂ DMAPA

DMAPA condensate

R–COOX Fatty acid FAME Veg oil

AEEA

H₂N–N–OH

AEEA condensate

NH₃ Ammonia

RCH₂CONH₂ Fatty amide — [H] → R–NH₂ Primary amine

MCA/NaOH

Epichlorhydrin + NaHSO₃

Acrylic acid + NaOH

Acrylic acid + NaOH

Scheme 24.4 A simplified chemistry for amphoteric viscoelastic surfactants.

Viscous aqueous or nonaqueous fluids are hydraulically injected into the wellbore to accomplish this function.

Once the rock is fractured, a "proppant" generally consisting of sand or gravel that is suspended in the viscous fracture fluid is placed into the fracture. These proppants prevent collapse of the fracture and provide improved flow of oil or gas into the wellbore. Traditionally the viscosifiers used have been hydratable polysaccharides such as galactomannoses (guars) and high molecular weight polyacrylamides (HPAM) cross-linked with zirconium, beryllium, and so on. Several problems have been associated with the use of cross-linked polysaccharides and HPAM as fracturing fluids. First, the process of hydration and cross-linking with transition metal ions to induce viscosity and yield stress is cumbersome as cross-linking takes time to react and requires labor and expensive blending/hydrating equipment. Second, controlling the viscosity with the cross-linking agent is highly concentration dependent. Third, recovering the cross-linked polysaccharides, once the fracturing operation is completed, is very tedious and involves oxidation or enzymatic hydrolysis of the hydrocolloid. Even then the recovery of the degraded polymer is rarely complete and leaves residue that prevents the ease of flow of oil and gas from the formation. Last, the use of heavy transition metal ions for cross-linking is an environmental hazard due to their acute toxicity.

In order to overcome the limitations of cross-linked hydrocolloids and their unfavorable environmental profile, the use of amphoteric/zwitterionic VES systems has become the choice of innovative technology in well stimulation [40, 42–44].

VES fluid systems that are now being used and that have shown very good results are the oleochemical based (oleic, stearic, behenic, and erucic) amphoteric surfactants [40, 42, 43] that have been optimized to form "wormlike" micelles (as opposed to the spherical micelles formed by most other surfactants). These VESs are unique in that they are highly viscoelastic with $G' > G''$ over a broad range of frequencies ranging from 0.01 to $100\,s^{-1}$ at concentrations as low as <1% by weight. This allows them to be used economically as well. Furthermore, these association structures are formed in a dynamic state with rapid shear recovery [44]. Coming in contact with hydrocarbons, the structures are readily altered. Under the latter conditions, the micelles rapidly revert to spherical micellar or monomer forms that do not build viscosity in the fracturing fluid. This allows almost complete recovery of these VES fluid systems once the fracture is completed, with uninhibited flow of oil and gas, further enhancing their environmental profile and safety over polymers in the recovery of oil and gas.

24.2.4.4 Other Novel Applications for VES

Apart from fracturing, the literature describes several novel applications for VES, ranging from friction reducers [45, 46], drift control agents for agrochemical adjuvants [47], cementing excavation [48], foam mobility control [49], enhanced oil recovery (EOR) [50], self-diverting acids in well stimulation [51], and home and personal care formulations [52, 53].

24.2.4.5 Conclusions

To meet the challenges of consumer demand for "sustainable" surfactants, the manufacture and use of oleo-based surfactants are rapidly growing. Within this context oleo-based amphoteric/zwitterionic VESs represent an emerging technology that allows significant opportunity for new and novel applications in home, personal care, and industrial markets.

24.3
Chemistry on Cellulose: Cellulose Acetate

24.3.1
What Is Cellulose Acetate?

Cellulose acetate (CA) is a family of different polymers. These polymers are synthetized by the reaction of cellulose with acetic anhydride (substitution of some hydroxyl groups of the cellulose backbone with acetate groups). The mainly produced and used CAs are:

- Cellulose triacetate (CTA): Almost all three hydroxyl groups of the cellulose glucose monomer are replaced by acetate groups (degree of substitution (DS) of 2.8–3.0).
- Cellulose diacetate (CDA): With a DS of 2.2–2.7 (Scheme 24.5).

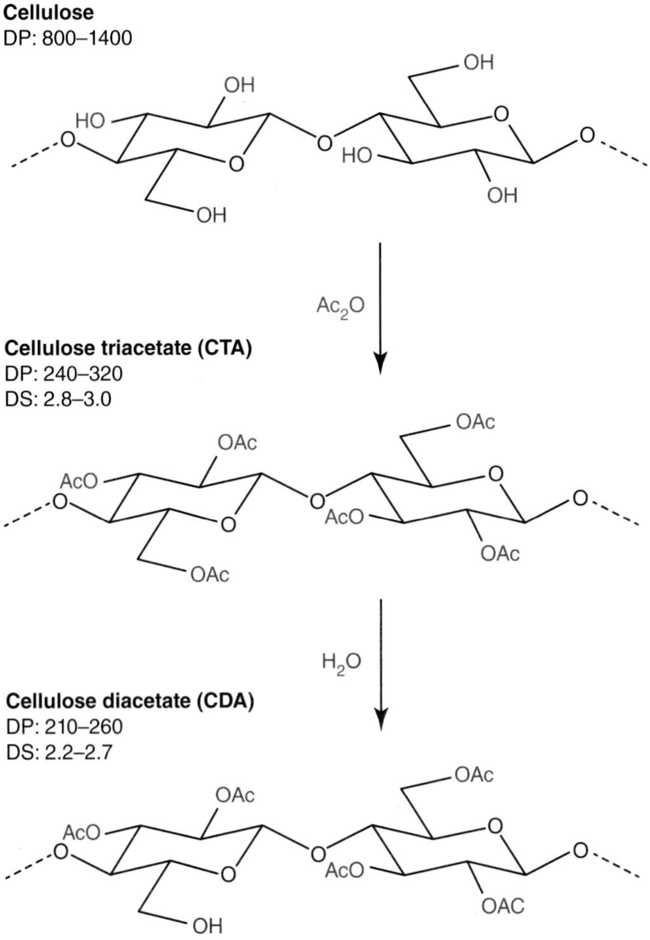

Cellulose
DP: 800–1400

Ac$_2$O

Cellulose triacetate (CTA)
DP: 240–320
DS: 2.8–3.0

H$_2$O

Cellulose diacetate (CDA)
DP: 210–260
DS: 2.2–2.7

Scheme 24.5 Simplified chemistry of cellulose acetate.

CAs are bio-based polymers, as the main raw material used is cellulose. The cellulose used is a high-quality cellulose (with an alpha cellulose content more than 96%). The mostly used cellulose raw material is high-quality wood pulp (also called *dissolving wood pulp*) which is sourced from sustainably managed forestry (SFI-certified forest).

CA polymers are also biodegradable [54].

A well-documented monograph on CA has been published in 2004, edited by Paul Rustemeyer [55], combining the academic expertise on physical and chemical properties of CA with the industrial knowledge in technology, product properties, and applications.

Figure 24.7 A. Eichengrün.

24.3.2
An History That Started More Than 100 Years Ago

CA was first mentioned in 1865 by Schützenberger [56], who synthetized it by heating cellulose in a sealed glass tube with acetic anhydride. In 1879, Franchimont [57] discovered that it could be produced at room temperature with sulfuric acid as catalyst.

Targeting the replacement of existing collodion and nitrocellulose, immense effort in industrial research was carried out at the end of the nineteenth century for the development of CTA for photographic films, fibers, and plastic material, but with very little success due to prohibiting costs, insufficient developed technologies, and so on.

The commercial breakthrough came with the discovery of CDA by Miles in 1904 [58] and Eichengrün in 1905 [59], as CDA is soluble in many available solvents like acetone, methyl acetate, or ethyl acetate and has good mechanical properties (Figure 24.7).

The first important application for CDA was in the field of coating for airplanes: a linen tissue coated with CDA gave a light-, water-, oil-, and petrol-resistant material with very flat surface.

24.3.3
Main Markets and Applications for CA

24.3.3.1 Textile
Since the coating market for plane dropped strongly after World War I, the development of new applications was a must. In England, the Dreyfus brothers successfully developed an innovative spinning process (dry spinning of CDA in acetone) for the company British Celanese Co. [60].

CDA textile fibers (known also as *artificial silk*, due to their natural touch and appearance) encountered a high commercial success and development.

Since World War II, CDA textile fibers demand declined, highly challenged by new oil-based man-made fibers such as polyamides or polyesters.

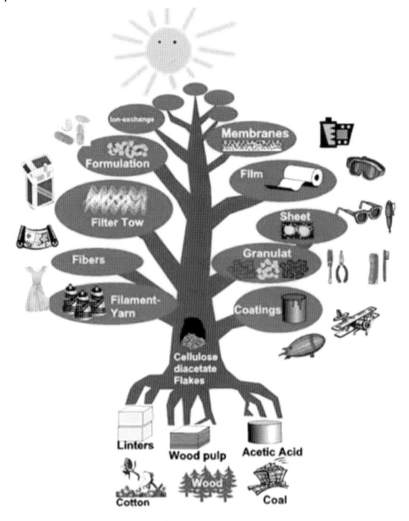

Figure 24.8 Cellulose acetate applications.

Today CA textile fibers keep its niche in the area of silk like women's apparel (Figure 24.8).

24.3.3.2 Filter Tow

Since the first introduction of cigarette filters in 1952 [61], CA filter tow has been the material of choice, with an almost linear growth which in 2014 stands at about 800000 metric tons. Today more than 97% of the world's cigarettes are made with CA filters. This success is linked to their high filtration performances and highly appreciated "taste signature." Solvay, one of the leaders in this industry, launched very recently an innovative fast degradable filter tow, Rhodia DE-Tow™, to reduce the environmental impact of bud's litters.

24.3.3.3 Plastic Applications

Highly transparent plastic with excellent clarity, good mechanical, natural touch, and feeling can be obtained by adding plasticizers to CDA and CTA. By injection molding or extrusion, plasticized CAs are today transformed in many products like tool handle, combs, barrettes, and spectacle frames. Solvay launched recently a new range of bio-based plastic, based on CA, Ocalio™, which is "the nature's plastic," bio-based, nontoxic, phthalate-free, with reduced CO_2 emissions and with excellent technical performances.

24.3.3.4 CA Films

Due to their high optical and mechanical performances and also dimensional stability, CA films first developed in many applications like photographic films. Today, CA films are also used in the growing market of LCD flat screen.

CA products are also present and developed in many other applications and markets like membrane technologies (as for water purification or desalinization), pharmaceutical applications (like control release matrix for actives), and so on.

24.3.4
Cellulose Acetate Today and Tomorrow

Today's worldwide CA market (and installed capacities) represents more than 1 million tons per year, Solvay (with its Solvay Acetow Global Business Unit) being one major actor.

It will continue to sustainably develop on its core businesses (filter tow, plastics, films, etc.).

CA is born more than 100 years ago, but it is clearly also a polymer for the future due to its specific high-performing functionalities: a biopolymer (bio-based, biodegradable with reduced CO_2 emissions) with good mechanical performances, high optical properties, natural touch and feeling, highly hydrophilic, superior filtration performances, and so on.

The acetylation technology is and will be also key for the future: it allows transforming natural raw material (like cellulose) into high-performance and easy-to-use materials. Beside CA, acetylated wood Accoya® has been introduced recently by Accsys with high success on the market. Accoya® is the wood solution for long-lasting external use (exceptional durability, high dimensional stability, nontoxic, long-lasting coating, not attractive for termites, etc.). Solvay Acetow is partnering with Accsys for the future business development of Accoya®.

24.4
Guars

24.4.1
Introduction

24.4.1.1 What Is Guar?

Native guar, or guar gum, is a biopolymer extracted from the seeds of *Cyamopsis tetragonolobus* (Leguminosae family), a plant that grows in semiarid zones such as Pakistan and India which is still the largest exporter of guar gum. Worldwide production of guar gum and guar gum derivatives jumped from ~200 000 tons in 2004 to ~450 000 tons in 2012, representing a value of $3200 millions, most of the increase being driven by the US shale gas drilling [62].

Guar pods contain five to nine seeds which are made of a hull (mainly cellulosics), a germ (mainly proteins), and the endosperm, which is in between the hull and the germ and whose function is to provide the germ with the energy needed during the germination phase. The endosperm contains the guar gum. It is isolated from the hull, and the germ in a two-step process and the resulting guar splits are either directly milled to produce native guar flour or chemically modified, washed, and milled, yielding in this case derivatized guar powders.

24.4.1.2 Why Is Guar Unique?

Polysaccharides that possess linear chains that can tightly pack together and develop hydrogen bonds are usually partly or fully crystalline and are not soluble in cold water: it is typically the case of cellulose, starch, as well as polymannose. Actually, polygalactomannans that have too few galactose side units to prevent the tight packing of polymer chains (cassia gum, locust bean gum, tara gum) are not soluble in cold water either [63].

As member of the polygalactomannans, guar is characterized by a linear polymannose backbone but contains enough randomly distributed galactose side units (Man/Gal ratio = (1.8 – 2)/1 Scheme 24.6) to avoid crystallinity. Thus guar is cold water soluble, which represents a significant processing advantage. For comparison purpose, guar is eight times more soluble than starch at ambient temperature.

It's now admitted thanks to multiangle laser light scattering analyses that guar is one of the highest Mw cold water-soluble polymers, with weight average Mws above 1 million g·mol^{-1} [64].

As a result of both its solubility at ambient temperature and its high Mws, guar is by definition a very good thickener, easy to process.

To conclude, if guar is by far the most produced galactomannan, it's not only because guar is easier and/or faster to grow compared to other galactomannans (locust bean gum, for instance) and relatively resistant to drought; it is also and mostly because guar combines these two key features – solubility in cold water *and* high Mws – making it a very good thickener, easy to process, as detailed later.

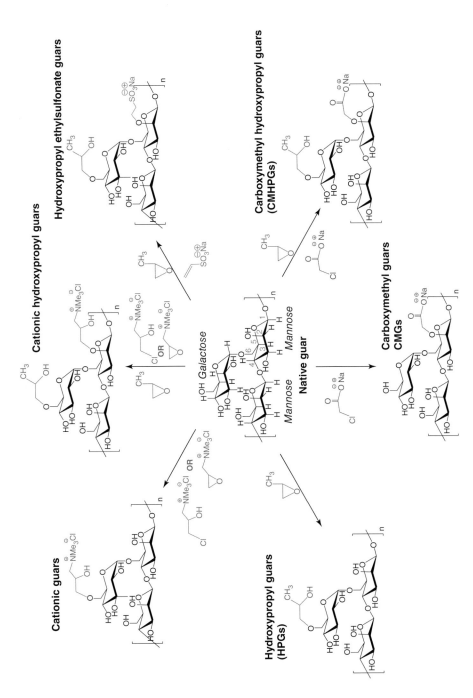

Scheme 24.6 Some chemistry of functionalization of guars.

24.4.2
Nonderivatized Guar: From Physicochemical Properties to Major Applications

Guar powders are easy to hydrate in water without any heating and yield very viscous solutions (at low shear rates) that are shear thinning. For these reasons, guar is a thickener of choice for cosmetic formulations such as shampoos and body washes.

With the viscosity being controlled by Mws, depolymerization is sometimes used to adjust the viscosity to the desired level. This is typically the case in food, where guar is used to thicken, stabilize, prevent ice crystal formation, and much more [65]. Among the most recent applications of guar, native or depolymerized, let's also cite its uses in nutraceuticals: due to its gel-forming properties, guar is now advertised to decrease cholesterol and glucose, which helps in weight loss and obesity prevention.

Beyond food and far ahead of cosmetics, the major use of native guars is in fracturing fluids in the oil and gas industry. Like most common polysaccharides (including other galactomannans and starches) but xanthan and succinoglycan, guar solutions don't display yield stress. However, guar *cis*-C-2 and C-3 hydroxyl groups are readily cross-linked by common metallic species such as boric acid, titanates, zirconates, and so on, generating yield stress in a well-controlled pH range which depends on the selected cross-linker. These gels are widely used to suspend proppants in fracturing fluids: the well-suspended proppants are transported into the fractures, and then, when the gel is chemically destroyed to remove the fracturing fluid, the proppants keep the fractures opened and the production can start.

For a more comprehensive list of nonderivatized guar applications, the reader is invited to read recent reviews [66].

24.4.3
Guar Derivatives: From Physicochemical Properties to Major Applications

Native guar is known to develop hydrogen bonds with lots of surfaces, for example, hydrated mineral surfaces or organic surfaces such as cellulose [67]. However, guar derivatization is a powerful lever to enhance the interaction between guar and surfaces or between guar and formulation ingredients, and so on. Last but not least, derivatizations usually create steric hindrance and weaken guar hydrogen bonds, which translate into faster guar hydration rates. This is the case for derivatizations shown Scheme 24.6.

At the industrial scale, volume-wise, the major derivatization is hydroxypropylation (reaction between propylene oxide and guar). As said earlier, hydroxypropylation increases guar hydration speed (as far as the hydroxypropylation is moderate) [68]. In addition, it makes guar solutions more compatible with salts (brines), and guar chains can still form gels with temporary cross-linkers (zirconates, titanates, etc.), widely used in fracturing fluids in the oil and gas industry. This application is by far the major application of HPGs, ahead of cosmetics (HPGs are often combined in body washes or sanitizers with xanthan to get an optimal viscosity and an optimal yield, while the "jelly" aspect that is typical of xanthan disappears) and ahead of agrochemicals.

Another important derivatization is the carboxymethylation (reaction between sodium monochloroacetate and guar). This derivatization is said to increase the guar thermal stability, and again, it's not detrimental to the formation of temporary gels with well-chosen cross-linkers.

Double derivatizations with propylene oxide and sodium monochloroacetate are also of industrial importance. The resulting products combine the features of both HPGs and CMGs and are widely used in oil and gas.

Cationization or hydroxypropylation combined with cationization has been performed at industrial scale for decades now. The resulting cationic guars are well known as *hair conditioning polymers* in shampoos and skin conditioning polymers in body washes. Regarding hair conditioning, the latest challenge was to provide hair conditioning to a wide variety of hair types, especially thin hair and damaged hair, but without using silicones which are perceived by the public as negatively affecting the environment and without sulfated surfactants which are perceived as irritating to the skin and scalp; this was recently achieved with a new grade of cationic guar [69].

Regarding EOR, microscopic insoluble species present in guar solutions have long been an obstacle to the valorization of guars in this field: these insoluble species significantly decrease the fluid injectivity in the formation. Recent studies showed that there is no insoluble when sulfonated guars are used, resulting in an optimal injectivity [70].

24.4.4
Challenges and Opportunities

Regarding the guar industry, as the industry demand for guar derivatives keeps increasing, the very first challenge for the years to come relates to current derivatization processes: the reactions are not quantitative; thus large quantities of solvents (either water or water/alcohol depending on processes) are needed to wash by-products and residuals away, and the washing steps are followed by a drying step that is very energy consuming.

The challenge will consist in increasing reaction yields and optimizing the washing processes. First progresses have been made at the lab scale and still need to be extrapolated at the industrial scale [71].

Regarding the uses of guar derivatives, considering that guar is a biodegradable, nonecotoxic, and renewable raw material, guar derivatives are investigated in a wide variety of applications. For example, guar derivatives appear as an alternative to nonionic surfactants in automatic dish washing, the rationale being to replace nonbiodegradable surfactants by nonecotoxic and readily biodegradable guar derivatives. In addition, guar derivatives are described as increasing plastic surface hydrophilicity, speeding up their drying, and facilitating their cleaning [72].

24.5
Vanillin

Vanillin (4-hydroxy-3-methoxybenzaldehyde) is the molecule present in natural vanilla bean extracts with the main organoleptic effect for the vanilla taste. However, a cured vanilla bean from, for example, the *Vanilla planifolia* orchids, contains no more than 2 wt% of vanillin, based on the dry matter. Only about 50 metric tons of pure vanillin can thus be extracted from the annual production of vanilla beans, representing about 0.3 wt% of worldwide vanillin demand. Besides the volume issue, the price of vanilla is high and volatile because the quality and the productivity of vanilla pods, exclusively hand cultivated and grown in tropical areas, are dependent of climatic events.

To fulfill consumer demand for natural flavors, alternative sources for natural vanillin are key to ensure availability, constant quality, and prices. For this purpose, Solvay introduced into the market more than 10 years ago Rhovanil® Natural, a vanillin with natural label. The European regulation EC 1334/2008, the reference for the European food industry, states that a flavoring substance can be labeled as natural in the final product only if the following three conditions are strictly met:

- The flavoring substance is naturally present in nature.
- The raw material is from a natural source.
- The process is traditional.

While vanillin is obviously present in nature, Rhovanil® Natural fulfills the last two conditions by being produced with the fermentation, a process known since the ancient age, of ferulic acid, an abundant phenolic phytochemical found in plant cell wall such as rice bran or maize.

Vanillin obtained from ferulic acid is currently the only vanillin recognized as a natural flavor by organizations such as the Directorate General for Competition, Consumer Affairs and Repression of Fraud (DGCCRF) in France and the Alcohol and Tobacco Tax and Trade Bureau (TTB) in the United States.

The ex-ferulic acid Rhovanil® Natural from Solvay displays a strong natural character, with a soft, sweet, and powdery note, and a touch of caramel. It is widely used in natural flavors, especially natural fruity flavors, with sweet, creamy, and dairy notes. These flavors can be used in all segments of the food and fragrance industry such as dairy, ice cream, beverages, chocolate, confectionery, bakery, perfumes, fragrance oils, and essential oils.

24.6
Summary and Conclusions

We have demonstrated by these examples the wide range of benefits that using renewable raw materials can bring to the chemical industry and finally the consumer. This is tightly linked to the very nature of renewable resources and their diversity and their unique properties. Being able to leverage the unique structure these raw materials exhibit is the basis for the design of highly specialized products. It could also offer a very competitive way to produce some of the molecules previously obtained through a petrochemical route and be a way to mitigate the risk linked to fossil oil volatility. The third facet of the benefits is the "bio" attribute that some products can bring, which itself could cover different characteristics the consumer would value (biocompatibility, biodegradability, biosourced content, etc.).

Solvay's strategy for a sustainable growth implies nevertheless to have a balanced use of renewable raw materials. Final application properties for the products are the main driver, and chemical processes should also take benefit. In that sense, using renewable resources cannot be an objective per se. LCA is thus a mandatory tool to be sure that our products are the results of a systemic analysis including of course the performance of our processes but also the benefits brought by our products to our customers.

Solvay as a leader for sustainable chemistry will continue and reinforce toward that direction.

References

1. G. Sienel, R. Rieth, K.T. Rowbottom, in Epoxides, *Ullmann's Encyclopedia of Industrial Chemistry*, Wiley-VCH Verlag GmbH: Weinheim, 2005, doi:10.1002/14356007.a09_531
2. L. Krähling, J. Kre, G. Jakobson, J. Grolig, L. Miksche, in Allyl compounds, *Ullmann's Encyclopedia of Industrial Chemistry*, Wiley-VCH Verlag GmbH: Weinheim, 2005, doi:10.1002/14356007.a01_425
3. McKetta, J.J. (1993) in *Chemical Processing Handbook* (ed J.J. McKetta), Marcel Dekker, New York, pp. 401–406.
4. (a) Carrà, S., Santacesaria, E., and Morbidelli, M. (1979) *Ind. Eng. Chem. Process Des. Dev.*, **18**, 424–427; (b) Carrà, S., Santacesaria, E., and Morbidelli, M. (1979) *Ind. Eng. Chem. Process Des. Dev.*, **18**, 428–433.
5. (a) Wenzl, H. (1970) *The Chemical Technology of Wood*, Academic Press, London, pp. 216–218; (b) Smith, P.B. (2012) *ACS Symp. Ser.*, **1105**, 183–196.
6. Ciriminna, R., Della Pina, C., Rossi, M., and Pagliaro, M. (2014) *Eur. J. Lipid Sci. Technol.*, Special Issue: Euro Fed Lipid Highlights, **116**, 1432–1439.
7. Gibson, G.P. (1931) *Chem. Ind.*, **20**, 949–975.
8. (a) Krafft, P., Gilbeau, P., Balthasart, D., and Daene, A. (2008) Aqueous composition containing a salt, manufacturing process and use (Solvay). WO Patent 2008152043 (Solvay); (b) Gilbeau, P. and Ward, B. (2009) WO Patent 2009095429 (Solvay).
9. Krafft, P., Gilbeau, P., Gosselin, B., and Claessens, S. (2005) Process for producing dicholoropropanol from glycerol, the glycerol coming eventually from the conversion of animal fats in the manufacture of biodiesel (Solvay). WO Patent 2005054167 (Solvay).
10. Carothers, W.H. (1938) Linear Polyamides and their production (Du Pont de Nemours). US Patent 2,130,523.
11. Kohan, M. (1995) *Nylon Plastics Handbook*, Hanser/Gardner Publications.
12. Baumann, F.E., Moser, K., Rohde-Liebenau, U., and Schmidt, F.G. (1998) *Polyamide Kunststoff Handbuch*, Hanser Verlag, pp. 769–802.
13. Naughton, F.C. (1974) *J. Am. Oil Chem. Soc.*, **51** (3), 65–71.
14. Mutlu, H. and Meier, M.A.R. (2010) *Eur. J. Lipid Sci. Technol.*, **112**, 10–30.
15. Häger, H. and Limper, J. (2009) Elements 28, Evonik Science Newsletter – Technical Leaflet.

16. Viot, J.F. (2011) Solvay Internal Communications and Technyl eXten© Technical Leaflet.

17. Magill, J.H. (1966) *J. Polym. Sci., Part A-2*, **4**, 243–265.

18. Wang, G. and Yan, D. (1998) *Chin. J. Polym. Sci.*, **16** (3), 241–252.

19. Xenopoulos, A. and Wunderlich, B. (1990) *J. Polym. Sci., Part B: Polym. Phys.*, **28**, 2271–2290.

20. Ramesh, C. (1999) Annual Technical conference of Society of Plastics Engineers. *Macromolecules*, **32**, 3721–3726.

21. Wunderlich, B. (1976) *Macromolecular Physics*, vol. 2, *5.1*, Academic Press Inc., p. 44.

22. Ruehle, D.A., Perbix, C., Castañeda, M., Dorgan, J.R., Mittal, V., Halley, P., and Martin, D. (2013) *Polymer*, **54**, 6961–6970.

23. Puglisi, C., Samperi, F., Di Giorgi, S., and Montaudo, G. (2003) *Macromolecules*, **36**, 1098–1107.

24. Samperi, F., Montaudo, M.S., Puglisi, C., Di Giorgi, S., and Montaudo, G. (2004) *Macromolecules*, **37**, 6449–6459.

25. Houska, C. Deicing salts – Recognizing the corrosion threat. http://www.imoa .info/download_files/stainless-steel/ DeicingSalt.pdf (accessed 29 September 2015).

26. Robeson, L.M. (2013) *Polym. Eng. Sci.*, **53** (3), 453–467.

27. Dunn, P. and Sansom, F. (1969) *J. Appl. Polym. Sci.*, **13**, 1641–1655.

28. Dunn, P. and Sansom, F. (1969) *J. Appl. Polym. Sci.*, **13**, 1657–1672.

29. Reimschuessel, A.C. and Kim, Y.J. (1978) *J. Mater. Sci.*, **13**, 243–252.

30. Burford, R.P. and Williams, D.R.G. (1979) *J. Mater. Sci.*, **14**, 2872–2880.

31. Burford, R.P. and Williams, D.R.G. (1988) *J. Mater. Sci. Lett.*, **7** (1), 59–62.

32. Mathew, B.A. and Normandin, M.G. (1993) SPE ANTEC 1993 Conference Proceedings, New Orleans, LA, pp. 196-202.

33. Nair, S.V. and Donovan, J.A. (2001) *J. Appl. Polym. Sci.*, **81**, 494–497.

34. Dhevi, D.M., Choi, C.W., Prabu, A.A., and Kim, K.J. (2009) *Polym. Compos.*, **30** (4), 481–489.

35. Isreaelachvili, J.N., Michell, D.J., and Ninham, W. (1976) *J. Chem. Soc., Faraday Trans. 1*, **72**, 1525; *Biochem. Biophys. Acta* (1977) **470**, 185.

36. Rosen, M.J. and Dahanayake, M. (2000) *Industrial Utilization of Surfactants, Principals and Practice*, 1st edn, AOCS Press, pp. 62–65.

37. Balzer, D. (1994) Aqueous viscoelastic surfactant solutions for hair and skin cleaning. US Patent 5965502 (Huels).

38. Klein, K. and Bator, P. (1981) *Drug Cosmet. Ind.*, **12**, 38.

39. Clapperton, R.M. and Fazia, D. (1993) Aqueous based liquid compositions. US Patent 6177396 B1 (Albright).

40. Dahanayake, M., Yang, J., Niu, J., Li, R.X., and Derian, P.J. (1997) Viscoelastic surfactant fluids. US Patents 6258859; 6831108; 6482866 (Rhodia).

41. Farmer, R.F., Doyle, A.K., Vale, G.D.C., and Gadberry, J.F. (1997) Method for controlling the rheology of a cationic aqueous fluid and gelling agent therefore. US Patent 6239183 B1 (Akzo Nobel).

42. Dahanayake, M. and Yang, J. (1997) Viscoelastic surfactants fluids and related methods of use. US Patents 6703352; 7238648 (Schlumberger).

43. Chen, Y., Lee, J.C., Dahanayake, M., Tillotson, R., and Colaco, A. (2002) Rheology enhancers. US Patent 7378378 (Schlumberger).

44. Rosen, M.J. and Dahanayake, M. (2000) *Industrial Utilization of Surfactants, Principals and Practice*, 2nd edn, AOCS Press, pp. 138–140.

45. Harswigsson, I. and Hellestan, M. (1996) *J. Am. Oil Chem. Soc.*, **73**, 921.

46. Savins, J.G. (1969) in *Viscous Drag Reduction* (ed C.S. Wells), Plenum Press, New York, pp. 183–212.

47. Dahanayake, M. and Yang, J. (1997) For use in fracturing a subterranean formation; as a drift control agent for agricultural formulations; and detergents. US Patent 6703352 (Schlumberger).

48. Deschaseaux, F., Prat, E., and Touzet, S. (2008) Use of betaines as foaming agents and foam drainage reducing agents. US Patent 2010140531 (Rhodia).

49. Dahanayake, M., Chabert, M., Morvan, M., and Sorin, D. (2011) Method for

mobility control in oil-bearing carbonate formations. US Patent 8,851,187 B2 (Rhodia).

50. Dahanayake, M. and Marchand, J.P. (2005) Enhanced oil recovery. US Patent 7789143 B2 (Rhodia).

51. Fu, D. (2009) Self-diverting preflush acid for sandstone. US Patent 20040009880 A1.

52. Dahanayake, M. and Douglas, A. (2008) New and novel highly viscoelastic surfactants. 6th CESIO, International World Congress and Exhibition.

53. Dahanayake, M. and Douglas, A. (2009) *Interaction and Synergism in Surfactants/Polymers in Boosting Performance in Personal Care Applications*, AOCS.

54. Puls, J., Wilson, S.A., and Hölter, D. (2011) *J. Polym. Environ.*, **19**, 152–165.

55. Rustemeyer, P. (ed) (2004) Cellulose acetate: properties and applications, in *Macromolecular Symposia*, vol. 208, Wiley-VCH Verlag GmbH, Weinheim.

56. Schützenberger, P. (1865) *C.R. Acad. Sci.*, **61**, 484.

57. A.P.N. Franchimont, Die Einwirkung von Essigsäureanhydrid mit Zusatz von Schwefelsäure auf die Cellulose *Berichte*,1899, **18**, 472.

58. Miles, G.W. (1904) Cellulose derivative and process for making the same. US Patent 838350.

59. Eichengrün, E.A., Becker, T., and H. Guntrum (1905) DRP Patent 252706.

60. British Celanese Manufacturing Co (1920) EP Patent 165 519.

61. Crawford, T.R. and Stevens, B.J. (1952) Tow for use in the production of tobacco smoke filters (Eastman Kodak Company). US Patent 2794239 (Eastman Kodak Company).

62. Yokose, K., Jebens, A., Lochner, U., and Yang, W. (2013) *Water-Soluble Polymers*, Specialty Chemicals Update Program, HIS Chemical.

63. Pollard, M.A. and Fischer, P. (2006) *Curr. Opin. Colloid Interface Sci.*, **11**, 184.

64. Dolnik, V., Gurske, W.A., and Padua, A. (2001) *Electrophoresis*, **22**, 707–719.

65. Mudgil, D., Barak, S., and Khatkar, B.S. (2014) *J. Food Sci. Technol.*, **51**, 409–418.

66. Labeau, M.P. (2012) in *Polymer Science: A Comprehensive Reference*, vol. 10 (eds M. Moeller and M. Matyjaszewski), Elsevier, Dusseldorf, pp. 195–204.

67. Eronen, P., Junka, K., Laine, J., and Osterberg, M. (2011) *Bioresources*, **6**, 4200–4217.

68. Cheng, Y., Prud'homme, R.K., Chik, J., and Rau, D.R. (2002) *Macromolecules*, **35**, 10155.

69. (a) Lizarraga, G., Mechineau, D., Lemos, D., and Mabille, C. (2013) Guar hydroxypropyltrimethylammonium chloride and uses thereof in hair treatment compositions (Solvay). WO Patent 2013/011122 (Solvay); (b) Mabille, C., Cazette, C., Lemos, D., and Vigneron, C.S. (2014) *SOFW*, **140** (9), 16–29.

70. Labeau, M.P., Degre, G., Ranjan, R., Henaut, I., Tabary, R., and Argillier, J.F. (2011) Flooding fluid and enhancing oil recovery method (Solvay). WO Patent 2011/055038 (Solvay).

71. Le-Thiesse, J.C., Lomel, S., and Gisbert, T. (2013) Method for preparing cationic galactomannans (Solvay). WO Patent 2013/050427 (Solvay).

72. (a)Labarre, D., Labeau, M.P., Lambert, F., Orizet, C., Lizarraga, G., and Geoffroy, V. (2012) Detergent composition with anti-spotting and/or anti-filming effects (Solvay). WO Patent 2012/042001 (Solvay); (b)Lambert, F., Orizet, C., and Serrurier, S. (2012) WO Patent 2012042000 (Solvay).

25
Biomass Transformation by Thermo- and Biochemical Processes to Diesel Fuel Intermediates

Daniele Bianchi, Carlo Perego, and Federico Capuano

25.1
Introduction

The decline of petroleum resources combined to the increasing demand of energy by emerging economies are driving a rapid expansion of biofuel production also boosted by government mandates and incentives for the greenhouse gas reduction [1–4].

Biomass can be directly used for the production of heat, steam, and electricity or can be converted by biological or thermal routes into a variety of liquid and gaseous energy carriers.

The main difference between renewable and petroleum feedstock chemical composition is the oxygen content, which ranges from 10 to 40% in biofeedstocks, while petroleum has essentially none.

Sulfur and nitrogen levels depend on the amino acid content of the starting biomass that is generally low in vegetable oils and lignocellulosic biomass and high in municipal organic waste and algal biomass.

Some of these properties are incompatible with the typical refinery operations, so these feeds require an intermediate transformation stage before entering the refinery processing (Figure 25.1). Only few processes can directly transform biomass into biofuels compatible with the existing transportation and fuel infrastructure, typically by fermentation of sugars, extracted from starch and sugar-based crops to produce ethanol or butanol.

In this chapter, the focus is directed to the technologies to produce advanced biofuels derived from plant- and algae-based material such as agricultural residues, forest resources, perennial grasses, woody energy crops, microalgae, municipal solid waste, urban wood waste, and food waste. First-generation biofuels, obtained from biomass in completion with food and feed application, will not be included.

In particular, processes providing liquid high-oxygenated intermediates (bio-oils), suitable for a further upgrading in a refinery cycle, will be discussed.

Chemicals and Fuels from Bio-Based Building Blocks, First Edition.
Edited by Fabrizio Cavani, Stefania Albonetti, Francesco Basile, and Alessandro Gandini.
© 2016 Wiley-VCH Verlag GmbH & Co. KGaA. Published 2016 by Wiley-VCH Verlag GmbH & Co. KGaA.

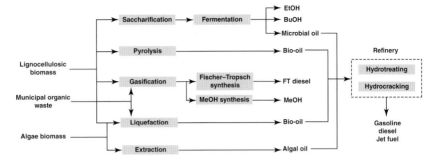

Figure 25.1 Technologies for biomass transformation into advanced biofuels.

25.2
Biological Processes

25.2.1
Microbial Oils

Vegetable oils and fats are at present the main source of renewable diesel. Because of their limited availability, they could only replace a small fraction of transportation fuel demand; however, the potential large supply of lignocellulosic biomass could supply a high percentage of the required biofuels if commercial processes were available to convert these feeds into triglyceride intermediates.

At this purpose the production of microbial oils by oleaginous microorganisms, also known as *single-cell oil*, has been investigated in order to find an alternative source of oils and fats [5].

Microbial oils, with chemical composition and energy value similar to plant and animal oils, could have many advantages, since they are not in competition with food and are independent of season and climate factors.

Among microorganisms, oil accumulation was only found in some yeasts, fungi, and a small number of algae.

A microorganism can be characterized as oleaginous when it accumulates lipids to more than 20% of its total cellular dry weight with a composition similar to vegetable oils. Among the 600 species of yeast, only 30 have been classified as oleaginous (OY). The typical OY genera so far identified include *Yarrowia, Candida, Rhodotorula, Rhodosporidium, Cryptococcus, Trichosporon, and Lipomyces* [6].

Lipids are accumulated in yeast as storage metabolites, usually in the form of triacylglycerols (TAGs). It is known that in most of the oleaginous yeasts, lipid production occurs when the organism is under a nutrient limitation, usually nitrogen, but with a large amount of carbon source still present in growing the medium [7, 8].

Under these conditions the supplied nitrogen is rapidly exhausted by the growing biomass, but the carbon source continues to be assimilated and selectively channeled into lipid synthesis. The produced oils are accumulated up to 80% of

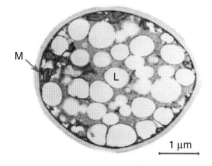

Figure 25.2 Oleaginous yeast cell – *Candida curvata* cell grown with limited nitrogen source. Total lipid content approximately 40%. M: mitochondrion and L: lipid bodies. Adapted from Holdsworth *et al.* [9].

Table 25.1 Lipid accumulation and fatty acid profiles of selected oleaginous yeasts.

Species[a]	Lipid accumulation (% dry weight)	Major fatty acid residues (% w/w)					
		C16:0	C16:1	C18:0	C18:1	C18:2	C18:3
Cryptococcus curvatus	58	25	Trace	10	57	7	0
Cryptococcus albidus	65	12	1	3	73	12	0
Candida sp. 107	42	44	5	8	31	9	1
Lipomyces starkeyi	63	34	6	5	51	3	0
Rhodotorula glutinis	72	37	1	3	47	8	0
Rhodotorula graminis	36	30	2	12	36	15	4
Rhizopus arrhizus	57	18	0	6	22	10	12
Trichosporon pullulans	65	15	0	2	57	24	1
Yarrowia lipolytica	36	11	6	1	28	51	1

a) Yeasts were grown on glucose as carbon source [10].

the cell dry weight, in discrete intracellular fat globular deposits, called the *lipid bodies* (Figure 25.2) [9]. These neutral lipids serve as energy source for the cell when required.

Table 25.1 illustrates the fatty acid profiles found in different yeast species together with the extent of lipid accumulation using a glucose-containing medium. The lipid composition remains similar between different species, being palmitic (C16:0), stearic (C18:0), oleic (C18:1), and linoleic (C18:2) as the main produced fatty acids. It has been suggested that such yeast oils could be suitable feedstocks for biodiesel production that mainly requires C16–C18 chains.

The fatty acid biosynthesis in yeasts begins with the enzyme-catalyzed formation of a three-carbon intermediate, malonyl-CoA, by carboxylation of acetyl-CoA. The elongation of fatty acid molecules progresses through the iterative addition of two-carbon units derived from malonyl-CoA with the elimination of CO_2. The process is generally complete when the chain length reaches 16 carbons (i.e., palmitate). However, other longer or shorter fatty acids may also be formed with different degrees of saturation, by effect of desaturase and elongase enzymes [7], whose combination determines the range of fatty acid structures [11].

The activity of these enzymes can be tuned by metabolic engineering strategies in order to modify the lipid profile and enhance the biomass and oil yields [12, 13].

The yeast *Yarrowia lipolytica* proved to be a versatile strain to produce non-specific lipids for biofuel applications or functional lipids for food and chemical applications [14].

The lipid synthesis can be directed toward the production of saturated fatty acids, mainly C16 [15], monounsaturated fatty acids (MUFAs), that are the best components for biodiesel because of the low-temperature fluidity and high oxidative stability [16] or toward the production of polyunsaturated fatty acids (PUFAs) [17] for nutraceutical applications. An engineered strain of *Y. lipolytica* has been engineered to produce up to 43% of its total lipids as ricinoleic acid [18].

A further development of lipid metabolic engineering could be a microorganism able to transform fatty acid into hydrocarbon fuel directly inside the cell, as pioneered by Lee and Choi [19]. The engineered strain is able to produce fatty acids with controlled chain length avoiding double bond formation and transforming them into hydrocarbons by reduction to fatty aldehyde followed by decarbonylation (Figure 25.3). The job done inside the cell includes the following macroscopic industrial stages: cultivation of oleaginous plant for oil production, oil recovery, hydrogen production, and its use for oil hydrodeoxygenation (HDO).

The theoretical metabolic mass yield (weight of product relative to sugar) for the biological production of lipids is lower than that calculated for ethanol (Table 25.2); however, this is partially balanced by the higher heating value of the oil that makes the overall energy output comparable for the two processes [20, 21].

Different cultivation methods, including fed-batch and continuous fermentation, have been used to increase the cell density of oleaginous microbes, greatly improving the performances of the overall process. OY fermentation can afford very high biomass productivity: a cell density of $185\,g\,l^{-1}$ in 84 h was described for fed-batch culture of *Rhodotorula glutinis* aerated with oxygen-enriched air [22].

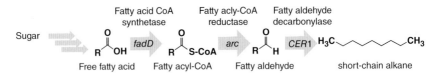

Figure 25.3 Metabolic pathway for microbial production of hydrocarbons. Adapted from Lee and Choi [19].

Table 25.2 Theoretical metabolic yields for lipids compared to ethanol.

	Metabolic mass[a] yield (%)	Carbon yield (%)	Oxygen content (%)	HHV[b] (MJ kg^{-1})	Energy yield[c] (%)
Ethanol	51	67	34.8	29.7	98
Fatty acid (palmitic acid)	35	67	12.5	43.3	89

a) Maximum stoichiometric yield in grams of product per gram of glucose.
b) HHV: high heating value.
c) Calculated on HHV basis relative to glucose [20].

A cell density of $153\,\mathrm{g\,l}^{-1}$ and a lipid content of 54% (w/w) were achieved using a fed-batch culture of *Lipomyces starkeyi* for 140 h [23]. The pilot-scale fed-batch cultures *Rhodosporidium toruloides* in a stirred-tank fermenter for 134 h resulted in dry biomass, lipid content, and lipid productivity of $106.5\,\mathrm{g\,l}^{-1}$, 67.5% (w/w), and $0.54\,\mathrm{g\,l}^{-1}\,\mathrm{h}$, respectively [24].

Oleaginous microorganisms, mostly yeast or fungi, have the capabilities of growing on a broad range of carbon sources, including the sugars obtained from the saccharification of lignocellulosic biomass, such as C5 sugars (xylose and arabinose) and C6 sugars (galactose, glucose, and mannose), thus allowing the lipid production using substrates which do not directly compete with food uses [25, 26].

In order to reduce the cost of microbial oil production, low-cost raw materials, such as wheat straw [27], sugarcane bagasse [28], rice straw [29], corn stover [30], and corncobs [31], have been used as sugar source. The most commonly used nitrogen sources are yeast extract, peptone, nitrate, and ammonium sulfate, but the combination of nitrogen sources is also commonly used.

Figure 25.4 shows the complete process flow for a microbial oil process starting from sugar. The main operations are fermentation, cell harvesting and concentration, cell disruption, solvent extraction, and solvent evaporation.

The yeast cell harvesting stage is simpler that the complex procedure required for the oleaginous microalgae biomass, because of the high cell concentration directly obtained from the fermentation stage ($>100\,\mathrm{g\,l}^{-1}$ cell dry weight) with respect to extremely low algal density typical of the raceway pond cultivation

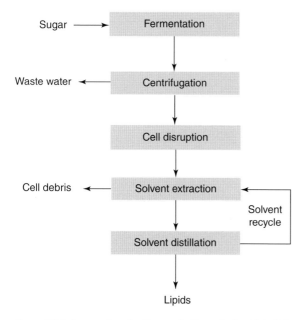

Figure 25.4 Process flow for the production of microbial oil from sugar.

($<1\,g\,l^{-1}$ cell dry weight). High-density yeast suspension (up to $400\,g\,l^{-1}$ cell dry weight) can be obtained by a standard centrifugation stage.

Since the oil is an intracellular product, the extraction is a key step of the process. Like any other biological material, yeast lipids are bound to biopolymers and membranes by several mechanical and chemical forces such as van der Waals or hydrophobic interactions, hydrogen, and ionic bonds of amino acids. These strong interactions must be weakened to make the oils accessible and achieve a quantitative product recovery [32].

Lipids bound to phospholipids membrane, are difficult to release and need a preliminary cell disruption treatment to break the rigid cell wall and thus favor more accessibility to extraction solvents without affecting the native state of the lipids.

Several oil extraction technologies are currently available to process the microbial biomass, in order to meet the requirement of being low cost, easy and safe to operate, and environmentally friendly, such as high-pressure homogenizer, bead mill, X-press, ultrasounds, and thermolysis [33–35].

Solvent extraction is a commonly used method for vegetable oil processing, and it is also used to extract lipids from microbial cells. Organic solvents should be insoluble in water, have a low boiling point, and be reusable. Current industrial solvents for microbial lipid extraction include propane, hexane, cyclohexane, gasoline distillation fractions containing heptane and/or octane, refinery low aromatic petroleum distillation fractions, methyl *tert*-butyl ether (MTBE), dimethylether (DME), and supercritical CO_2 [36–40].

Simultaneous extraction and acid-catalyzed transesterification with methanol allow to obtain lipids directly as fatty acid methyl ester (FAME) biodiesel but requires an additional and energy-intensive cell drying stage [41].

A number of major oil companies are currently running research projects in order to scale up the microbial oil production.

In October 2012, BP and DSM announced a joint agreement to advance the development of a technology for conversion of sugars into renewable diesel. The concept of this technology has been demonstrated at pilot scale [42, 43].

Neste Oil has commissioned Europe's first pilot plant dedicated to producing microbial oil from agricultural and forestry residues in 2012 at Porvoo. The goal is to develop technology capable of yielding commercial volumes of microbial oil for use as an alternative feedstock for NExBTL renewable diesel production [44, 45].

In 2014, eni has demonstrated up to pilot scale a proprietary technology for the production of yeast oils from lignocellulosic biomass, suitable as a cofeeding feedstock for the Ecofining green diesel process [34].

25.2.2
Algal Oils

Microalgae are unicellular microorganism species, typically found in aquatic systems, that can be classified as oleaginous due to their capability to accumulate intracellular lipids. Microalgae are photosynthetic organisms, autotrophically

Table 25.3 Microalgae photosynthetic efficiency in outdoor cultivation.

	Solar efficiency (%)
Total solar radiation (TSR) at sea level	100
PAR (photosynthetic active radiation, suitable radiation)	45
Maximum (theoretical) photosynthetic efficiency (PE) on PAR	27
Maximum (theoretical) PE on TSR	12
Maximum PE under optimal conditions outdoors	5
Maximum PE under real conditions	2.5
Actual best average in industrial plants	1.5

Adapted from Tredici [47].

growing using light as the energy source and assimilating CO_2 as the carbon source, according to the photosynthesis pathway.

In addition, many species of microalgae can grow in dark systems; in this case, the cells lose chlorophyll and shift their metabolism from photoautotrophic to heterotrophic pathway, using organic substrates as carbon and energy sources [46].

25.2.2.1 Photoautotrophic Cultivation Technology

In almost all microalgae photoautotrophic production systems, the energy source is the solar radiation. Table 25.3 reports the steps and the thermodynamic limits for the photosynthetic efficiency, corresponding to the effective fraction of solar energy that can be exploited from microalgae [47].

The main advantages of the photoautotrophical growth are the relatively simple reactor layout and management and the capability to use potentially cheap substrate for growing, such as the free energy coming from sun and the CO_2 supplied from industry flue gas. The main drawbacks are the low concentration of cell in cultures and the relatively low areal productivity that rise production costs [48]. The photoautotrophic reactors for outdoor microalgae production can be classified as open and closed systems.

Among the open systems, the raceway paddle wheel ponds are the most common reactors for algae cultivation. A raceway pond is a shallow artificial pond where algae, CO_2-rich water, and nutrients are pumped around the racetrack by a motorized paddle wheel. The ponds are usually kept shallow (20–30 cm depth) because the algae need to be exposed to sunlight, and sunlight can only penetrate the pond water to a limited depth. The algae culture will grow continuously, and part of the algae will be removed during the growing process.

Open culture systems allow relatively low-cost production but are subject to contamination. Algae produced in open systems include *Spirulina*, *Chlorella*, *Dunaliella*, *Haematococcus*, *Anabaena*, and *Nostoc* [49].

The maximum values of productivity find literature range from 80 to 100 tons per hectare per year of algal dry biomass [50]. R&D efforts are focused on increasing the productivity by improving fluid dynamics and pond layout in order to maximize the solar penetration on culture [51].

Contamination-free monoseptic culture of most microalgae requires a fully closed culture system or a photobioreactor. These devices provide a protected environment safe from contamination by other microorganisms and with culture parameters easy to control. They also prevent evaporation, reduce water use, lower CO_2 losses due to outgassing, and permit higher cell concentration, up to $1-2$ g of biomass per liter with respect to $0.3-0.5$ g of biomass per liter typical of open ponds. Considering the areal productivity (produced biomass per effective occupied land), the photobioreactor productivity is around two times higher than the open ponds, but the capital cost are four times higher than the latter [48].

Although photobioreactor culture is expensive relative to open culture, it is useful for culturing high-value algae products.

Various kinds of photobioreactors have been developed such as vertical tubular, horizontal tubular, and plate and plastic bag systems [49, 52], mainly used for aquaculture feed, human food, nutraceuticals, and cosmetic applications.

Several microalgae strains are able to accumulate a large quantity of intracellular lipids and can be used for biofuel applications. According to the literature, the maximum value of microalgae biomass production ranges between 80 and 100 tons per hectare per year [50] with a lipid concentration ranging between 30 and 60% of dry microalgae biomass [53–55], corresponding to a lipid production of 30–50 tons per hectare per year. This amount is much higher than that achieved with the most utilized oleaginous energy crops, as shown in Table 25.4 [56].

The main sections of a typical process for oil production from microalgae are described in Figure 25.5. After cultivation in open raceway ponds, microalgae are harvested, concentrated, and then processed with different technologies to produce bio-oil by wet solvent extraction process, dry extraction process [57], or thermal treatments [58].

There are several techniques used for algae biomass harvesting, depending on cell dimensions and culture water characteristics (e.g., the high ionic salt concentration in seawater strongly inhibits the flock formation in flocculation processes) [59]. The most common methods are filtration, sedimentation, flotation,

Table 25.4 Lipid production of microalgae compared with most common oleaginous crops.

Energy crop	tons$_{oil}$ per hectare per year
Corn	0.15
Soybean	0.41
Sunflower	0.87
Rapeseed	1.08
Palm	3.62
Microalgae	30–50

Adapted from Molina *et al.* [56].

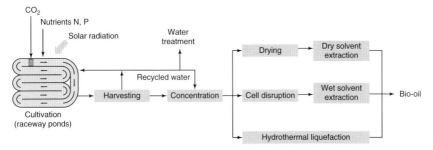

Figure 25.5 Microalgae cultivation and algal oil downstream processes.

and centrifugation. Typically the algal concentration after harvesting rises up to $40-60\,\mathrm{g\,l^{-1}}$ [56].

In order to meet the high biomass concentration required by oil extraction processes, preferably ranging from 100 to $250\,\mathrm{g\,l^{-1}}$, the harvested algal slurry needs a further concentration step the can be achieved by filtration, flocculation, and centrifugal recovery [56].

Figure 25.5 shows three different options for oil production: dry solvent extraction, wet solvent extraction, and thermal treatment. The dry extraction process requires a previous dewatering step to lower the water content of the biomass below 10%. High extraction yields (near 95% of total lipid content) are obtained using nonpolar solvent-like hexane or octane [60]. Suitable drying methods include spray drying, drum drying, freeze-drying, and sun drying [61]. In general, these techniques are too energy-intensive for fuel application and are commonly used for high-value products (>$1000 per ton).

In the case of wet solvent extraction methods, a water miscible polar solvent (e.g., acetone) is used. A preliminary cell disruption step is needed to reach high extraction yields up to 90% [62, 63].

The algal oil recovered by solvent extraction is nonequivalent to vegetable oil, since it contains approximately $40-70\%$ of mono-, di-, and triglyceride and free fatty acid (FFA) components, suitable for biofuel production, with the rest of lipids coming from cell membranes, mainly consisting in phospholipids and glycolipids [64].

In thermal downstream processes, the concentrated algal biomass (typically $10-20\%$ dry weight) is subjected to high-temperature $(250-300\,^{\circ}\mathrm{C})$ and pressure $(50-100\,\mathrm{bar})$ treatment in the absence of oxygen, as described in the following hydrothermal liquefaction section of the chapter. Around 40% of the organic components of the cells are converted in crude bio-oil. The product has high concentration of metals, heteroatoms, and molecules not suitable for biofuel production and needs a further upgrading treatment.

The critical issues for the industrial development of a biofuel process from microalgae are mainly the low culture concentration, which means high downstream energy costs, and the still low areal productivity, which pushes the production cost out of the market level, around 13 US$ per gallon of algal oil [65].

Another critical point is the crude oil composition: as previously discussed, the algal oils contain a significant fraction of products not suitable for biofuel production and high concentration of metals and heteroatoms that can rapidly deactivate the catalysts used in the upgrading processes.

In order to overcome this critical drawback, many companies are involved in microalgae process development for biofuel production, as listed in the following [66].

Cellana (United States), Aurora Biofuels (United States), and Muradel (Australia) are currently operating microalgae cultivation pilot plants.

Sapphire Energy has built a 120 ha demonstration algae cultivation plant in Columbus, NM, for the production of a so-called green crude, derived from a thermal treatment of algae biomass and compatible with the current refinery infrastructures.

Neste Oil (Finland) has signed a commercial agreement with Cellana to partially replace vegetable oils in its proprietary hydrotreating process NExBTL

Eni (Italy) has built in 2010 a pilot plant including a series of 50, 200, 1000, and 2000 m^2 open ponds with the integrated downstream facilities including filter press, centrifuge, crossflow membrane filtration, and thermal treatment sections.

25.2.2.2 Heterotrophic Cultivation Technology

A feasible alternative for phototrophic cultivation of microalgae is the exploitation of their heterotrophic growth capacity in the absence of light, replacing the fixation of atmospheric CO_2 of autotrophic cultures with organic carbon sources dissolved in the culture media [67].

In general, heterotrophic cultivation is cheaper and simpler to construct in facilities than autotrophic cultivation systems, since it can be carried out in traditional stirred tank bioreactors or fermenters. Oil accumulation occurs under growing conditions similar to these previously described for oleaginous yeasts, typically under nitrogen starvation. Suitable carbon sources are glucose, glycerol, and acetate. These systems provide a high degree of growth control and also lower harvesting costs due to the high cell densities achieved, typical of fermentation processes.

Solazyme, San Diego, CA, has developed a biotechnology platform to transform a wide variety of sugars into renewable oils for fuel and chemical applications, using heterotrophic microalgae, engineered to produce tailored lipids, including short-chain-length fatty acids (C10–C14), high saturated chains, and high oleic (C18:1) products. Productivity of over 80% oil within each individual cell has been achieved at commercial scale.

Solazyme is manufacturing products at three large-scale facilities, including a 2000 tons per year integrated plant in Illinois, a 20 000 tons per year plant in Iowa, and a100 000 tons per year facility in Brazil, in partnership with Bunge Global Innovation.

25.2.3
Microbial and Algal Oils Upgrading to Fuels

The most common way of upgrading vegetable oils to fuels is the transesterification of triglycerides with methanol to produce FAMEs, also called *biodiesel*. Both acid and base catalysts could be used in the transesterification process; however, base catalysts are generally preferred for their superior activity (approximately 4000 times higher than that of acid catalysts). Sodium methoxide is the most widely used biodiesel catalyst. Waste cooking oils, grease, and animal fats can also be used as feedstocks because of their availability and low cost, but the process has to deal with FFAs and water impurities [68].

While FAME has many desirable qualities, such as high cetane, there are other issues associated with its use such as low energy content, cold weather freezing, low storage stability, high solvency of seals, gaskets, and other fuel system components, formation of deposits at injector tips, and high NO_x emissions with respect of mineral diesel [69]. For these reasons current European fuel standards allow up to 7% volume FAME (B7) blends.

As an alternative, traditional petrochemical refinery operations such as hydrotreating and catalytic cracking can be applied with modifications to produce hydrocarbon-like drop-in biofuels from triglyceride-based feedstocks, as shown in Figure 25.6 [1].

The hydrotreating of vegetable oils and fats, as well as microbial and algal oils, uses hydrogen to remove the oxygen from the triglyceride molecules producing a paraffinic diesel fuel.

The oxygen is removed via three competing reactions: HDO, decarbonylation, and decarboxylation. In addition, all olefinic bonds are saturated, resulting in a product consisting of only *n*-paraffins. Hydrogen requirements are variable depending on the degrees of unsaturation on the fatty acid chains. Straight-chain paraffins display a high-cetane but poor cold-flow properties that need to be

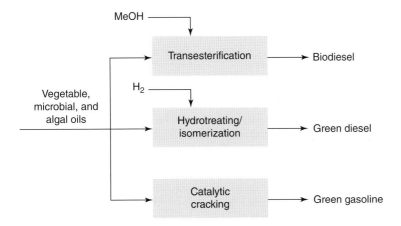

Figure 25.6 Technologies for lipid-based feedstock conversion to biofuels.

adjusted in a further isomerization stage in order to meet ASTM diesel specifications. Hydrogenated vegetable oil (HVO) shows better fuel and blending properties than biodiesel [1].

Several companies (Neste Oil, BP, PetroBras, Syntroleum/Tyson Foods, ConocoPhillips, Haldor Topsoe, Nippon Oil, Axens, and eni/UOP) have developed proprietary hydrotreating technologies to produce HVO, green diesel, and green jet.

Neste Oil has developed a process to produce diesel fuel, marketed as NExBTL fuel and is the world's leading producer of renewable diesel with online capacity of 2 million metric tons per year, distributed between three facilities located in Porvoo (Finland), Rotterdam (Holland), and Singapore [70]. Eni and UOP have jointly developed the integrated deoxygenation/isomerization Ecofining process to produce a high cetane and low sulfur content green diesel product with >98% yield, calculated on volumetric basis [71]. Eni started up an Ecofining commercial plant with 500 000 metric tons per year capacity in Venice refinery in 2014, and a twin facility is under construction at Gela Refinery (Sicily).

Fluid catalytic cracking (FCC) is the most widely used process in petrochemical refineries for the conversion of the heavy fraction of crude oil (vacuum gas oil (VGO)) into gasoline and other hydrocarbons and can be also adapted to upgrade triglyceride-based feedstocks.

A pretreatment unit is usually required to remove catalytic poisons such as alkali metals, water, and solids. The conditioned feed can be coprocessed with crude mineral oil (VGO) to produce gasoline or green olefins, depending on the process conditions and catalysts. Zeolite catalysts have been mainly used for this reaction, including HZSM-5, beta-zeolite, and USY.

Catalytic cracking offers significant advantages over hydrotreating since no hydrogen is required and the process operates under atmospheric pressure. However, poor yields to gasoline fraction can be achieved due to the formation of cracked $C2-C4$ fractions and heavier products, such as light cycle oil (LCO), clarified slurry oil (CSO), and coke. UOP has investigated the vegetable oil processing by FCC, reporting the following product distribution: $C2-C4$ (13.9%), gasoline (45.5%), LCO (11.4%), CSO (13.1%), and coke (4.5%), similar to that obtained from mineral VGO [72].

25.3
Thermal Processes

Thermochemical conversion processes include three main categories: gasification, pyrolysis, and hydrothermal liquefaction, also called *hydropyrolysis*. These processes differ in the operating temperature, reaction time, and the amount of required oxygen.

They can be applied to a wide range of feedstocks, including lignocellulosic and waste biomass, for the production of solid, liquid, and gaseous fuels.

Compared to biological processes, they are relatively simple; can convert the whole biomass, including lignin; and do not need a complex pretreatment and saccharification preliminary stage.

The gaseous and liquid products of these thermochemical processes are mostly comprised of oxygenated species that can be directly used for heat and power generation but need to be further processed to produce fuel suitable for automotive applications [73].

This review focuses on the production of liquid intermediates (bio-oils) by pyrolysis and liquefaction technologies.

25.3.1
Pyrolysis

Pyrolysis is a thermal decomposition of biomass carried out in the absence of oxygen. Gas, liquid, and char are the three major products of pyrolysis, with different relative distributions depending on the reaction temperature and the residence time.

As shown in Table 25.5, low process temperatures and long residence times (torrefaction) favor the production of charcoal. Medium temperature and short residence times (fast pyrolysis) favor the production of liquid oils, while at high temperature gas is the dominant product (gasification) [74].

Fast pyrolysis occurs in the temperature range of 400–700 °C with fast heating rate ($500 \, °C \, s^{-1}$), low pressure (1–5 bar), and short solid residence time (0.5–10 s). In fast pyrolysis process biomass decomposes to generate vapors, aerosol, and some solid residue. After rapid cooling and condensation of vapors and aerosol, a dark brown liquid (bio-oil) is formed [75–78].

The main gas-phase components are H_2, CO, CH_4, and C_2H_5. Under fast pyrolysis conditions, in the absence of oxygen, the carbon fraction in the solid product does not react and leaves the process as a char material.

Biomass must be grinded to less than 2–10 mm particles before being introduced into the pyrolysis reactor in order to ensure high heat transfer rates at the reaction interface.

A preliminary drying pretreatment of the feedstock, to lower the moisture content below 10 wt%, is required in order to minimize the water in the produced bio-oil.

Table 25.5 Product distribution obtained by different pyrolysis modes from wood.[a]

Mode	Temperature (°C)	Residence time	Liquid (%)	Solid (%)	Gas (%)
Torrefaction	~290	~10–60 min	0	80	20
Intermediate	~500	~10–30 s	50	25	25
Fast	~500	~1 s	75	12	13
Gasification	~750–900		5	10	85

a) Weight yields calculated on dry wood basis [74].

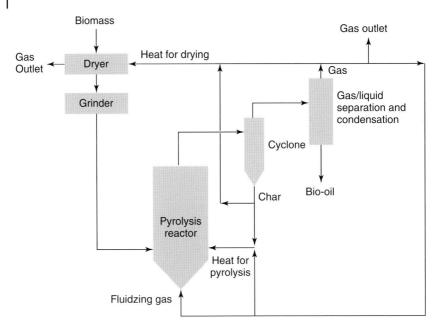

Figure 25.7 Fast pyrolysis reactor system. Adapted from Bridgwater and Peacocke [77].

Different reactor technologies have been developed for fast pyrolysis such as fluidized bed, circulating fluid bed, ablative, and vacuum pyrolyzers [79].

Figure 25.7 shows the scheme of a typical pyrolysis process, including drying and grinding of the biomass, pyrolysis reactor, cyclones for the char separation, vapor condensation, and bio-oil and gas separation sections. Gas and char fractions account to 5% and 25% of the biomass energy content, respectively, and can be combusted to provide heat to the main reaction chamber [77].

A wide variety of different wood and waste biomass feedstocks can be converted by pyrolysis processes, as reported by UOP, with different bio-oil yields: hardwood (70–75%), softwood (70–80%), hardwood bark (60–65%), softwood bark (55–65%), corn fiber (65–75%), bagasse (70–75%), and wastepaper (60–80%) [80].

Pyrolysis–bio-oils have an oxygen content similar to that of the starting biomass (35–40%), a higher moisture content (up to 30%), and a lower heating value ($17\,MJ\,kg^{-1}$) with respect to mineral fuel oil ($43\,MJ\,kg^{-1}$) as shown in Table 25.6. Due to the high organic acid content, bio-oils show an acidic pH (2–3) and can be corrosive for common construction materials, especially at elevated temperature and high-water content [81].

Bio-oils are complex mixtures containing more than 400 individual compounds, including short-chain organic acids (formic, acetic, propionic), alcohols (methanol, ethanol), esters, ketones (acetone), aldehydes (acetaldehyde, formaldehyde), miscellaneous oxygenates (hydroxyacetaldehyde), furans (5-hydroxymethylfurfural, furfural), phenols, guaiacols (isoeugenol, eugenol, 4-methyl guaiacol), and syringols (2,6-dimethoxyphenol) [83–85].

Table 25.6 Properties of bio-oils and upgraded bio-oils from wood.

Property	Fast pyrolysis oil	Liquefaction oil	Heavy fuel oil	Hydrodeoxygenated bio-oil
Moisture content (wt%)	15–30	5.1	0.1	0.001–0.008
pH	2.5	—	—	—
Specific gravity	1.2	1.1	0.94	0.79–0.92
Elemental composition (wt%)				
Carbon	54–58	73	85	85–89
Hydrogen	5.5–7.0	73	11	10.5–14.1
Oxygen	35–40	16	1	0.0–0.7
Nitrogen	0–0.2	0–7.0	0.3	0.0–0.1
Ash	0–0.2	—	0.1	—
Higher heating value (MJ kg^{-1})	16–19	34	40	42–45
Viscosity (50 °C) (cP)	40–100	15 000	180	1.0–4.6

Adapted from Czernik and Bridgwater [82] and Huber *et al.* [2].

Since fast pyrolysis is a nonthermodynamically controlled process, the resulting bio-oils are chemically unstable and undergo progressive physicochemical modifications with time and temperature. During storage, the chemical composition of the bio-oil moves toward the thermodynamic equilibrium, resulting in changes in viscosity, molecular weight, and solubility of its components [86, 87].

The bio-oil stability can be improved by instantly treating the hot pyrolysis vapor with a catalyst, before the condensation stage [88, 89].

There are two approaches to catalytic pyrolysis: *in situ* and *ex situ*. For *in situ* catalytic pyrolysis, the catalyst is introduced in the pyrolysis reactor together with the feedstock. For *ex situ* catalytic pyrolysis, the catalyst is placed in a reactor, separated from the pyrolyzer, and only contacted with pyrolysis vapors after the removal of char and other solid particulates.

Both *in situ* and *ex situ* pyrolysis can convert the oxygenated and heavy compounds into stable products in the range of transportation fuels.

In both cases, coke formation on the catalyst is unavoidably high, and the rapid deactivation requires frequent and continuous regeneration treatments.

Acid zeolites like ZSM-5 have been extensively studied for application in catalytic fast pyrolysis, due to their strong acidity, shape selectivity, and ion exchange capacity.

HZSM-5 can remove oxygenated compounds and increase the aromatic product content in the bio-oil by dehydration, decarboxylation, and oligomerization reactions leading to a mixture of gasoline, diesel fuel, heating oil, and renewable chemicals including benzene, toluene, and xylenes [90, 91]. Basic metal oxides, like CaO, are also effective by reducing the acids and heavy products and increasing the formation of hydrocarbons by ketonization and aldol condensation reactions [92].

ZnO-catalyzed pyrolysis produces bio-oils with greatly improved stability and low viscosity after aging treatment, without any yield reduction with respect to the reference process [93].

Numerous pyrolysis technologies have developed up to commercial scale throughout the world, using a variety of biomass feedstocks and different reactor designs.

Ensyn has developed the Rapid Thermal Processing (RTP) technology based on an FCC reactor design. Ensyn Corp.'s manufacturing facility in Renfrew, Ontario, consumes about 70 tons per day of wood fiber to produce 3 million gallons per year of a liquid fuel known as renewable fuel oil (RFO) [94]. Currently there are six commercial biomass-processing facilities in operation, based on RTP technologies [95].

Ensyn and UOP have established a strategic alliance related to the commercial deployment of Ensyn's RFO advanced cellulosic fuels.

Dynamotive Energy Systems has built a 100 and a 200 tons per day plant in Canada which are currently in operation, based on its proprietary fast pyrolysis process.

Fortum has commissioned in 2013 a plant to produce 50 000 tons per year of bio-oil in Joensuu (Finland).

KIOR has built in 2012 a 45 000 tons per year facility in Columbus, Mississippi, based on a proprietary catalytic pyrolysis technology for the production of renewable crude. The plant is now out of operation due to technical and economical problems.

Empire BV, owned by BTG BioLiquids, has started in 2014 the construction of a pyrolysis facility with a capacity of 20 million liters per year of renewable oil in Hengelo, Netherlands.

25.3.2
Hydrothermal Liquefaction

Hydrothermal liquefaction is a thermal treatment carried out at high pressure (50–200 bar), low temperatures (250–350 °C), and longer reaction time (15–90 min) than that typical of fast pyrolysis [96–98].

The reaction proceeds in the presence of water, consequently drying processes of the feedstock are not required, and high moisture content biomass, such as aquatic biomass, garbage, and organic sludge can be directly fed to the process.

The liquefaction mechanism comprises the following consecutive steps: (i) depolymerization of the biomass to form water-soluble monomers, (ii) degradation of monomer by dehydration, deamination, and decarboxylation reactions, (iii) recombination of the reactive fragments to form a water-insoluble bio-oil, and (iv) further polymerization at prolonged reaction time to form char.

The bio-oil yields are strictly depending on the feedstock chemical composition. Liquefaction of oleaginous wet microalgae biomass affords bio-oil yields ranging from 20 to 64 wt%, calculated on dry feedstock basis [99, 100]. Lower yields are reported for wood biomass, typically producing bio-oil (20–45 wt%), water-soluble products (9–17 wt%), and solid residue (20–30 wt%) [101, 102]. The bio-oil productivity reported for disposed activated sewage sludge and digested sludge is 31.4 and 11.0 wt%, respectively [103].

The bio-oil produced by liquefaction has a lower oxygen content and therefore higher energy content than fast pyrolysis-derived oils, as shown in Table 25.6. Physically, the liquefaction oil is much more viscous but less dense than pyrolysis oil. The chemical composition is also different due to the condensation reactions of the light fragments, resulting in a more hydrophobic product with less dissolved water. Liquefaction of feedstock with high protein content (e.g., organic fraction of domestic wastes, microalgae, sewage sludges) produces bio-oil with nitrogen content ranging from 1 to 9% [104].

The main nitrogen-containing species detected in bio-oils are fatty acid amides and heterocyclic aromatic compounds containing up to three nitrogen atoms [105–107].

A number of catalysts have been used in liquefaction processes with the aim to reduce the heteroatom content, improve the bio-oil yield, and minimize the char formation, including TiO_2, $CaCO_3$, Na_2CO_3, Al_2O_3, Ni, CuO, KNO_3, Na_2S, $ZnCl_2$, and $FeSO_4$ [108, 109].

There are many reports about pilot-scale development of liquefaction processes, but no commercial plants have been yet built. Battelle (United States) has developed the sludge-to-oil reactor system (STORS) process for primary sewage sludge liquefaction. Shell Co., Biofuel Co., BTG, and TNO (Netherlands) have developed the hydrothermal unit, HTU, process for the liquefaction of domestic, agricultural, and industrial residues. Eni (Italy) has developed up to pilot plant a multifeedstock liquefaction technology integrated with the bio-oil upgrading by cofeeding to a heavy oil hydrotreating refinery process [110].

25.3.3
Bio-Oil Upgrading

Bio-oils can be used directly as a fuel oil for combustion in a boiler or a furnace without any upgrading. However, some properties affecting bio-oil fuel quality, such as the high oxygen content, high solid content, high viscosity, and chemical instability, are out of specifications for conventional diesel and gasoline replacement.

As in the case of vegetable and microbial oils, two general routes for bio-oil upgrading have been considered: HDO and zeolite cracking [78, 111–113].

HDO treatment of bio-oils is typically carried out at moderate temperatures (300–600 °C) under high hydrogen pressure, in the presence of heterogeneous catalysts.

The most used catalysts are traditional hydrodesulfurization (HDS) catalysts, based on sulfided CoMo and NiMo, or metal catalysts based on Pd, Pt, and Ru [114].

During HDO processes, the oxygen is removed as H_2O and CO_2 and the C–C and C–O double bonds are saturated. The reaction conditions are usually set in order to minimize the hydrogenation of aromatic rings that would increase H_2 consumption.

As shown in Table 25.6, the hydrotreating of wood-derived pyrolysis oil shows an oxygen content lower than 1% and a heating value increase from 22.6 up to $42.3-45.3\,MJ\,kg^{-1}$. The typical upgraded bio-oil yields are in the $21-65\,wt\%$ range [111].

Bio-oils can also be upgraded using zeolite catalysts (e.g., HZSM-5), typically operating at $350-500\,°C$ under atmospheric pressure. These processes have an economic advantage over hydrotreating since hydrogen is not required; however, higher amounts of coke and char are produced (typically $26-39\,wt\%$); lower yields to liquid products are achieved (typically $14-23\,wt\%$), and extensive carbon deposition results in very short catalyst lifetimes. The oil upgraded by zeolite cracking is also affected by a lower H/C ratio (typically $0.3-1.8$) with respect to hydrotreated bio-oil (typically $1.3-2.0$) and to the reference crude mineral oil (typically $1.3-2.0$), due to the high aromatic content [111].

Pacific Northwest National Laboratory and National Renewable Energy Laboratory have developed an integrated technology for fast pyrolysis oil upgrading, whose flow sheet is reported in Figure 25.8 [115].

The bio-oil stream from the pyrolysis reactor is stabilized and deoxygenated by catalytic hydrotreating in a multistage process with increased severity in each subsequent stage, in order to reduce overall coking. The process includes (i) a stabilization stage, carried out at $140\,°C$, 1200 psia over Ru/C catalysts, to remove reactive carbonyl and olefin functional groups; (ii) a mild HDO stage, carried out at $180\,°C$, 2000 psia over sulfided CoMo catalysts; and (iii) a severe HDO stage,

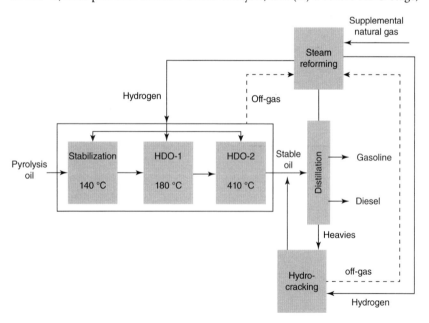

Figure 25.8 Multistage bio-oil hydrotreating process. Adapted from Jones *et al.*, 2013 [115].

carried out at 410 °C, 2000 psia over sulfided CoMo catalysts, to convert oxygenated compounds to hydrocarbons and saturate some of the aromatics.

The overall hydrotreatment yields are stable hydrocarbon oil (44%), gas (13%), and water (48%), based on dry pyrolysis oil weight [116, 117].

The hydrocarbon product is then fractionated into gasoline blendstock, diesel blendstock, and heavies.

The heavy fraction (with boiling point above 350 °C) is sent to the hydrocracker to completely convert the oil to gasoline and diesel components. The final gasoline/diesel ratio depends on the feed and the process conditions [118].

The off-gas streams from the hydrotreating and hydrocracking stages, containing noncondensable hydrocarbons (methane, ethane, propane, butane) and carbon dioxide, are sent to a steam reforming unit together with natural gas to produce the amount of hydrogen required for the process.

UOP has also developed a combined hydrotreating/hydrocracking technology that produces naphtha range (21%) and diesel range (21%) hydrocarbons, based on raw bio-oil. The carbon recovery is 45%, based on the starting wood biomass [119].

25.4
Conclusions

The processing of biomass by biochemical, thermal, and catalytic technologies will play an increasing role for the future development of advanced biofuels. One of the main goals to make the transition from first-generation to second-generation biofuels more economically convenient is the utilization of existing infrastructure in order to lower capital costs.

The final upgrading of oxygenated intermediate products such as microbial and algal oils and bio-oils from thermal treatments requires high amounts of hydrogen with a negative impact on the operation costs and overall process emissions. It is likely that in the future this hydrogen could be produced by using renewable energy sources such as the sun, wind, or biomass.

Future work should focus on improvement of the reaction pathways to integrate process operation, reactor, and catalyst design into an efficient biorefinery system for the production of biofuels and chemicals.

References

1. Huber, G.W. and Corma, A. (2007) *Angew. Chem. Int. Ed.*, **46**, 7184–7201.
2. Huber, G.W., Iborra, S., and Corma, A. (2006) *Chem. Rev.*, **106**, 4044–4098.
3. Melero, J.A., Iglesias, J., and Garcia, A. (2012) *Energy Environ. Sci.*, **5**, 7393.
4. Perego, C. and Ricci, M. (2012) *Catal. Sci. Technol.*, **2**, 1776–1786.
5. Wynn, J.P. and Ratledge, C. (2005) in *Bailey's Industrial Oil and Fat Products*, 6th edn, vol. 6 (ed. F. Shahidi), John Wiley & Sons, Inc., pp. 121–153.
6. Ageitos, J.M., Vallejo, J.A., Veiga-Crespo, P., and Villa, T.G. (2011) *Appl. Microbiol. Biotechnol.*, **90**, 1219–1227.

7. Ratledge, C. (2004) *Biochimie*, **86** (11), 807–815.

8. Ratledge, C. and Wynn, J.P. (2002) *Adv. Appl. Microbiol.*, **51**, 1–51.

9. Holdsworth, J.E., Veenhuis, M., and Ratledge, C. (1988) *J. Gen. Microbiol.*, **134**, 2907–2915.

10. Beopoulos, A., Nicaud, J.M., and Gaillardin, C. (2011) *Appl. Microbiol. Biotechnol.*, **90**, 1193–1206.

11. Hashimoto, K., Yoshizawa, A.C., Okuda, S., Kuma, K., Goto, S., and Kanehisa, M. (2008) *J. Lipid Res.*, **49**, 183–191.

12. Ledesma-Amaro, R. (2015) *Eur. J. Lipid Sci. Technol.*, **117**, 141–144.

13. Stephanopolos, G., Tai, M., Chakraborty, S., (2013) Engineered Microbes and Methods for Microbial Oil Production, US Patent US2013/143282 A1, Applicant: Massachusetts Institute of Technology.

14. Beopoulos, A., Mrozova, Z., Thevenieau, F., Le Dall, M.T., Hapala, I., Papanikolaou, S., Chardot, T., and Nicaud, J.M. (2008) *Appl. Environ. Microbiol.*, **74** (24), 7779–7789.

15. Wang, J., Zhang, B., and Chen, S. (2011) *Process Biochem.*, **46**, 1436–1441.

16. Cao, Y.J., Liu, W., Xu, X., Zhang, H.B., Wang, J.M., and Xian, M. (2014) *Biotechnol. Biofuels*, **7** (59), 1–11.

17. Xue, Z., Sharpe, P., Hong, S., Yadav, N.S., Xie, D., Short, D.R., Damude, H.G., Rupert, R.A., Seip, J.E., Wang, J., Pollak, D.W., Bostick, M.W., Bosak, M.D., Macool, D.J., Hollerbach, D.H., Zhang, H., Arcilla, D.M., Bledsoe, S.A., Croker, K., McCord, E.F., Tyreus, B.D., Jackson, E.N., and Zhu, Q. (2013) *Nat. Biotechnol.*, **31**, 734–740.

18. Beopoulos, A., Verbeke, J., Bordes, F., Guicherd, M., Bressy, M., Marty, A., and Nicaud, J.M. (2014) *Appl. Microbiol. Biotechnol.*, **98** (1), 251–262.

19. Choi, Y.J. and Lee, S.Y. (2013) *Nature*, **502**, 571–574.

20. Davis, R., Tao, L., Tan, E.C.D., Biddy, M.J., Beckham, G.T., Scarlata, C., Jacobson, J., Cafferty, K., Ross, J., Lukas, J., Knorr, D., and Schoen, P. (2013) Process Design and Economics for the Conversion of Lignocellulosic Biomass to Hydrocarbons: Dilute-Acid and Enzymatic Deconstruction of Biomass to Sugars and Biological Conversion of Sugars to Hydrocarbons. NREL Technical Report NREL/TP-5100-60223, NREL.

21. Dugar, D. and Stephanopoulos, G. (2011) *Nat. Biotechnol.*, **29**, 1074–1078.

22. Pan, J.G., Kwak, M.Y., and Rhee, J.S. (1986) *Biotechnol. Lett*, **8**, 715–718.

23. Yamauchi, H., Mori, H., Kobayashi, T., and Shimizu, S. (1983) *J. Ferment. Technol.*, **61**, 275–280.

24. Li, Y., Zhao, Z., and Bai, F. (2007) *Enzyme Microb. Technol.*, **41**, 312–317.

25. Galafassi, S., Cucchetti, D., Pizza, F., Franzosi, G., Bianchi, D., and Compagno, C. (2012) *Bioresour. Technol.*, **111**, 398–403.

26. Huang, C., Chen, X., Xiong, L., Chen, X., Ma, L., and Chen, Y. (2013) *Biotechnol. Adv.*, **31**, 129–139.

27. Yu, X.C., Zheng, Y.B., Dorgan, K.M., and Chen, S.L. (2011) *Bioresour. Technol.*, **102**, 6134–6140.

28. Tsigie, Y.A., Wang, C.Y., Truong, C.T., and Ju, Y.H. (2011) *Bioresour. Technol.*, **102**, 9216–9222.

29. Huang, C., Zong, M.H., Wu, H., and Liu, Q.P. (2009) *Bioresour. Technol.*, **100**, 4535–4538.

30. Huang, X., Wang, Y.M., Liu, W., and Bao, J. (2011) *Bioresour. Technol.*, **102**, 9705–9709.

31. Huang, C., Chen, X.F., Xiong, L., Chen, X.D., and Ma, L.L. (2012) *Bioresour. Technol.*, **110**, 711–714.

32. Jacob, Z. (1992) *Crit. Rev. Biotechnol.*, **12** (516), 463–491.

33. Geciova, J., Bury, D., and Jelen, P. (2002) *Int. Dairy J.*, **12**, 541–553.

34. Franzosi, G., Pizza, F., Cucchetti, D. Bianchi, D., Compagno, C.M., Galafassi, S., (2013) Process for The Production of Lipids from Biomass, US Patent US2013/0289289 A1, Applicant: eni S.P.A.

35. Espinosa-Gonzalez, I., Parashar, A., and Bressler, D.C. (2014) *J. Biotechnol.*, **187**, 10–15.

36. Malm, A., Tanner, R., Hujanen, M., (2012) Method for Lipid Extraction from Biomass, US Patent US

2012/0110898 A1, Applicant: Neste Oil Oyj.

37. Aaltonen, O., Jauhiainen, O., Hujanen, M., (2012) Method for recovery of oil from biomass, US Patent US 2012/0116105 A1, Applicant: Neste Oil Oyj.

38. Catchpole, O., Ryan, J., Zhu, Y., Fenton, K., Grey, J., Vyssotski, M., and MacKenzie, A. (2010) *J. Supercrit. Fluids*, **52**, 34–41.

39. Matyash, V., Liebisch, G., Kurzchalia, T.V., Shevchenko, A., and Schwudke, D. (2008) *J. Lipid Res.*, **49**, 1137–1146.

40. Hegel, P.E., Camy, S., Destrac, P., and Condoret, J.S. (2011) *J. Supercrit. Fluids*, **58** (1), 68–78.

41. Liu, B. and Zhao, Z. (2007) *J. Chem. Technol. Biotechnol.*, **82**, 775–780.

42. Hartig, S. (2012) Value creation in bio-energy. Carnegie Biofuels Seminar, Copenhagen.

43. Apt, K., Borden, J., Hansen, J., Sellers, M., Behrens, P.W., (2014) Microorganisms with altered fatty acid profiles for renewable materials and biofuel production, World Patent WO2014138732 (A2), Applicant: BP Corporation North America Inc.

44. (2012) Neste Oil Annual Report 2012, p. 50.

45. Holnback, M., Lehesto, M., Koskinen, P., Selin, J.F., (2011) Process and Microorganisms for Production of Lipids, US Patent US 2011/0294173 A1, Applicant: Neste Oil Oyj.

46. Droop, M.R. (1974) in *Algal Physiology and Biochemistry* (ed. W.D.P. Stewart), University of California Press, California, pp. 530–560.

47. Tredici, M. (2010) *Biofuels*, **1** (1), 143–162.

48. Jorquera, O., Kiperstok, A., Sales, E.A., Embiruçu, M., and Ghirardi, M.L. (2010) *Bioresour. Technol.*, **101** (4), 1406–1413.

49. Chisti, Y. (2006) *Environ. Eng. Manage. J.*, **5** (3), 261–274.

50. Chisti, Y. (2007) *Biotechnol. Adv.*, **25** (3), 294–306.

51. Mendoza, J.L., Granados, M.R., de Godos, I., Acién, F.G., Molina, E., Banks, C., and Heaven, S. (2013) *Biomass Bioenergy*, **54**, 267–275.

52. Tredici, M.R. (1999) in *Encyclopedia of Bioprocess Technology: Fermentation, Biocatalysis and Bioseparation*, vol. 1 (eds M.C. Flickinger and S.W. Drew), John Wiley & Sons, Inc., New York, pp. 395–419.

53. Huang, G., Chen, F., Wei, D., Zhang, X., and Chen, G. (2010) *Appl. Energy*, **87**, 38–46.

54. Gouveia, J. and Oliveira, A.C. (2009) *J. Ind. Microbiol. Biotechnol.*, **36**, 269–274.

55. Liu, Z.Y., Wang, G.C., and Zhou, B.C. (2008) *Bioresour. Technol.*, **99** (11), 4717–4722.

56. Molina, G.E., Belarbi, E.H., Acién Fernández, F.G., Robles, M.A., and Chisti, Y. (2003) *Biotechnol. Adv.*, **20**, 491–515.

57. Lam, M.K. and Lee, K.T. (2012) *Biotechnol. Adv.*, **30**, 673–690.

58. Patil, V., Tran, K.Q., and Giselrød, H.R. (2008) *Int. J. Mol. Sci.*, **9** (7), 1188–1195.

59. Bilanovic, D., Shelef, G., and Sukenik, A. (1988) *Biomass*, **17**, 65–76.

60. Halim, R., Gladman, B., and Danquah, M.K. (2011) *Bioresour. Technol.*, **102** (1), 178–185.

61. Ben-Amotz, A. and Avron, M. (1989) in: *Algal and Cyanobacterial Technology* (eds R.C. Cresswell, T.A. Rees, and N. Shah), Longman, London, pp. 90–114.

62. Sathish, A. and Sims, R.C. (2012) *Bioresour. Technol.*, **118**, 643–647.

63. Chen, M., Liu, T., Chen, X., Chen, L., Zhang, W., Wang, J., Gao, L., Chen, Y., and Peng, X. (2012) *Eur. J. Lipid Sci. Technol.*, **114** (2), 205–212.

64. Schlagermann, P., Göttlicher, G., Dillschneider, R., Rosello-Sastre, R., and Posten, C. (2012) *J. Combust.*, **2012**, 1–14.

65. US Department of Energy (2014) Bioenergy Technology Office Multi-Year Program Plan, July 2014, pp. 2–42.

66. Singh, J. and Gu, S. (2010) *Renewable Sustainable Energy Rev.*, **14**, 2596–2610.

67. Perez-Garcia, O., Escalante, F.M.E., de-Bashan, L.E., and Bashan, Y. (2011) *Water Res.*, **45**, 11–36.

68. Janaun, J. and Ellis, N. (2010) *Renewable Sustainable Energy Rev.*, **14**, 1312–1320.

69. Knothe, G. (2010) *Prog. Energy Combust. Sci.*, **36**, 364–373.

70. (2014) Neste Oil Annual Report 2014, p. 21–25.

71. Holmgren, J., Gosling, C., Marinangeli, R., Marker, T., Faraci, G., and Perego, C. (2007) *Hydrocarbon Processes*, 67–71.

72. Marinangeli, R., Marker, T., Petri, J., Kalnes, T., McCall, M., Mackowiak, D., Jerosky, B., Reagan, B., Nemeth, L., Krawczyk, M., Czernik, S., Elliott, D., Shonnard, D. (2005) Opportunities for Biorenewables in Oil Refineries Final Technical Report. DOE Report No. DE-FG36-05GO15085.

73. Chiaramonti, D., Oasmaa, A., and Solantausta, Y. (2007) *Renewable Sustainable Energy Rev.*, **11**, 1056–1086.

74. Bridgwater, A.V. (2012) *Biomass Bioenergy*, **38**, 68–94.

75. Mohan, D., Pittman, C.U. Jr., and Steele, P.H. (2006) *Energy Fuels*, **20**, 848–889.

76. Bridgwater, A.V. (2004) *Therm. Sci.*, **8**, 21–49.

77. Bridgwater, A.V. and Peacocke, G.V.C. (2000) *Renewable Sustainable Energy Rev.*, **4**, 1–73.

78. Xiu, S. and Shahbazi, A. (2012) *Renewable Sustainable Energy Rev.*, **16**, 4406–4414.

79. Scott, D.S., Majerski, P., Piskorz, J., and Radlein, D. (1999) *J. Anal. Appl. Pyrolysis*, **51**, 23–37.

80. Nair, P. (2010) *Alternative Fuels and Power from Renewable Feedstocks Alternative Feedstocks, Energy Frontiers International*, Denver, Colorado.

81. Oasmaa, A., Elliott, D.C., and Korhonen, J. (2010) *Energy Fuels*, **24**, 6548–6554.

82. Czernik, S. and Bridgwater, A.V. (2004) *Energy Fuels*, **18**, 590–598.

83. Mullen, C.A. and Boateng, A.A. (2008) *Energy Fuels*, **22**, 2104–2109.

84. Marsman, J.H., Wildschut, J., Mahfud, F., and Heeres, H.J. (2007) *J. Chromatogr. A*, **1150**, 21–27.

85. Bertero, M., Puente, G., and Sedran, U. (2012) *Fuel*, **95**, 263–271.

86. Diebold, J.P. (2000) A Review of the Chemical and Physical Mechanisms of the Storage Stability of Fast Pyrolysis Bio-Oils. NREL Report NREL/SR-570-27613.

87. Oasmaa, A. and Kuoppala, E. (2003) *Energy Fuels*, **17**, 1075–1084.

88. Ruddy, D.A., Schaidle, J.A., Ferrell, J.R., Wang, J., Moens, L., and Hensley, J.E. (2014) *Green Chem.*, **16**, 454–490.

89. Wan, S. and Wang, Y. (2014) *Front. Chem. Sci. Eng.*, **8** (3), 280–294.

90. Fostera, A.J., Jaeb, J., Chengb, Y.T., Huberb, G.W., and Loboa, R.F. (2012) *Appl. Catal., A*, **423**, 154–161.

91. Carlson, T.R., Vispute, T.P., and Huber, G.W. (2008) *ChemSusChem*, **1**, 397–400.

92. Lu, Q., Zhang, Z.F., Dong, C.Q., and Zhu, X.F. (2010) *Energies*, **3**, 1805–1820.

93. Nokkosmäki, M., Kuoppala, E., Leppämäki, E., and Krause, A.J. (2000) *J. Anal. Appl. Pyrolysis*, **55**, 119–131.

94. Kryzanowski, T. (2015) *Logging Sawmilling J.*, 1–4.

95. (2014) *Bioenergy Insight*, **5** (6), 58–59.

96. Tekin, K., Karagöz, S., and Bektas, S. (2014) *Renewable Sustainable Energy Rev.*, **40**, 673–687.

97. Toor, S.S., Rosendahl, L., and Rudolf, A. (2011) *Energy*, **36**, 2328–2342.

98. Elliott, D.C., Hart, T.R., Schmidt, A.J., Neuenschwander, G.G., Rotness, L.J., Olarte, M.V., Zacher, A.H., Albrecht, K.O., Hallen, R.T., and Holladay, J.E. (2013) *Algal Res.*, **2** (4), 445–454.

99. Barreiro, D.L., Prins, W., Ronsse, F., and Brilman, W. (2013) *Biomass Bioenergy*, **53**, 113–127.

100. Elliott, D.C., Bille, P., Ross, A.B., Schmidt, A.J., and Jones, S.B. (2015) *Bioresour. Technol.*, **178**, 147–156.

101. Xu, C. and Lad, N. (2008) *Energy Fuels*, **22**, 635–642.

102. Zhong, C. and Wei, X. (2004) *Energy*, **29**, 1731–1741.

103. Liu, J., Zhang, X., and Chen, G. (2012) *J. Sustainable Bioenergy Syst.*, **2**, 112–116.

104. Leow, S., Witter, J.R., Derek, R.,Vardon, D.R., Brajendra, V.,Sharma, B.K., Guest, J.S., and Strathmann, T.J. (2015) *Green Chem.*, **17**, 3584–3599.

105. Chiaberge, S., Leonardis, I., Fiorani, T., Cesti, P., Reale, S., and De Angelis, F. (2014) *Energy Fuels*, **28**, 2019–2026.

106. Leonardis, I., Chiaberge, S., Fiorani, T., Spera, S., Battistel, E., Bosetti, A., Cesti, P., Reale, S., and De Angelis, F. (2013) *ChemSusChem*, **6**, 160–167.

107. Chiaberge, S., Leonardis, I., Fiorani, T., Bianchi, G., Cesti, P., Bosetti, A., Crucianelli, M., Reale, S., and De Angelis, F. (2013) *Energy Fuels*, **27**, 5287–5297.

108. Yuan, X., Xie, W., Zeng, G., Tong, J., and Li, H. (2008) *Int. J. Biotechnol.*, **10** (1), 35–44.

109. Karagoz, S., Bhaskar, T., Mutoa, A., Sakata, Y., Oshiki, T., and Kishimoto, T. (2005) *Chem. Eng. J.*, **108**, 127–137.

110. Bosetti, A., Bianchi, D., Moscotti, D.G., Pollesel P., (2013) Integrated process for the production of biofuels from solid urban waste, European Patent EP 2 749 626 A1, Applicant: eni S.P.A.

111. Mortensen, P.M., Grunwaldta, J.D., Jensena, P.A., Knudsenc, K.G., and Jensena, A.D. (2011) *Appl. Catal., A*, **407**, 1–19.

112. Choudhary, T.V. and Phillips, C.B. (2011) *Appl. Catal., A*, **397**, 1–12.

113. Elliott, D.C. (2007) *Energy Fuels*, **21**, 1792–1815.

114. Wang, H., Male, J., and Wang, Y. (2013) *ACS Catal.*, **3** (5), 1047–1070.

115. Jones, S., Meyer, P., Snowden-Swan, L., Padmaperuma, A, Tan, E., Dutta, A., Jacobson, J., and Cafferty, K. (2013) Process Design and Economics for the Conversion of Lignocellulosic Biomass to Hydrocarbon Fuels. NREL Report NREL/TP-5100-61178.

116. Susanne, J., Valkenburg, C., Walton, C., Elliott, D., Holladay, J., Stevens, D., Kinchin, C., and Czernik, S. (2009) Production of Gasoline and Diesel from Biomass via Fast Pyrolysis, Hydrotreating and Hydrocracking: A Design Case, Pacific Northwest National Laboratory Report PNNL-18284.

117. Jones, S.B. and Male, J.L. (2012) Production of Gasoline and Diesel from Biomass via Fast Pyrolysis, Hydrotreating and Hydrocracking: 2011 State of Technology and Projections to 2017, Pacific Northwest National Laboratory Report PNNL-22133.

118. Christensen, E.D., Chupka, G.M., Luecke, J., Smurthwaite, T., Alleman, T.L., Iisa, K., Franz, J.A., Elliott, D.C., and McCormick, R.L. (2011) *Energy Fuels*, **25**, 5462–5471.

119. Holmgren, J., Marinangeli, R., Nair, P., Elliott, D., and Bain, R. (2008) Hydrocarbon Processing, 95–103.

26
Food Supply Chain Waste: Emerging Opportunities

Katie Privett, James H. Clark, Mehrdad Arshadi, Apostolis Koutinas, Nicholas Gathergood, Piergiuseppe Morone, and Rafael Luque

26.1
Introduction

Society has been benefiting from the manufacturing/chemical industry closely related to petroleum refining since the twentieth century. However, chemical industries have recently been challenged by several important issues for its long-term development. These include limited reserves of feedstocks (fossil-derived resources), escalating costs of oil, rise of public awareness on human and environmental safety, growth of legislations affecting the industry, and related others.

Biorefineries have been recently perceived as an alternative and more environmentally friendly possibility to move away from petrochemically derived products we have relied upon the past 60+ years. In the IEA Bioenergy Task 42, a thorough definition of biorefinery was introduced as "sustainable processing of biomass into a spectrum of marketable products (food, feed, materials, chemicals) and energy (fuels, power, heat)" [1]. In this promising paradigm, analogous chemicals, materials, fuels, and related end products can be potentially produced from renewable feedstocks, biomass, and waste using environmentally friendly technologies (e.g., microwave irradiation, continuous flow processes, multistep fractionation, etc.) as illustrated in Figure 26.1 [2].

Accordingly, biomass and residues including food waste, agriculture residues, and biomass are prospective, suitable raw materials in future biorefineries. Valorization of these types of waste can provide not only a solution to waste disposal and management but also potential replacement of products derived from the petrochemical industry.

Food supply chain waste (FSCW) is a key example of a waste that can be turned into a resource. FSCW comprises of wastes throughout the food production and use chain, including agricultural residues (e.g., wheat straw), processing wastes (e.g., citrus peel), and post-consumer wastes (e.g., catering surplus). In most cases, these wastes are produced in concentrated regions and are easily separated pre-disposal, making them a convenient resource and source of multiple products

Chemicals and Fuels from Bio-Based Building Blocks, First Edition.
Edited by Fabrizio Cavani, Stefania Albonetti, Francesco Basile, and Alessandro Gandini.
© 2016 Wiley-VCH Verlag GmbH & Co. KGaA. Published 2016 by Wiley-VCH Verlag GmbH & Co. KGaA.

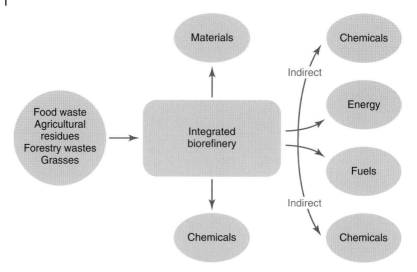

Figure 26.1 The integrated biorefinery as a mixed feedstock source of chemicals, energy, fuels, and materials [2].

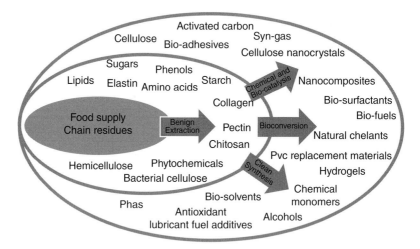

Figure 26.2 Components derived from food waste and their applications. (Reproduced with permission from the Royal Society of Chemistry.)

(Figure 26.2). Reusing FSCW has two broad environmental impacts. Firstly, redirection of organic waste away from landfill reduces Greenhouse Gas (GHG) emissions and water and soil contamination associated with landfill leachate. Secondly, valorization creates a new set of biobased chemicals, materials, and fuels that could directly substitute traditional petroleum-based products, reducing dependence on fossil fuels and boosting the bioeconomy.

FSCW valorization has developed across a range of disciplines of academia and has caught the attention of industry and policymakers too [3]. An important consideration in applied science such as this is the necessity of integration and comparison of different approaches. In light of this, networks and special interest groups have arisen throughout the academic community whose aim is to gather a critical mass of researchers and industrialists to tackle the problem together. The COST Action TD1203 on FSCW valorization (EUBIS) is one of these networks, comprising over 200 researchers from 120 institutions, both academic and industrial, from across Europe and the world. The Action is structured under four Working Groups (WGs) – pretreatment and extraction, bioprocessing, chemical processing, and technical and sustainability assessment/policy analysis. This chapter will be based around these themes, utilizing case studies from within and across these WGs and initiatives worldwide, aiming to provide a broad and general overview of some emerging opportunities in FSCW valorization.

26.2
Pretreatment and Extraction

26.2.1
Extraction

Once a waste has been identified and collected, the first priority is to derive value in the simplest means possible. Some FSCWs already contain complex chemicals with good market value, which would require multiple steps to synthesize in a laboratory. This valuable "shortcut" to complex chemicals should be exploited via their extraction, before allowing the material to undergo any further transformation. Some of the more valuable extracts are bioactive compounds including flavonoids and phenolics which can exhibit antioxidant, antibacterial, or anticancer properties. Present in very low concentrations, their isolation at a purity that would allow for medical uses is often unfeasible. However, if food-safe techniques are used, these compounds can be marketable in functional foods and nutraceuticals.

26.2.2
Microwaves

Extraction methods should be mild enough in most cases to be able to separate out compounds without inducing any chemical changes on the final products. Importantly, methods complying with food industry standards and/or utilized in food industry technology are preferable – many of the valuable products can be used as food additives and nutraceuticals if the extract is uncontaminated. It is also prudent to consider the impacts on the remaining material as preservation and enhancement of further value is vital to ensure potential for future processing under the biorefinery concept.

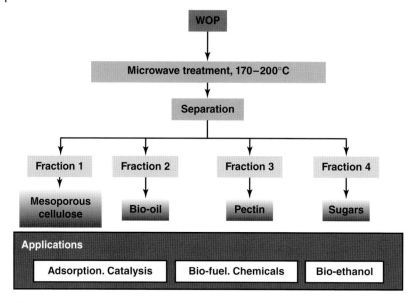

Figure 26.3 Waste orange peel (WOP) valorization to useful end products using microwave-assisted extraction technologies. (Reproduced from reference [4]. Copyright Wiley VCH Verlag.)

There is a wide range of green extraction techniques that are being applied to food waste including supercritical CO_2, subcritical water, ultrasounds, and microwaves. Microwaves have become increasingly popular – their efficient, rapid, and tuneable heating facilitates extraction of chemicals often using waste's inherent water content as solvent. Citrus peel research at the University of York has determined a process by which selective extraction of D-limonene, pectins, and flavonoids was found to be possible (Figure 26.3) using only microwave-assisted hydrothermal extraction and ethanol solvent extraction, eliminating the environmentally harmful hot acid extraction of pectin currently used commercially. All products are food grade and the remaining sugar-rich fraction can be used as a fermentation feedstock [4]. Microwaves can also be used for pretreatment – Gaziantep University's Food Engineering Department has succeeded in using them for drying olive pomace, yielding a cocoa butter alternative from an enzymatic transformation [5].

The variable and varied composition of mixed waste such as urban biowaste restricts extraction opportunities, and valorization of these materials is often limited to conversion to fuels. Researchers in Italy have discovered recalcitrant lignin-like fractions of urban biowastes have the potential to be cost-effective sources of soluble biobased substances. These have been extracted via microwave and conventional heating and have a wide range of applications, for example, flocculants, emulsifiers, plastics, platform chemicals, feed supplements, fertilizers, and plant biostimulants [6].

26.2.3
Pretreatment

Pretreatment methods are often necessary for the effective valorization of FSCWs. These processes can impact how long the materials can be stored, how easily they can be transported and transferred from vessel to vessel, and convenience of extraction and further chemical/biological processing. Thermomechanical pre-processing methods can have an impact on chemical components in the waste material, which coupled with intentional extractions can have positive impacts on the quality of the material for further processing, for example, for energy generation or biomaterial applications [7].

Pretreatments can prepare biomass for further chemical or biological processing by breaking up recalcitrant fractions allowing access to smaller molecules that are easier to work with. High-pressure CO_2 and ionic liquids are good examples, resulting, respectively, in structural changes that can enhance glucan and xylan accessibility in fractions for enzymatic hydrolysis or in selective fractionation of biomass into high purity fractions of cellulose, hemicellulose, and lignin which can be converted separately [8].

26.3
Bioprocessing

FSCW is an organic resource that is generally susceptible to biochemical processes, which comprise any transformations mediated by enzymes or microorganisms. Research groups at international level have been developing bioprocesses at lab and pilot scale, aiming to produce fuels, chemicals, biopolymers, food additives, and high value-added products, with biorefining of remaining fractions also envisaged.

26.3.1
Food Additives and Functional Foods

As well as extracting food additives from FSCW, the material can be used as a feedstock for fermentations yielding a wide range of chemicals for use in food and nutraceuticals. Fermentation of lignocellulosic FSCWs can yield products such as prebiotic oligosaccharides, natural sweeteners, and phenolics, produced by combining hydrothermal pretreatment and enzymes or microbes acting on a range of wastes [9]. Enzyme-mediated production of soluble fibers at the University of Buenos Aires can also produce texture modifiers and antioxidants from globe artichoke, beetroot, plums, butternut, quince, and papaya waste [10].

Production of fungal enzymes using starch-rich food waste (bakery waste) by solid-state fermentation (SSF) has also been recently reported, particularly in view of future generation of biofuels, platform chemicals, and materials that will be subsequently illustrated in coming sections [11–14]. With the high glucoamylase yield ($253.7 \pm 20.4\,U\,g^{-1}$) of the crude enzyme extract, the application of bakery waste for the production of glucoamylase revealed a great potential for its production in industrial scale [15].

26.3.2
Biofuels

Transformation into biofuels such as biogas has been the most common FSCW valorization option. However, practicalities of biogas from FSCW such as system stability, cost minimization, and need for pretreatment still need addressing. One potential expansion is for bioenergy generation from wastewaters, the organic content of which leads to massive environmental issues if released. At DUTH, recent developments in valorization of cheese whey wastewater have included a two-stage biogas system creating high methane yields under high organic loading rates, which has been demonstrated as stable at both lab and full scale [16]. By separating acidogenesis and methanogenesis in anaerobic digestion, H_2 can be produced in the first stage, creating an added energy value source, with the second stage yielding alcohols and useful chemicals such as lactic, acetic, and butyric acid [17]. This two-stage process has also been used to valorize olive mill pulp and wastewater for production of H_2, methane, and higher value microbial products such as biopolymers [18].

Complex anaerobic ecosystems are being used to produce biogas and bio-H_2 from wheat straws and sunflower stalks as well as chemicals such as acetate, butyrate, propionate, and other dark fermentation metabolites (lactate, ethanol). Physicochemical or biological pretreatment can be applied to enhance the conversion of waste into energy or to favor the production of some metabolites [19]. Optimization and enhancement of biogas and H_2 production at the University of Leon focuses on combining fermentative valorization of agro-industrial wastes and microbial electrolysis cells for treatment of low load liquid effluents, resulting in a significant increase in H_2 yields when treating cheese whey [20]. The group are also determining the best reactor configuration for different agro-wastes (energy crops, corn stover, and potato wastes), with results demonstrated at pilot scale [21].

Oleaginous waste, such as used cooking oil, containing triglycerides and relatively high quantities of free fatty acids (FFAs), can be applied to the production of biodiesel (fatty acid methyl esters and fatty acid ethyl esters, FAMEs and FAEEs) via simultaneous transesterification of glycerides and the esterification of FFAs with methanol or ethanol in the presence of a range of solid acid catalysts [22, 23]. In this regard, a new family of sulfonated carbonaceous materials denoted as Starbons were able to promote the aforementioned chemistries for waste oils containing a large concentration of FFAs (>10 wt%) at mild temperatures both under conventional heating and microwave irradiation [23]. Quantitative yields to biodiesel could be obtained after 12 h at 80 °C (conventional heating) and even reduced to 30 min under microwave irradiation. Additional to these relevant efforts toward biodiesel production using solid acid catalysts, an integrated

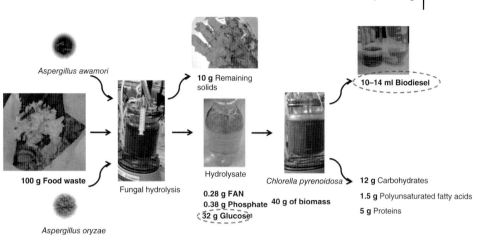

Aspergillus awamori

100 g Food waste

Fungal hydrolysis

Aspergillus oryzae

10 g Remaining solids

Hydrolysate

0.28 g FAN
0.38 g Phosphate   40 g of biomass
32 g Glucose

Chlorella pyrenoidosa

10–14 ml Biodiesel

12 g Carbohydrates
1.5 g Polyunsaturated fatty acids
5 g Proteins

Figure 26.4 Oleaginous food waste to valuable products: concept and quantities. Reproduced with permission of Elsevier, 2013, ref. [24].

concept to valorize lipid-containing food waste to biodiesel was recently envisaged by Lin *et al.* (Figure 26.4) [24].

Food waste was firstly converted into crude hydrolytic products by enzymatic hydrolysis and then fractions (fungal biomass and hydrolysates) were sequentially separated for further processing [11–15, 24]. Lipid-rich fungal biomass was converted into biodiesel by transesterification, while the hydrolysate rich in glucose, free amino acid (FAN), and phosphate can be an appealing substrate for microbial biomass cultivations to further process organic feedstocks into value-added products [11–15, 24]. Polyunsaturated fatty acids (PUFAs) were also produced in relevant quantities with further applications in cholesterol-reducing agents and nutraceuticals.

Apart from these lab-scale food waste valorization processes, some food waste-based industrial-scale applications have emerged in recent decades worldwide. A biodiesel plant, using waste oil and grease as feedstock, was set up and put into production in November 2013 by ASB Biodiesel (Hong Kong) Limited in Hong Kong. The adopted BioDiesel International (BDI) technology provided an interesting feasibility to process multi-feedstocks containing a high level (over 20%) of FFAs for biodiesel production [25]. Not to mention the economic value, 257 000 tons of CO_2 emission can be avoided every year and there is 3.6% of potential reduction of CO_2 in HK's transport sector due to the prevention of waste disposal and substitution of fossil fuel [25]. Similar initiatives are also available in Europe which have been developed in recent years [26].

26.3.3
Platform Chemicals and Biopolymers

A promising number of possibilities are available to convert FSCW into useful platform chemicals and biopolymers.

Lactic acid is a highly useful bioderived platform chemical, and its fermentative production from FSCW has been widely investigated [27]. At Agrartechnik *Potsdam*-Bornim Potsdam-Bornim, fermentation processes are being designed to produce lactic acid from several feedstocks (e.g., bagasse, oilseed meal, whey, coffee residues, waste bread, vegetable residues, slaughterhouse waste, etc.) [28]. To demonstrate that this technology is viable at full scale, a multifunctional pilot plant facility has been built here, and testing on substrate preparation, microbial transformation, and substance separation are being carried out, as well as technological optimization of fermentation and downstream processing steps [29]. The group at BOKU has developed a theoretical future biorefinery on wheat bran, producing food, lactic acid (via on-site cellulolytic enzyme production), and energy, with a theoretical maximum product efficiency under optimized conditions of around 80% (30% food, 39% lactic acid, 6% convertible energy) [30]. Moreover, by coupling organosolv treatment with enzymatic hydrolysis, conversion of the carbohydrate fraction into their monomers can reach 60%, and with the addition of hydrothermal treatment, conversion efficiency can reach 75% [31].

Butyric acid is another interesting platform chemical produced from FSCW bioprocessing from pretreated and hydrolyzed wheat straw using an adapted strain of *Clostridium tyrobutyricum* with enhanced tolerance [32]. By applying *in situ* butyric acid removal (using reverse electro-enhanced dialysis), 100% fermentation became possible in a continuous operation mode. The obtained butyric acid yield has been optimized to achieve 92% of the maximum theoretical yield, and the developed process has been verified in a 20-l pilot-scale bioreactor [32]. Furthermore, initial techno-economic analysis of the overall process scheme, from wheat straw to 99% (w/w) butyric acid based on experimental data and theoretical figures, suggested feasible economic viability of the process [33].

26.4
Chemical Processing

26.4.1
Combining Chemical and Bioprocessing

Bioprocessing can be complemented with chemical processing options to yield more complex chemicals. At WUR's Biobased Products laboratory, the unstable galacturonic acid obtained via enzymatic hydrolysis of the pectin fraction of sugar beet pulp can be catalytically oxidized to the corresponding aldaric acid, yielding the versatile platform chemical galactaric acid [43]. This diacid has potential applications as a sequestering agent, polymer building block, and feedstock to produce furan-2,5-dicarboxylic acid, a biobased alternative to terephthalic acid in various polyesters [34]. A combination of high hydrostatic pressure and enzyme transformation can also be applied to FSCWs such as okara from soymilk production and orange pulp. This combination leads to improved characteristics of the materials, including an impact on composition and properties of dietary fiber, oligosaccharides, carotenoids, polyphenols, and peptides [35].

26.4.2
Chemical Processing and the Biorefinery

In current food industry practice, coriander seeds are processed by hydrodistillation to produce essential oils. Coriander seeds contain vegetal and essential oils, and so a biorefinery cascade process has been developed to apply to spent seeds – single-screw/double-screw extrusion (virgin vegetal oil expression), hydrodistillation (essential oil obtention), and extraction cake thermopressing (agromaterials design) [36]. Vegetal and essential oils contain >70% petroselinic acid and linalool, respectively, which are target molecules for flavor chemistry and lipochemistry [37]. Following coriander vegetal oil's recent classification as a Novel Food by the European Food Safety Agency, a regional initiative was set up to recultivate and process coriander in southwest France. Alongside food applications, promising chemical applications include exploitation of petroselinic acid's unique structure as a niche building block for novel bioactives (e.g., sophorolipids analogs) [38].

Biodiesel production from waste sources such as used cooking oil is a key example of chemical processing in a biorefining context – the by-product, glycerol, is a valuable commodity in its own right and subsidizes the cost of biodiesel production. However, researchers in Slovenia and Bulgaria are working together to perfect greener catalytic conversions of glycerol to mono-, di-, and triacetyl glycerol, useful in cosmetics, medicines, and as a monomer for the production of biodegradable polyesters (mono- and di-) or as a biodiesel additive (tri-). Currently the process involves using corrosive homogeneous catalysts that cause technical and environmental complications – the team utilize safer, recoverable heterogeneous catalysts for efficient conversion [39].

26.5
Technical and Sustainability Assessment and Policy Analysis

The technical aspects of FSCW valorization explored so far have great potential to affect a future bioeconomy built on waste, but in order to assess the feasibility of particular products and processes, each option needs to be subjected to a techno-economic evaluation, a social acceptance review, and a policy analysis to understand if it would be marketable. This requires collaboration between the research scientists identifying techniques and economists, chemical engineers, social scientists, and policy experts. This cross-cutting approach is supported by networks such as EUBIS, and so presented below are case studies from some of these collaborations.

26.5.1
Assessing a Market

The market for a particular product is not just based on the theoretical changeout of an established product for a new alternative – often, social network architecture will define whether or not a new product is accepted into a particular marketplace. One example, explored by a group of Italian researchers, is the potential development of the bioplastics technological niche that uses secondary feedstock (biowaste) instead of dedicated crops. By investigating structural characteristics of the Italian bioplastics network and determining how information is passed throughout this system, the team could investigate how easily a new product could find acceptability by assessing the opinions of key influential actors within the network. The empirical investigation provided evidence that the architectural structure of the Italian bioplastics producers network offers great opportunities for development of a technological niche based on biowaste valorization. However, the system appears weak as far as expectations are concerned, as these are generally low and, more critically, are low for those actors occupying central positions in the network. This shortcoming could jeopardize the niche development process, if no appropriate policy actions are undertaken [40].

26.5.2
Cost Analysis of Biorefining

Process flowsheets in biorefineries need to undergo techno-economic evaluation and life cycle assessment to evaluate the feasibility of industrial implementation for the production of a range of chemicals, materials, and fuels from one/many inputs. At the Agricultural University of Athens, a process of converting confectionary waste streams into microbial oil for subsequent biodiesel production was evaluated. The cost analysis for the whole biorefinery, including the conversion to biodiesel, was compared with biodiesel plants that use conventional feedstocks (e.g., vegetable oils) or microalgae. For a plant capacity of 10 000 tons of microbial oil, a unitary cost of $3.4 per kg was calculated. The investigation found that the designed plant is greatly affected by the price of the raw material (glucose-based medium) and the productivity of the fermentation process from oleaginous yeasts [41].

26.5.3
Sustainability Transition Patterns

The need for a transition toward sustainability, including safe and sustainable energy production and food security, is widely acknowledged. Multilevel perspectives are often used to investigate sustainability transition patterns – a team of researchers are looking closely at how pressure exerted from the landscape level contributes to the opening up of windows of opportunities allowing sustainable transition to occur. Using biorefineries as a case study, the group developed a conceptual model that provides a framework to understand the endogenous role of the landscape in the sustainability transition process and the conditions that may encourage opening of windows of opportunities for a technological niche to compete with the incumbent technological regime. Ultimately, this study could support policymakers in pursuing specific strategies to control the opening/closure of these windows of opportunities and to manage the interactions among levels more effectively [42].

26.6
Conclusions and Outlook

Technologies that can be potentially applied in the biorefinery process are developing rapidly, potentially being able to provide more possibilities for biorefineries while enhancing their productivity. Valorization of the biomass/waste can not only convert wastes into a variety of value-added products (biofuel, chemicals, and materials) but also release an important environment burden resulting from the disposal/management of these waste sources. Abundant biomass/food waste residues, rich in organic compounds, can be economically and ecologically feasible to be valorized as a low-cost, renewable feedstocks for the biorefinery industry.

The opportunities for expansion of products and value from FSCW highlighted in this chapter are just a small sample from a massively expanding area. With awareness of the issues surrounding waste rising and the potential profits to be made from treating waste as a resource becoming more widely discussed, FSCW valorization will play an important role in the future expansion of the bioeconomy.

Acknowledgments

The authors would like to acknowledge the support of the COST Action TD1203 (EUBIS), especially the network members whose research has been showcased in this chapter, and the teams supporting the Working Group leaders during the construction of this chapter.

References

1. Sonnenberg, A., Baars, J.,Hendrickx, P. (2013) IEA Bioenergy Task 42 Biorefinery, p. 28

2. Clark, J.H., Luque, R., and Matharu, A.S. (2012) *Annu. Rev. Chem. Biomol. Eng.*, **3**, 183–207.

3. House of Lords Science and Technology Select Committee (2014) Waste or Resource? Stimulating a Bioeconomy. 3rd Report of Session 2013-2014. The Stationary Office Limited, London.

4. (a) Pfaltzgraff, L. (2014) The study & development of an integrated & additive-free waste orange peel biorefinery. PhD thesis. University of York.;

(b) Balu, A.M. (2012) *ChemSusChem*, **5**, 1694–1697.

5. (a) Göğüş, F. and Maskan, M. (2001) *Nahrung*, **45** (2), 129–132; (b) Çiftçi, O.N., Göğüş, F., and Fadıloğlu, S. (2010) *J. Am. Oil Chem. Soc.*, **87**, 1013–1018.

6. (a) Baxter, M., Edgar, A., Montoneri, E., and Tabasso, S. (2014) *Ind. Eng. Chem. Res.*, **53**, 3612–3621; (b) Orazio, S., Montoneri, E., Patanè, C., Rosato, R., Tabasso, S., and Ginepro, M. (2014) *Sci. Total Environ.*, **487C**, 443–451; (c) Vargas, A.K.N., Savarino, P., Montoneri, E., Tabasso, S., Cavalli, R., Bianco Prevot, A., Guardani, R., and

Roux, G.A.C. (2014) *Ind. Eng. Chem. Res.*, **53**, 8621–8629.

7. (a) Brlek, T., Pezo, L., Voća, N., Krička, T., Vukmirović, Ð., Čolović, R., and Bodroža – Solarov, M. (2013) *Fuel Process. Technol.*, **16**, 250–256; (b) Brlek, T., Bodroža – Solarov, M., Vukmirović, Ð., Čolović, R., Vučković, J., and Lević, J. (2012) *Bulg. J. Agric. Sci.*, **18** (5), 752–758; (c) Attard, T. (2015) Supercritical CO_2 extraction of waxes as part of a holistic biorefinery. PhD thesis. University of York.

8. (a) da Costa Lopes, A.M., João, K.G., Bogel-Lukasik, E., Roseiro, L.B., and Bogel-Lukasik, R. (2013) *J. Agric. Food. Chem.*, **61** (33), 7874–7882; (b) Morais, A.R.C., da Costa Lopes, A.M., and Bogel-Lukasik, R. (2014) *Chem. Rev.*, **115** (1), 3–27; (c) Morais, A.R.C., Mata, A.C., and Bogel-Lukasik, R. (2014) *Green Chem.*, **16** (9), 4312–4322.

9. (a) Kiran, E.U., Akpinar, O., and Bakir, U. (2013) *Food Bioprod. Process.*, **91**, 565–574; (b) Akpinar, O. and Usal, G. (2015) *Food Bioprod. Process.*, **95**, 272–280. doi: 10.1016/j.fbp.2014.11.001 (c) Akpinar, O., Sabanci, S., Levent, O., and Sayaslan, A. (2012) *Ind. Crops Prod.*, **40**, 39–44; (d) Akpinar, O., Levent, O., Bostanci, S., Bakir, U., and Yilmaz, L. (2011) *Appl. Biochem. Biotechnol.*, **163**, 313–325.

10. (a) Fissore, E.N., Rojas, A.M., Gerschenson, L.N., and Williams, P.A. (2013) *Food Hydrocolloids*, **31** (2), 172–182; (b) Fissore, E.N., Domingo, C.S., Pujol, C.A., Damonte, E.B., Rojas, A.M., and Gerschenson, L.N. (2014) *Food Funct.*, **5** (3), 463–470.

11. Pleissner, D., Kwan, T.H., and Lin, C.S.K. (2014) *Bioresour. Technol.*, **158**, 48–54.

12. Karmee, S.K. and Lin, C.S.K. (2014) *Lipid Technol.*, **26** (9), 206–209.

13. Zhang, A.Y., Sun, Z., Leung, C.C.J., Han, W., Lau, K.Y., Li, M., and Lin, C.S.K. (2013) *Green Chem.*, **15**, 690.

14. Leung, C.C.J., Cheung, A.S.Y., Zhang, A.Y.Z., Lam, K.F., and Lin, C.S.K. (2012) *Biochem. Eng. J.*, **65**, 10–15.

15. Lam, W., Pleissner, D., and Lin, C.S.K. (2013) *Biomolecules*, **3**, 651–661.

16. Stamatelatou, K., Giantsiou, N., Diamantis, V., Alexandridis, C., Alexandridis, A., and Aivasidis, A. (2012) Anaerobic digestion of cheese whey wastewater through a two stage system. 3rd International Conference on Industrial and Hazardous Waste Management, Chania, Greece, September 12–14, 2012.

17. (a) Stamatelatou, K., Antonopoulou, G., Tremouli, A., and Lyberatos, G. (2011) *Ind. Eng. Chem. Res.*, **50** (2), 639–644; (b) Antonopoulou, G., Stamatelatou, K., Venetsaneas, N., Kornaros, M., and Lyberatos, G. (2008) *Ind. Eng. Chem. Res.*, **47**, 5227–5233.

18. (a) Koutrouli, E., Kalfas, H., Gavala, H.N., Skiadas, I.V., Stamatelatou, K., and Lyberatos, G. (2009) *Bioresour. Technol.*, **100**, 3718–3723; (b) Ntaikou, I., Kourmentza, C., Koutrouli, E.C., Stamatelatou, K., Zampraka, A., Kornaros, M., and Lyberatos, G. (2009) *Bioresour. Technol.*, **100** (15), 3724–3730; (c) Kourmentza, C., Ntaikou, I., Lyberatos, G., and Kornaros, M. (2015) *Int. J. Biol. Macromol.*, **74**, 202–210.

19. (a) Quemeneur, M., Bittel, M., Trably, E., Dumas, C., Fourage, L., Ravot, G., and Steyer, J.P. (2012) *Int. J. Hydrogen Energy*, **37**, 10639–10647; (b) Monlau, F., Trably, E., Kaparaju, P., Steyer, J.P., and Carrere, H. (2015) *Chem. Eng. J.*, **260**, 377–385.

20. Moreno, R., Escapa, A., Cara, J., Carracedo, B., and Gómez, X. (2015) *Int. J. Hydrogen Energy*, **40** (1), 168–175.

21. (a) Molinuevo-Salces, B., García-González, M.C., González-Fernández, C., Cuetos, M.J., Morán, A., and Gómez, X. (2010) *Bioresour. Technol.*, **101** (24), 9479–9485; (b) Cuetos, M.J., Gómez, X., Martínez, E.J., Fierro, J., and Otero, M. (2013) *Bioresour. Technol.*, **144**, 513–520.

22. Arancon, R.A., Barros, H.R. Jr., Balu, A.M., Vargas, C., and Luque, R. (2011) *Green Chem.*, **13**, 3162.

23. Luque, R. and Clark, J.H. (2011) *ChemCatChem*, **3**, 594–597.

24. Pleissner, D., Lam, W.C., Sun, Z., and Lin, C.S.K. (2013) *Bioresour. Technol.*, **137**, 139–146.

25. ASB Biodiesel (Hong Kong) Limited ASB Biodiesel (Hong Kong) Ltd. – About Us,

http://www.asb-biodiesel.com/glance.php (accessed 13 March 2015).

26. BDI BDI – BioEnergy International AG – References BioDiesel, http://www.bdi-bioenergy.com/en-references&uscore;biodiesel-104.html (accessed 13 March 2015).

27. Dusselier, M., Van Wouwe, P., Dewaele, A., Makshina, E., and Sels, B.F. (2013) *Energy Environ. Sci.*, **6**, 1415–1442.

28. Pleissner, D. and Venus, J. (2014) in *Green Technologies for the Environment*, vol. 1186 (eds S.O. Obare and R. Luque), American Chemical Society, Washington, DC, pp. 247–263.

29. Walsh, P. and Venus, J. (2013) Method for producing L(+)lactic acid using bacillus strains. EP Patent 2013/059184; WO/2013/164423.

30. Tirpanalan, Ö., Reisinger, M., Smerilli, M., Huber, F., Neureiter, M., Kneifel, W., and Novalin, S. (2015) *Bioresour. Technol.*, **180**, 242–249.

31. Reisinger, M., Tirpanalan, Ö., Prückler, M., Huber, F., Kneifel, W., and Novalin, K. (2013) *Bioresour. Technol.*, **144**, 179–185.

32. Baroi, G.N. (2015) Fermentative Butyric acid production from lignocellulosic hydrolysate with in-situ products removal. PhD thesis. Aalborg University.

33. Baroi, G.N., Skiadas, I.V., Westermann, P., and Gavala, H.N. (2015) *Waste Biomass Valorization* **6**, 317–326. doi: 10.1007/s12649-015-9348-5

34. (a) Knoop, R.J.I., Vogelzang, W., van Haveren, J., and van Es, D.S. (2013) *J. Polym. Sci., Part A: Polym. Chem.*, **51**, 4191; (b) Cruz-Izquierdo, A., van den Broek, L.A.M., Serra, J.L., Llama, M.J., and Boeriu, C.G. (2015) *Pure Appl. Chem.*, **6**, 1640.

35. (a) Mateos-Aparicio, I., Redondo, A., and Villanueva, M.J. (2013) in *Dietary Fiber: Sources, Properties and Their Relationship to Health* (eds D. Betancur-Ancona, L. Chel-Guerrero, and M.R. Segura-Campos), Nova Publishers Inc., New York; (b) Mateos-Aparicio, I., Redondo, A., and Villanueva, M.J. (2011) *J. Sci. Food Agric.*, **92**, 697–703; (c) Mateos-Aparicio, I., Mateos-Peinado, C., Jimenez-Escrig, A., and Rupérez, P. (2010) *Carbohydr. Polym.*, **82**, 245–250; (d) Mateos-Aparicio, I., Mateos-Peinado, C., and Rupérez, P. (2010) *Innovative Food Sci. Emerg. Technol.*, **11**, 445–450.

36. Uitterhaegen, E. (2015) Coriander oil – extraction, applications and biologically active molecules. MSc thesis. University of Gent.

37. Sriti, J., Talou, T., Faye, M., Vilarem, J., and Marzouk, B. (2011) *Ind. Crops Prod.*, **33** (3), 659–664.

38. Metzger, J.O. and Bornscheuer, U. (2006) *Appl. Microbiol. Biotechnol.*, **71**, 13–22.

39. Popova, M., Szegedi, A., Risti, A., and Novak Tušar, N. (2014) *Catal. Sci. Technol.*, **4**, 3993.

40. Morone, P., Tartiu, V.E., and Falcone, P. (2015) *J. Cleaner Prod.*, **90**, 43–54.

41. Koutinas, A., Chatzifragkou, A., Kopsahelis, N., Papanikolaou, S., and Kookos, I. (2014) *Fuel*, **116**, 566–577.

42. Dellino, G., Meloni, C., Morone, P., and Tartiu, V. Working paper on "Sustainability transitions in multipurpose biorefineries: an agent-based conceptual framework". presented a two international conferences: (a) 15th International Conference of the International Joseph A. Schumpeter Society (ISS), hosted by the Friedrich Schiller University Jena, Germany, July 27–30, 2014; (b) 5th Anniversary of the International Sustainability Transitions (IST) Conference hosted by the Utrecht University, The Netherlands, August 27–29, 2014.

43. van der Klis, F., Frissen, A.E., van Haveren, J., and van Es, D.S. (2013) *ChemSusChem*, **6**, 1640.

Index

Chemicals and Fuels from Bio-Based Building Blocks, First Edition.
Edited by Fabrizio Cavani, Stefania Albonetti, Francesco Basile, and Alessandro Gandini.
© 2016 Wiley-VCH Verlag GmbH & Co. KGaA. Published 2016 by Wiley-VCH Verlag GmbH & Co. KGaA.